Local Content for the International Petroleum Industry

LOCAL CONTENT FOR THE INTERNATIONAL PETROLEUM INDUSTRY

EDITED BY

EDUARDO G. PEREIRA

TONJE GORMLEY

PennWell®

Disclaimer

The recommendations, advice, descriptions, and the methods in this book are presented solely for educational purposes. The author and publisher assume no liability whatsoever for any loss or damage that results from the use of any of the material in this book. Use of the material in this book is solely at the risk of the user.

PennWell Corporation
1421 South Sheridan Road
Tulsa, Oklahoma 74112-6600 USA

800.752.9764
+1.918.831.9421
sales@pennwell.com
www.pennwellbooks.com
www.pennwell.com

National Account Executive: Barbara McGee Coons

Publisher: Matthew Dresher
Managing Editor: Stephen Hill
Production Manager: Sheila Brock
Production Editor: Tony Quinn
Cover Designer: Charles Thomas

Library of Congress Cataloging-in-Publication Data
Names: Pereira, Eduardo G., 1984- editor. | Gormley, Tonje, 1974- editor.
Title: Local content for the international petroleum industry / Eduardo G. Pereira and Tonje Gormley, editors.
Description: Tulsa : PennWell, [2018] | Includes bibliographical references and index.
Identifiers: LCCN 2017027854 | ISBN 9781593704100
Subjects: LCSH: Petroleum industry and trade.
Classification: LCC HD9560.5 .L63 2017 | DDC 338.8/8722338--dc23

Printed in the United States of America

1 2 3 4 5 22 21 20 19 18

To India and Sienna for their patience

—*Tonje P. Gormley*

CONTENTS

INTRODUCTION

Petroleum products are nonrenewable resources that have the potential to increase state revenues. If the exploitation of petroleum resources and the resulting revenues flowing to the state are managed prudently, petroleum activities may contribute to sustainable development for the host country and its people.[1] The prospect of increased state revenues is the incentive most host countries have to initiate the exploration for petroleum resources. Host countries are therefore also looking at ensuring that petroleum activities contribute to domestic economies beyond direct state revenues, through increasing the local content in the domestic petroleum industry. There is no universal definition of the term *local content* in petroleum activities. The term typically refers to the added value that petroleum activities may bring to a host country other than the direct revenues. Such revenues may be obtained through the sales of government shares of petroleum produced or through the taxes, service fees, or other dues collected from companies carrying out upstream petroleum activities.

Typically, the host county will model policies with a goal of facilitating and promoting participation of state-owned or privately owned domestic oil companies in petroleum activities. Such policies often encourage the use of domestically produced goods, the use of domestic service providers, and requirements for industry training and employment for nationals. In order to speed up the development of local content, many host countries want their domestic companies to learn from the international companies. Requirements are therefore often established for international companies to enter into joint ventures with domestic companies, transfer know-how and technology to governmental or private domestic entities, or carry out joint research and development projects.

This book provides an overview of how various host countries have approached local content development. It examines how the policies for local content development may be modeled. It provides examples of what a host country could or should take into consideration in modeling adequate local content policies, along with how such policies may be implemented through statutory, regulatory, or contractual requirements. Moreover, we hope to illuminate how host countries today can learn from their own experiences or from those of other host countries. As we will see, there is no one-size-fits-all model for successful local content development. Although local content policies and local content requirements are rather similar throughout the world, the level of successful implementation varies from country to country.

Ownership, Management, and Cooperation

The expectation that exploitation of petroleum resources will lead to increased state revenues is linked to the host countries' ownership of petroleum resources. Pursuant to an established public international law principle, permanent sovereignty of natural resources is vested in the state in which the resources are located, on behalf of its people.[2] Often, this public international law principle is also translated into the national law of the host countries.[3] It is common for host countries to adopt constitutional or statutory rules clearly establishing that title to natural resources is vested in the state and that the state has the exclusive right to resource management. As such, host countries are free to exploit their petroleum resources and dispose of any petroleum produced, which in turn can generate revenues.

Nevertheless, host countries often wish for the involvement of international oil companies in developing and maintaining their petroleum industry. The reasons for this may be multifaceted. Typically, host countries that are new to petroleum activities do not possess the required technical competence or even the financial resources to prudently and efficiently explore, develop, and produce petroleum. International oil companies, however, have access to the capital, technical know-how, and infrastructure required. The international petroleum business is very risky and capital intensive. Even where there are commercial discoveries of petroleum, there is a long lead time from when investments are first made until profits can be collected. Therefore, even if a host country has built competence and has access to financial means, it may not wish to assume the sole risk for petroleum activities. Therefore, host countries often need or wish to engage international oil companies in their domestic petroleum industry. It might even be a policy goal to create competition in the domestic petroleum industry in a bid to obtain more efficient and cost-effective projects. As such, despite the distinct domestic nature of exploitation of petroleum resources, it typically involves a degree of cooperation between the host country and international oil companies.

In order to maintain control of the manner in which petroleum activities are carried out, however, host countries need to ensure that the international oil companies are adhering to the directions that are specified. Host countries will thus require that oil companies that wish to carry out petroleum activities will have to either enter into a contract with or obtain a license from the host country. This right is typically limited to a given area, for a limited period of time, and on certain conditions. Those carrying out petroleum activities will also often have to abide by a wider set of rules embedded in laws and regulations. Although the level of state control will vary, such rules will typically be aimed to ensure that the host country controls the manner in which exploration, development, production, and decommissioning is carried out. If exploration leads to development and production, the host country as owner of the resources will require payment from the oil companies, either in cash or in kind. The components that constitute "government take" in a host country vary. The host country, however, will collect direct revenues from the petroleum activities in the form of sales of the government's share of petroleum produced and the collection of taxes, service fees, or other dues.

Local Content Policy Development: A Balancing Act

Most host governments are keen to maximize the contribution that petroleum activities may make to their domestic economies. This political goal typically materializes in efforts to ensure that the economic rent is captured[4] by way of adequate fiscal models and in efforts to develop local content. As with any other political goal, host governments will model policies in order to achieve the goal. As petroleum resources are nonrenewable, and a given host country's resource base might be limited or uncertain, it is important that the policies are effective from the start. This is especially true for host countries where the petroleum resources are assumed to be of a limited extent. Experience shows that there are certain important factors that a host government would be well advised to consider when modeling local content development policies.

When aiming to develop beneficial local content policies, the petroleum sector cannot be viewed in isolation. Rather, the local content policies must be aligned with the other economic development policies of the host country.[5] The ministry responsible for petroleum activities should therefore consult and cooperate with other ministries with interfacing areas of responsibility to ensure that the policy goals are viable and suitable for the host country. For instance, if there is no focus on—or even intent to focus on—providing education within engineering, geology, or other technical disciplines in the host country, problems will ensue. A policy aimed at maximizing employment of nationals in the petroleum industry in specialized positions will be dependent on such training to prepare its national workforce. Without this provision, the policy will eventually fail. While one may require that the international companies train and educate nationals, to achieve ambitious employment goals in this specialized industry, adequate educational institutions are required in order to fully succeed in the long-term economy.

Policy makers also must duly consider the timing and the general market conditions for establishing domestic oil companies and petroleum-related goods and services. This aspect alone raises many questions. Other questions will arise related to preferential rights for participation of domestic oil companies or the establishment of one or more state-owned national oil companies. If NOCs are to be established, what should their respective roles be? What are the short- and long-term objectives? If domestic, privately owned oil companies are to enjoy preferential rights to enter and operate within the industry, how many domestic oil companies can the industry support? Should there be exceptions from award requirements as to financial strength, technical competence, and experience? And after the contracts are awarded, how will the national companies' share of the costs be covered? Should they receive financial aid from public institutions, or instead be carried by the international oil companies?

Policies for preferential treatment for providers of goods and services aimed at assisting them to enter and stay in the market give rise to other questions. Should the policies for local content development be general (i.e., targeting all types of petroleum-related businesses) or more specific as to the types of goods and services targeted? The petroleum industry today is well developed. It is technology heavy, specialized, and capital intensive. Local companies and their employees will have to compete against an established international industry that is already being served by a well-developed network of global goods and service providers. These service companies are often strong players in both a financial and technical sense. It is prudent to question whether the global petroleum industry has room for yet another company providing a certain service before time, money, and effort are spent to build such competence domestically. Depending on the global market and the overall competitiveness of the domestic businesses, a requirement to expand the local content requirements, for instance in a certain service category, might not be prudent for a relatively short-lived domestic petroleum industry. The question of sustainability arises.

Another aspect to consider is how much time is needed to establish domestic providers of specialized goods and services. Policy makers should consider if there are areas in which local content can be developed more quickly than others. Available human resources and the domestic level of industry knowledge, skills, and entrepreneurial spirit will play a role in how fast a specific domestic competency might be built. Host governments should therefore examine whether the host country has human resources and industrial capacity in a certain area that could be easily transferred to serve the petroleum industry, or whether it would be necessary to start from scratch. Without some level of existing, transferable competency, it may take a long time to establish domestic players that can compete in this market.

Host governments are well advised to be realistic in setting their policy goals. There are two main approaches for local content policies and requirements. Some host countries take a rigid, prescriptive approach, and others take a more goal-oriented and dynamic approach. Countries that choose the latter model will have to assess the competitiveness and competency of the domestic companies when choosing target levels for local content goods, services, and labor. In fact, it may be easy to argue for the selection of international companies rather than domestic companies. It is also difficult for host governments to supervise and enforce such requirements. Countries that choose the first model, on the other hand, must question the overall economic benefits of an overly ambitious and rigid local content policy. This is especially true if the local content policies and requirements are not realistic for a country's context. It can be very challenging for the international companies to comply with rigid and unrealistic requirements. In turn, this may complicate compliance, decrease efficiency of petroleum operations, increase costs, and even impair health and safety. Altogether, such requirements may therefore impair the investment climate in the host country. Still, as seen in chapter 16, the Nigerian local content level increased after the introduction of an ambitions and rigid local content act, setting very specific goals. But it is unclear how sustainable the local businesses will be over time, how much added value these requirements are giving the Nigerian economy as a whole, and how the local content requirements will affect the investment levels in the Nigerian petroleum industry.

Host governments should also be sensitive to the long-term effects of maintaining preferential rights for domestic companies within a specific industry. There might be unfortunate effects on the domestic economy as a whole. The obvious disadvantage is that businesses that rely on preferential rights might struggle to be competitive internationally. One may question how long a petroleum-related business will survive if it cannot expand its operations internationally. If the domestic companies that are enjoying preferential rights will be short-lived, and thus will not contribute to the domestic economy over the long-term period, one may question whether it is worth building them up in the first place. This is especially something to consider if the use of domestic goods and services is found to have a negative effect on costs and efficiency of the petroleum activities. Moreover, preferential arrangements can lead to an increase in the salary levels in the petroleum industry. Other domestic businesses that do not enjoy preferential treatment, but that are exposed to the full effects of international competition, could then struggle to cope with this higher salary level.[6]

To achieve local content policy goals, it is also important that the host government considers its own capacity and competence. A policy for institutional capacity building should be prepared alongside, or as a part of, local content development. After all, experience shows that host countries are best served by leaving supervision and enforcement to governmental entities. The host government should have a plan to hire adequate human resources for its regulatory agencies, with the technical and commercial competence to supervise, evaluate compliance, and enforce the local content requirements.

All in all, a local content policy that is not sufficiently debated, prepared, and balanced against other national policy goals and the host country's specific context at all times may not have the desired overall impact over the long term.

This brings us to another sensitive issue, namely that of a government's political robustness to make and implement unpopular decisions. Modeling effective local content development policies is challenging in terms of maximizing the contribution that petroleum activities make to the domestic economy. A long-term perspective is necessary, often beyond the reign of any given government. Therein lies the problem. It is not only governments that expect to see great domestic economic benefits from petroleum activities. In particular, in poorer countries, the people expect job creation. As we have seen, however, job creation can be challenging. In some instances it is not sustainable or even economically sound to try to build an entire national industry from scratch. At the very least, it will take some time to get there. For this reason and others, some commentators state that local content requirements can be a trade-off for profitability. Furthermore, this might well be against the overall political goal of maximizing the contribution to the domestic economy. Thus, having the political strength to think long term and choose the "right" priorities can be a challenge. In certain host countries, it possibly is not even politically viable.[7] Concerning local content policy (LCP), Tordo et al. aptly observe:

> [O]ur review of countries' experience suggests that LCP design is not solely or principally guided by considerations of economic efficiency and the need to address market failures. On the contrary, it would seem that political imperatives are key drivers, and economic efficiency is often an afterthought.[8]

Transparency in policy making and providing information to the people may be useful tools for a host government in the management of people's expectations. But as we will see in this book, the ways in which policies are presented to the public vary. As an example, Ghana's Ministry of Energy issued a separate document "Local Content and Local Participation in Petroleum Activities—Policy Framework" on February 26, 2010. This policy document was subsequently implemented in the Petroleum (Local Content and Local Participation) Regulations, 2013 (L.I. 2204), passed on November 19, 2013. Norway, on the other hand, expresses policies in regular reports from the Ministry of Petroleum and Energy to the Parliament. Policy considerations are also presented and discussed in preparatory works of statutory laws (white papers). Other countries do not issue specific documents on policy.

Local Content Policies Implemented: Requirements in Law, Regulations, and Contracts

We noted above that local content development policies are often implemented by requirements that facilitate and promote the participation of state-owned and/or privately owned domestic oil companies in petroleum activities. They also promote the use of domestically produced goods, the use of domestic service providers, and the training and employment of nationals. It was mentioned that many host countries have established requirements for joint ventures, transfer of technology, and joint research and development for these purposes. There is variety, however, in how such statutory, regulatory, and contractual requirements are modeled. At a macro level, we encounter many of the same issues as discussed above in modeling such requirements. Moreover, difficult issues will materialize in detailing the requirements.

Reflecting the differences in policy models as described above, there are, broadly speaking, two main models for drafting rules on giving domestic providers of goods and services preferential rights. Some countries have adopted flexible rules that state, for instance, that where domestic goods and services of a similar price and quality are available, preferential rights shall be given to domestic providers. Such rules typically set a threshold as to price, e.g., if the domestic goods or services are not more than 10% of the price of international goods. Availability and quality are also addressed. It can be challenging to set these thresholds. Other countries, such as Nigeria and Ghana, set very prescriptive, detailed requirements as to the achievement of certain local content percentages in a given area of business within a specified period. This model is certainly more challenging for the prudent drafter, who will have to evaluate what an adequate level for local content is for each specific service. In order to be useful, the levels should at least reflect the competence and capacity within that service in the host country at any given time. The former model is more dynamic in the sense that the door is left open for increased involvement of the domestic providers of goods and services as they step up their efforts in terms of availability, quality, and price. We noted above that requirements based on discretionary evaluations may make it all too easy for companies to find arguments to choose international providers, even when the domestic companies are competitive. However, we also noted that the more prescriptive and rigid approach may not always be optimal either. If there are no providers of goods and services of a specific kind available, it is virtually impossible for the international oil companies to comply with a prescriptive requirement demanding a certain percentage of local content. This may be viewed as a shift in responsibilities. In many host countries, it is normally a governmental task to enable the development of new businesses. The prescriptive model, however, will effectively leave the international companies to deal with the lack of domestic competence in the sense that there are requirements to use something that might not exist. This may lead to undesired effects, such as a high level of noncompliance or even corruption.

Another basic question that must be addressed in laws, regulations, and contracts is the definition of what constitutes a "local" company. Does it suffice for an international player to establish a local company or branch in the host country, or is a joint venture with a domestic company or full or part ownership by a national company required? Simply requiring the establishment of a local branch of an international company can be sufficient for job creation, provided that there are restrictions on the employment of expatriates. However, many host countries prefer to go one step further, namely to require the establishment of joint ventures with domestic companies, often combined with a requirement for transfer of know-how and technology.

In this manner, domestic businesses are learning from their international counterparts in order to be able to run similar businesses on their own at a later stage. This may lead to something more than job creation. It can pave the ground for sustainable activities even on a long-term perspective. However, as discussed above, sustainability might be an issue. In any event, we note that merely requiring local ownership does not in itself build industrial competence. Very few may actually benefit from such requirements. Moreover, we also see that there might be pitfalls; commentators note that requirements for joint ventures can lead to corruption.[9]

Requirements also vary regarding the training and employment of nationals. A typical challenge for a host country with a nascent petroleum industry is that its people lack the skills required for the more specialized jobs. However, it is important not to forget that even this complex and specialized industry requires a certain degree of unskilled labor, even if use of such labor is decreasing due to mechanization. A typical requirement is for companies to employ qualified nationals in lieu of expatriates when possible. In order to increase the local content in terms of national employees, host countries also require that companies train nationals, i.e., prepare them to work in skilled positions in the petroleum industry. This can be done in the form of on-the-job training, contributing to universities or educational institutions, providing courses, or even supporting nationals to study abroad. Again, some countries, such as Ghana, take the more prescriptive approach of requiring companies to comply with certain minimum levels of employment of nationals at junior or mid-level positions and to prepare succession plans for nationals to assume positions held by expatriates. Needless to say, the road to increase the level of competence among nationals to make them employable in the petroleum industry is longer and tougher in countries with a low level of education and poor educational systems. The reverse is true in some countries. Lebanon, for instance, might even have vast national competence in the diaspora and potentially could hire nationals who are already employed in the international petroleum industry. Here, the issue of local content in the form of employment might rather be to ensure that competent nationals are encouraged to come home to help build the industry.

As a part of enablement of the domestic companies and nationals alike, requirements to carry out research and development in the host country, in collaboration with domestic companies, can be important. Norway benefited much from requiring international companies to carry out a certain percentage of their research and development activities in Norway in collaboration with Norwegian partners. For Norway, the timing was lucky, too.

At the time there was a need for improvement of offshore technology, and Norwegians were part of developing it. Requiring transfer of technology is also a part of the enablement of domestic businesses and nationals. Without the requisite technology, it is difficult to succeed in the petroleum industry. As an example, we note that Iran, a country that has for a very long time been isolated from great parts of the world due to international sanctions, is focusing on technology. Now that Iran can again welcome foreign investment, it is highly focused on establishing joint ventures and obtaining access to new and modern technology.

Much can be said in detailing local content requirements, and this book provides many examples. Nonetheless, to be beneficial, local content rules must be complied with. The primary responsibility for compliance rests, of course, with the companies. However, in order to ensure a high level of compliance, the host government must be able to supervise and enforce the legal framework for local content development. To carry out these tasks efficiently, the relevant laws, regulations, and contracts must establish clarity of roles and responsibilities.[10] The laws, regulations, and contracts must also provide the authorities with adequate tools for supervision and enforcement. Most countries require the submission of local content plans, regular reporting on compliance, and the use of tenders for purchases exceeding a certain price threshold. Governments also typically ensure they have the right to carry out audits and inspections. The level of government interference in procurement varies; however, we see that some countries require preapproval of purchases and contracts. Nonetheless, the manner in which supervision is modeled should be carefully considered. Compliance with a high level of complex reporting requirements takes time and money to administer, both for the companies and the authorities. Requirements for frequent approvals might delay projects, and compliance might become an administrative burden for all those involved, including the government.

The tools chosen for enforcement should also be chosen with care. High penalties might be efficient for "scaring" companies into compliance, but at what price? Indeed, companies might be more prone to accept corruption in order to avoid harsh punishments. Host countries should instead consider how they might actually assist companies with compliance, for instance by building databases of qualified or competent domestic companies. Alternately, the incentives for compliance might be more indirect, perhaps through consideration of historical levels of compliance in tenders or licensing rounds.

So, much burden rests on the host government to set realistic, adequate policy goals given its specific context and to implement them adequately. The responsibility of the companies is equally clear; they should actively seek to comply with local content requirements. However, what is not so often discussed is the role of civil society in assisting successful and sustainable local content development. We note that without a degree of entrepreneurial spirit in business communities and the civil society at large, sustainable local content development is a struggle. There really is no such thing as a "free lunch."

In This Book

It is against this background that we deem that an analysis of international experience from a policy, regulatory, and a compliance point of view would be interesting. This text offers an analysis of the development of local content in a number of different host countries. We examine the host countries' policy goals, how these are implemented in legislation and contracts, and the factual framework conditions for development of local content in each country (i.e., industrial capacity, existing infrastructure, the competency of the human resources base, and the short- and long-term goals of the host country). Included are observations on how, if at all, these countries are succeeding in developing the wider benefits from the petroleum industry and how companies tackle compliance with local content requirements. In the concluding chapter, we summarize some of the pitfalls, challenges, and possible success factors and comment on how emerging petroleum countries may benefit from past experiences, from both a regulatory and compliance point of view.

Among the host countries analyzed, the reader will find some major producing countries as well as young or nascent petroleum provinces. One may divide these countries in three separate groups, as seen in table I–1.

Hopefully, practical examples from experienced jurisdictions (i.e., Group A) could provide valuable inputs for other petroleum provinces. Each jurisdiction in our book was given a specific content framework to facilitate comparisons. The outcome of our research is interesting. The reader will note that the policy goals and implementing measures are quite often very similar. It is rather the factual contexts of each host country, as well as the manner in which measures have been implemented, that vary.

Table I–1. Groups of host countries

	Background	Jurisdiction – Examples
Group A	This group is the core of our book as they represent relevant producers of oil and/or gas and they should have higher experience with local content to certain extent.	We consider the following countries as part of Group A: Angola, Algeria, Brazil, Canada, China, Egypt, Iran, Libya, Nigeria, Norway, and Qatar.
Group B	This group represent emerging countries with recent discoveries and they are yet to test their local content measures.	We consider the following countries as part of Group B: Ghana and Mozambique.
Group C	This group represent petroleum provinces with higher hope for large hydrocarbon discoveries but they are yet to find and/or produce large-scale hydrocarbon discoveries.	We consider the following countries as part of Group C: Cyprus, Kenya, Lebanon, and East-Timor.[11]

The reader will note that the host countries in Group A vary in their experiences with local content development. The common denominator is that these are all countries with long or even very long histories of petroleum activity. Some of these countries failed at first and then progressed with more successful policies, such as Brazil and Nigeria. Some were quite successful from the beginning, such as Norway. Others appear to still be working on fine-tuning or even rethinking their direction when it comes to local content development. The host countries in Group B are similar; both countries have recently made large hydrocarbon discoveries and they are trying to maximize the benefit from those resources. They are eager to increase local content, but they do not yet have extensive practices and experience to understand the success or failure of their policies. Group C has limited experience, but the resource countries are already being challenged with the management of expectations in the population. The combination of categories A, B, and C should therefore provide interesting perspectives for all types of host countries. We hope you enjoy reading this text.

Notes

1. The impact that exploitation of petroleum resources has made on the Norwegian society within a reactively short period of time is a good example of how those resources may benefit a host country. Since the commencement of the petroleum activities in the North Sea in the 1960s, Norway has moved from being a postoccupancy state with an economy mainly based on farming, shipping, and fishery to a country with sound financial growth and a welfare state largely funded by revenues from petroleum exploitation.

2. The principle was developed through a series of UN resolutions on this and related issues that emerged from the late 1950s up until the 1970s, including the Declaration on Permanent Sovereignty over Natural Resources, Resolution 1803 (XVIII) of December 14, 1962, UN Doc A75217 (1962), 2ILM (1963), 223; the Declaration on the Establishment of a New International Economic Order, Resolution 3202 (S-VI) of May 1, 1974 UN Doc. A/5217 (1962), 2ILM (1974), 715; and Charter of Economic Rights and Duties of States, Resolution 3281 (XXIX) of December 12, 1974, UN Doc. A/9631 (1974), 14 ILM (1975), 215.
3. This is especially true where host countries have a dualistic approach to the application of international law.
4. The term *economic rent* can be defined as "Excess profits above the return on capital needed for investment to go on, so that in principle, changes in the size of the economic rent do not influence the willingness to invest in the particular industry, as long as return on capital is satisfactory, compared with the second best investment option adjusted for risk." (Øystein Noreng, *Crude Power: Politics and the Oil Market* [London: I.B. Tauris, 2007], 178.) In the present context, the statement that host countries wish to "capture the economic rent" means that while taking the volatility and fluctuations in petroleum prices duly into account, host countries will seek to obtain as much revenue as possible from the exploitation of petroleum resources, without the international oil companies losing interest in investing in the host country.
5. Silvana Tordo, Michael Warner, Osmel E. Manzano, and Yahya Anouti, *Local Content Policies in the Oil and Gas Sector* (Washington, DC: International Bank for Reconstruction and Development, World Bank, 2013), xiii, doi: 10.1596/978-0-8213-9931-6, or http://documents.worldbank.org/curated/en/549241468326687019/pdf/789940REVISED000Box377371B00PUBLIC0.pdf.
6. For more detailed discussion of Norway's regulation and development of its resources and local content policies, including the observations made by the Norwegian government in the late 1970s, see chapter 17.
7. A related issue is that not many countries have the political stability to remain consistent through changes in governments and government policies. It takes time to build competence relevant to the petroleum industry, and frequent shifts in policy clearly affect its potential impact.
8. Silvana Tordo et al., *Local Content Policies in the Oil and Gas Sector*, xiii.
9. Malachi Brown and Augustine Evangel, "Petro-Economy and Corruption in Nigeria: A Legal Examination," *IOSR Journal of Humanities and Social Sciences*, 16, no. 2 (September–October 2013), doi: 10.9790/0837-16295102.
10. Moreover, as mentioned above, the relevant authorities must be able to carry out their responsibilities in this respect. For this very reason, it is important that the host government focuses on institutional capacity building alongside local content development. Thus, we see that the training obligations of nationals can also extend to include training of government officials.
11. I.e., the Timorese Exclusive Area and not the framework in place for the Joint Development Zone that is established in a disputed area shared by Timor-Leste and Australia.

1

ALGERIA

Waniss A. Otman

Introduction

The People's Democratic Republic of Algeria is located in North Africa. It is made up of 48 provinces and more than 1,500 communes. With a population estimated in 2015 at more than 40 million, it ranks as the 33rd most populous country in the world. In terms of area, it is the 10th largest country globally, and the largest country in Africa.[1] The capital, Algiers, has more than 2.5 million inhabitants. The other major cities in terms of number of inhabitants are Sétif, Oran, Constantine, and Annaba.

The country has an area of nearly 2.4 million square kilometers. With a coastline of 998 km, it shares a border with Libya of 989 km; Mali, 1,359 km; Mauritania, 460 km; Morocco, 1,900 km; Niger, 951 km; Tunisia, 1,034 km; and the Western Sahara, 41 km. The Sahara covers 84% of the country, the steppe, 8.5%, and the coastal fringe (mountains and plains), 7.5%. The climate is mainly Mediterranean in the northern and coastal regions, continental in the hinterland, and tropical in the south, with large variations in temperature between day and night.

Politically, for most of the 1990s, chaos ruled. Around 200,000 people were killed in a civil war that resulted when Algerian generals refused to accept an Islamist victory in the 1991 election. However, for the past 15 years, President Abdelaziz Bouteflika has kept the peace. The Arab Spring has largely passed Algeria by.[2]

Today, at least on paper, Bouteflika's regime is still dominant, although he is now 80 years old. He was reelected in 2014 despite appearing only once, and then in a wheelchair. Behind him the cabal of security and army men who make up *le pouvoir*, as Algeria calls those who pull the strings, are fighting among themselves over succession issues.[3]

Macroeconomic Issues

Algeria's economy remains dominated by the state, a legacy of the country's socialist postindependence development model.

Overview

According to the International Monetary Fund (IMF) Country Report No. 14/341 of December 2014,[4] economic activity in Algeria picked up in 2014, with real gross domestic product (GDP) growth projected to reach 4.0%, following 2.8% growth in 2013. The hydrocarbon sector is expected to expand for the first time in eight years, while nonhydrocarbon growth remains supportive.

Inflation has decelerated sharply to 2.1%, thanks in part to tighter monetary policy. Algeria continues to enjoy substantial external and fiscal buffers, but threats to macroeconomic stability are growing. For the first time in nearly 15 years, the current account is expected to record a deficit.

Deficits are projected to widen over the medium term, as strong domestic hydrocarbon consumption and lower oil prices weigh on exports. Imports continue to grow, driven by public spending. As the *Economist* reported in September 2015, with 40 million people, Algeria faces a far more immediate squeeze. It ran a trade deficit of $8 billion in the seven months leading up to August 2015 (roughly 7% of GDP when annualized), compared with a $4 billion surplus in the same period last year. Its currency has lost a quarter of its value against the dollar.

The export base is undiversified, and foreign direct investment (FDI) is hampered by restrictions on ownership. For the years 2015 and 2016, the share of total exports from hydrocarbons is estimated to amount to 96.8% and 96.5%, respectively.[5]

The fiscal deficit is expected to widen to more than 7% due to lower hydrocarbon revenue, a sharp increase in capital expenditure, and continued high current spending. Nonhydrocarbon revenues are below their potential, the wage bill is high, and subsidies and transfers are costly, amounting to about 26% of GDP. Fiscal savings are expected to decline for the second consecutive year.

Although Algeria has enjoyed macroeconomic stability, faster and more inclusive growth is necessary to provide enough jobs for the country's youthful population. Public investment efficiency is low, and private sector growth is hindered by a cumbersome business climate, an underdeveloped financial sector, and limited international integration. Finally, rigidities in the labor market and skills mismatches reduce the impact of economic growth on job creation.

The IMF Country Report also identified a range of key macroeconomic issues as detailed in the following.

Although inflation has subsided, longstanding vulnerabilities are becoming increasingly apparent. Economic growth is holding up well, inflation has fallen below the central bank's target, and fiscal and external buffers remain large. But the economy's dependence on the hydrocarbon sector has led to serious vulnerabilities that are now coming to light. Stagnating hydrocarbon production and declining oil prices, combined with increasing domestic consumption of hydrocarbon products, are depressing export receipts. Public spending continues to increase, fueling imports and a loss of competitiveness, and placing fiscal policy on an increasingly unsustainable path. As a result, the current account was expected to record a deficit in 2014 for the first time in more than 15 years. Unemployment remains high among youth and women.

Algeria also needs to consolidate macroeconomic and financial stability. A full-fledged fiscal rule, combined with an ambitious consolidation effort over the medium term, would help put the fiscal stance on a sustainable path. The decline in inflation experienced since 2012 is a positive outcome that needs to be maintained through continued efforts to absorb liquidity and a more active use of interest rates. The exchange rate policy should avoid deviations of the dinar from its equilibrium level and should support the deepening of foreign exchange markets. Implementation of the Financial Sector Assessment Program (FSAP) recommendations should be pursued, notably to strengthen financial sector stability.

An export-oriented strategy is needed to ensure external sustainability. Export diversification is imperative to ensure external sustainability and reduce vulnerability to fluctuating oil prices. Diversification policies should focus on maintaining economic stability and improving competitiveness, increasing openness to trade and capital flows (particularly FDI), and creating a more export-friendly business climate. Additional reforms are needed to increase hydrocarbon exports, including reforms to increase production and phase out implicit subsidies on energy and electricity, which are fueling domestic consumption.

Algeria's significant growth potential needs to be more fully realized. Reforms are needed to reduce the dominance of the public sector and transform the private sector into an engine for growth, supported by a deeper financial sector. To increase employment, Algeria will need better-functioning labor markets, a workforce equipped with more relevant skills, and more effective labor market programs. In this regard, a more aggressive and legally mandated local content policy for the oil and gas industry than that which exists at present would be advantageous. Related to this, industry/government initiatives to provide training for young Algerians in the oil and gas sector could help to significantly reduce unemployment, particularly in the under-30 segment of the labor market.

Significantly, the report stressed that Algeria's external position is highly vulnerable to a sustained decline in oil prices. Despite efforts to diversify, Algeria's external position is determined largely by the performance of hydrocarbon exports, which accounted for 98% of merchandise exports in 2013.[6]

Table 1–1 highlights estimated and projected economic and financial indicators for the years 2012 to 2019.

Table 1–1. Algeria: Selected economic and financial indicators, 2012–2019

	Estimated		Projected					
	2012	2013	2014	2015	2016	2017	2018	2019
OIL AND GAS SECTOR								
Hydrocarbon production (in tons of oil equivalent)	142	137	138	139	142	144	147	149
HYDROCARBON EXPORTS								
Of which: liquid petroleum exports (in millions of barrels/day)	1.3	1.2	1.2	1.1	1.1	1.1	1.1	1.1
Of which: natural gas exports (in billions of m^3)	51.7	46.3	44.9	45.5	46.3	47.1	47.9	48.6
Crude oil export unit value (US$/bbl)	110.8	109.5	104.1	89.0	90.2	90.1	89.7	89.5
Share of hydrocarbons in total exports (in percent)	98.4	98.4	97.5	96.8	96.5	96.2	95.9	95.6
NATIONAL INCOME AND PRICES								
GDP at constant prices	3.3	2.8	4.0	3.9	4.1	4.0	4.0	4.0
Hydrocarbon sector	–3.4	–5.5	0.4	1.3	1.7	1.9	1.6	1.6

	Estimated		Projected					
	2012	2013	2014	2015	2016	2017	2018	2019
Other sectors	7.1	7.1	5.5	5.0	4.9	4.7	4.7	4.6
Consumer price index (period average)	8.9	3.3	3.0	4.0	4.0	4.0	4.0	4.0
EXTERNAL SECTOR*								
Exports, f.o.b. (percent change)	−1.6	−10.2	−6.3	−14.7	2.1	0.8	0.4	0.6
Hydrocarbons	−1.5	−10.3	−7.2	−15.4	1.8	0.5	0.1	0.3
Nonhydrocarbons	−6.0	−8.9	44.3	11.1	9.9	9.7	8.4	7.3
Imports, f.o.b. (percent change)	14.9	6.6	5.0	0.9	5.3	1.1	2.0	2.5
Current account balance (in percent of GDP)	5.9	0.4	−4.0	−7.7	−8.2	−7.8	−7.5	−7.5
MONEY AND CREDIT								
Net foreign assets (percent change)	7.4	1.9	0.5	−3.5	−5.4	−6.8	−8.7	−11.5
Credit to the economy	15.1	19.9	21.7	12.0	10.8	10.2	8.6	8.3
Money and quasi-money	10.9	8.4	14.5	9.0	5.7	4.3	3.0	3.0
Velocity of broad money (GDP/M^2)	1.5	1.4	1.3	1.2	1.3	1.3	1.4	1.4
IN PERCENT OF GDP								
Saving-investment balance	5.9	0.4	−4.0	−7.7	−8.2	−7.8	−7.5	−7.5
National savings	36.2	35.0	34.1	32.1	32.4	32.5	32.9	33.0
Government	9.1	9.9	6.9	4.4	5.1	5.6	5.9	6.0
Nongovernment	27.1	25.1	27.1	27.7	27.4	26.9	27.0	27.0
Investment	30.3	34.6	38.0	39.8	40.7	40.3	40.3	40.5
Government	14.1	11.4	13.9	13.9	13.3	12.7	12.4	12.1
Nongovernment	16.2	23.2	24.1	25.9	27.4	27.6	27.9	28.4
CENTRAL GOVERNMENT FINANCE								
Overall budget balance (deficit –)	−5.0	−1.5	−7.0	−9.5	−8.2	−7.1	−6.6	−6.1
Total revenue	39.7	35.9	33.5	31.8	31.6	31.2	31.1	30.9
Total expenditure (incl. net lending)	44.7	37.4	40.5	41.3	39.8	38.3	37.6	37.0

	Estimated		Projected					
	2012	2013	2014	2015	2016	2017	2018	2019
CENTRAL GOVERNMENT FINANCE (in percent of nonhydrocarbon GDP)								
Total revenue	60.5	51.2	46.7	42.4	41.8	41.1	40.3	39.5
Hydrocarbon	39.6	31.7	27.1	22.8	22.0	20.9	19.8	18.8
Nonhydrocarbon	20.9	19.5	19.7	19.6	19.8	20.2	20.5	20.7
Total expenditure	66.7	52.5	55.6	54.3	52.1	50.0	48.4	46.9
Current expenditure	45.2	36.2	36.2	35.8	34.5	33.2	32.3	31.4
Capital expenditure	21.5	16.3	19.4	18.5	17.6	16.8	16.1	15.5
Nonhydrocarbon primary balance	–46.8	–33.5	–36.6	–35.1	–32.6	–30.0	–28.0	–26.3
NH primary balance, net of one-off spendings	–41.9	–33.1	–36.6	–35.1	–32.6	–30.0	–28.0	–26.3
Nonhydrocarbon balance	–47.1	–33.8	–36.9	–35.5	–32.9	–30.3	–28.3	–26.6
MEMORANDUM ITEMS								
GDP (in billions of dinars at current prices)	16,115	16,569	17,646	18,422	20,010	21,725	23,141	24,641
NHGDP (in billions of dinars at current prices)	10,579	11,601	13,843	12,645	15,106	16,497	17,847	19,272
GDP (in billions of US$ current prices)	208	209	211	208	217	225	240	256
GDP capita per (in US$)	5,542	5,508	5,446	5,273	5,405	5,510	5,770	6,041
Exchange rate (DZD per US$)	77.6	79.4	—	—	—	—	—	—
Crude oil exports (in millions of barrels/day)	0.7	0.6	0.6	0.6	0.6	0.6	0.6	0.6
Nonhydrocarbon exports (percent of total exports)	1.6	1.6	2.5	3.2	3.5	3.8	4.1	4.4
Gross official reserves (end of period)	191	194	188	173	157	142	127	110
In months of next year's imports of goods and services	35	34	32	28	25	23	21	17
Gross government debt (in percent of GDP)	10.0	8.3	8.8	9.0	8.9	8.8	8.7	8.6

Source: IMF Country Report No. 14/341. December 2014. https://www.imf.org/external/pubs/ft/scr/2014/cr14341.pdf.
*In US dollars terms.

Local Content Policy and Baseline

The regulatory framework concerning local content policy in Algeria has been shaped by several programs in the last 20 years. In 2001, a development strategy called the Economic Recovery Program (PSRE), 2001–2003, was launched. This was followed in the period 2004 to 2009 by a new development program called the Economic Growth Support Program (PSCE). In 2010, a new 2010–2014 Public Investment Program (PIP) was structured within the framework of a development strategy originally launched in 2001.[7]

The Five-Year Public Investment Program, 2010–2014

The strategic objectives were to provide the country with basic modern infrastructure and make these facilities a prime asset to encourage FDI and national investments. Development of human resources was also seen as a key objective to strengthen the foundations of sustainable industrial development and economic growth. Economic growth in Algeria remains largely dependent on public spending. The third objective was to develop production capacities through public investments, particularly those related to industry and agriculture.

The PIP was adopted by the Council of Ministers on May 24, 2010. It provided for a total investment of US$286 billion dollars. The investment broke down into US$130 billion for the completion of major projects and US$156 billion for new projects.

The PIP focused on six main areas: human development, the construction of basic infrastructure, the improvement of public services, economic development, the fight against unemployment, and scientific research and new communication technologies.

The key infrastructure projects under the PIP included a major expansion of the rail network, the completion of the East–West motorway, and an additional 1,486 km of road. These projects benefited all the economic sectors, including the hydrocarbon sector. Nonetheless, the focus on human development and the fight against unemployment are components that could have meaningfully linked the oil and gas (O&G) industry to these initiatives.

For example, the national education sector received US$11 billion to build 3,000 primary schools, 1,000 colleges, 850 secondary schools, and 2,000 residential schools and cafeterias. If curriculum changes and new courses related to the oil and gas industry had been required, or at least encouraged, such major public spending could have positively impacted the O&G industry.

Similarly, the higher education sector was also allocated around US$11 billion to provide 322,000 new university teaching posts, 161,000 student accommodation places, and 22 campus restaurants. The education and professional training sector also received approximately US$2.3 billion to set up 27 vocational education institutes, 104 vocational training and apprenticeship centers, 58 residential schools, 134 extensions, 21 semi-residential schools, 39 libraries, and 1,900 training centers. Funding included support for 160,000 training positions for 221 institutes for more specialized vocational training.

Had there existed "joined-up thinking" between certain ministries, e.g., Ministère de l'Energi (Energy Ministry),[8] Ministère de l'Education Nationale (Education Ministry), and the Ministère du Travail, de l'Emploi et de la Sécurité Sociale (Ministry of Work, Employment and Social Security), then at least part of the funding under the PIP could have been earmarked specifically for the O&G industry. Funds could have been directed toward setting up apprenticeship programs for students to gain industry engineering skills, such as those of the United Kingdom,[9] Norway,[10] and Denmark.[11]

Through the PIP, the industry and energy sector received US$29 billion. Of this, US$10.2 billion was for the development of the chemical industry. US$3.8 billion was for the building of new power plants, to bring total capacity to more than 4,000 MW, and US$13 billion was designated to relaunch and modernize public industrial enterprises. Finally, US$2 billion was allocated to support small- and medium-sized enterprises (SMEs), i.e., 200,000 SMEs providing productive employment opportunities. Again, a more focused approach to the oil and gas industry, Algeria's major export industry, could have led to significant enhancement of needed workforce skills.

The 2010–2014 PIP also included US$3.25 billion for the scientific research and information technologies sector.[12] This was aimed at the development of a knowledge-based economy through support for scientific research and the widespread provision of education and use of IT throughout the national education system and the public sector. Again a more nuanced approach to the oil and gas industry could have delivered both jobs and particular skills for the oil and gas sector.

Industrial Knowledge

Overview of Algerian industry

According to the Oxford Business Group,[13] after Algeria's independence in 1962, the new government adopted a state-led strategy aimed at industrial diversification and self-sufficiency. Foreign imports were restricted, and the state channeled oil revenues into heavy industry.

By the late 1970s, large state-owned enterprises dominated most commercial and industrial sectors, including construction materials, machinery, and textiles. Due largely to the inefficiency of these state enterprises, in the 1980s policymakers attempted to recalibrate the industrial strategy through more privatization.

However, due to the lack of private-sector funding, privatization did not work, and a state-owned industrialized strategy has been in place since the early 2000s. Progress has been slow, evidenced by the fact that although industry contributed to 18% of GDP in the 1980s, it was only 5% in 2013.

Thus, although today the oil and gas industry contributes 97% of export receipts and roughly one-third of GDP, it employs only a fraction of the workforce. This is largely because, unlike countries such as Norway, there are insignificant legal requirements mandating the employment of Algerians or the use of Algerian-made equipment in the oil and gas industry.

In terms of industrial diversification, fundamental factors work in Algeria's favor, including a large and well-educated workforce, expanding transport network, and, crucially, a cheap energy supply. One obvious area that would further strengthen Algeria's manufacturing industry would be to focus on producing oil and gas equipment.

Refining and petrochemicals

As the Oxford Business Group further noted, given the economy's reliance on oil and gas, downstream processing activities are a cornerstone of local value-added activity. The government is seeking to boost such activity through the construction of five new refineries in the period 2013–2017. The facilities will have a combined fuel processing capacity of 30 million tonnes a year, which would more than double Algeria's current capacity of 22 million tonnes. Raising the country's refinery threshold will help further reduce import spending; Algeria imported AD 344.5 billion (€3.2 billion) in energy products and lubricants in 2013, according to the national Customs Directorate.

Fertial, a joint venture between Spain's Grupo Villar Mir (66%) and the Algerian firm Asmidal (34%), also produces phosphate fertilizers for the domestic market. Algeria used an average of 12.7 kg of fertilizer per ha of arable land in 2009–2013, well below Morocco (39.1 kg) and Tunisia (40.4 kg). The government subsidizes 20% of the cost of fertilizers, and demand is expected to grow as Algeria works to increase agricultural output.

Steel

Again, according to the Oxford Business Group, Algeria's reindustrialization strategy and the availability of relatively cheap energy mean that the country has considerable scope to scale up production of steel to meet demand from its ambitious building programs in public housing, roads, and railway networks, and of course components such as steel tubulars and casing for its oil and gas industry. Algeria imported €1.37 billion in steel products in 2013, down 11% from €1.54 billion in 2012, according to the Ministry of Finance, but steady demand in the coming years will weigh on import spending.

Although currently the El Hadjar plant in Annaba dominates steel production with a nameplate capacity of 2 million tonnes per year, several new plants are in the pipeline, including a €551.3 million steel plant developed in partnership with the Turkish group Tosyali Holding, adding 1.25 million tonnes of annual capacity. A second facility is under construction, a joint venture between SIDER (51%) and Qatar Steel International and Qatar Mining (49%), with an initial overall annual capacity of 2 million tonnes, including steel rebar and wiring rod.

Evolution of the Algerian Petroleum Industry

Early history

The existence of oil had been known in Algeria since the early Amazigh (Berber) kingdom of antiquity.[14] It was not until World War I, however, that the strategic importance of oil was recognized. In 1917, for example, three concessions near Ain Zeft and Tliouanet in Algeria were awarded to British companies, as part of the British government's strategic search for oil. The French colonizers became serious about Algerian oil, however, relatively late. The Bureau de Recherches de Pétrole (BRP) was not established until 1945, with 100% ownership by the French government. The entity was charged with the "establishment of a national programme for oil exploration in Metropolitan France and in Algeria, the countries of the Protectorate, the country under Mandate and in French Colonies."[15]

In 1946 BRP established, jointly with the French colonial government in Algeria, an oil exploration company, the Société Nationale de Recherche de Pétrole en Algérie (SN-Repal). As a direct result, a huge oil field was eventually discovered at Hassi-Messaoud in June 1956, and an equally impressive gas field was discovered at Hassi R'Mel in November 1956.

It is worth noting that it was French planners, geologists, and engineers who made these discoveries and carried out their commercial exploitation. According to Mohamed Sassi,

> Efforts to put technical teaching in place, initiated during the interwar period, were accelerated after the end of the Second World War. France had a large training center (the first of its kind) beginning in 1950. It trained oil industry specialists under the direction of the Institut Français du Pétrole (IFP), which contained an application school, the École Nationale Supérieure du Pétrole et des Moteurs (a higher national school for oil and engines). It trained engineers in four specialized fields: a) geological and geophysical prospecting; b) drilling and exploitation of oil layers; c) refining, chemical engineering, and scientific research; and d) engines and applications to oil derivatives.[16]

The Évian Accords of 1962 between France and Algeria transferred political independence to Algeria, but also attempted to safeguard existing French and other foreign rights. On paper, at least, "in controlling most of the companies operating in the country...the French government had actually maintained its pre-independence advantages intact."[17] In the years that followed independence, the Algerian government nationalized foreign oil and gas interests, notably those of the French in 1971.

Algerian oil and gas reserves and production

Algerian oil and gas reserves and production can be traced from 1984 to 2014 in the timelines in figures 1–1 and 1–2. Both petroleum and natural gas proved reserves show a dramatic increase beginning in the mid-1990s. While production in natural gas saw a similar increase in the 1990s (fig. 1–3), the increase in petroleum production was more incremental, peaking in late 2008 (fig. 1–4).

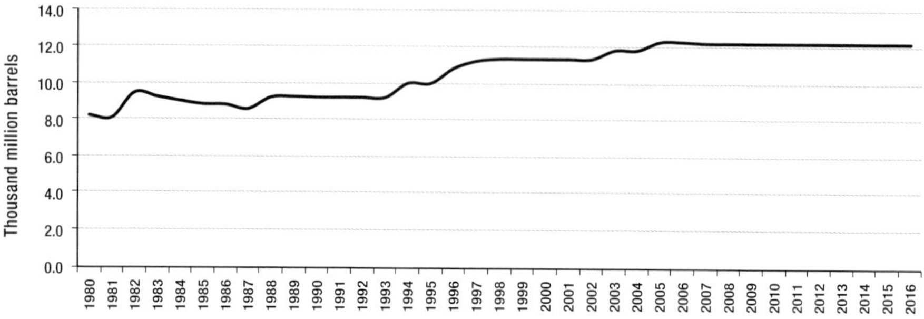

Fig. 1–1. Algerian oil reserves, 1980–2016
Source: BP Statistical Review of World Energy 2015.

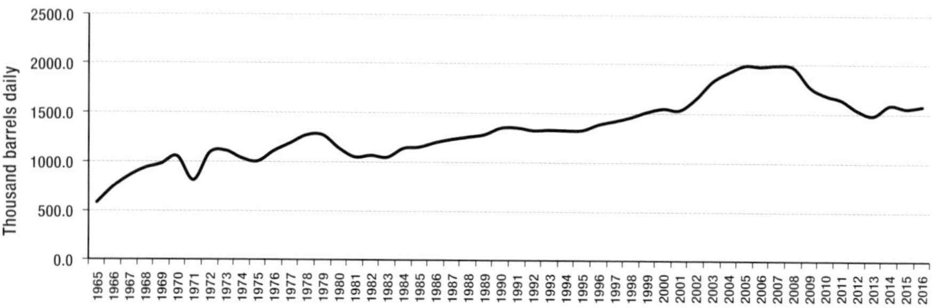

Fig. 1–2. Algerian gas reserves, 1965–2016
Source: BP Statistical Review of World Energy 2015.

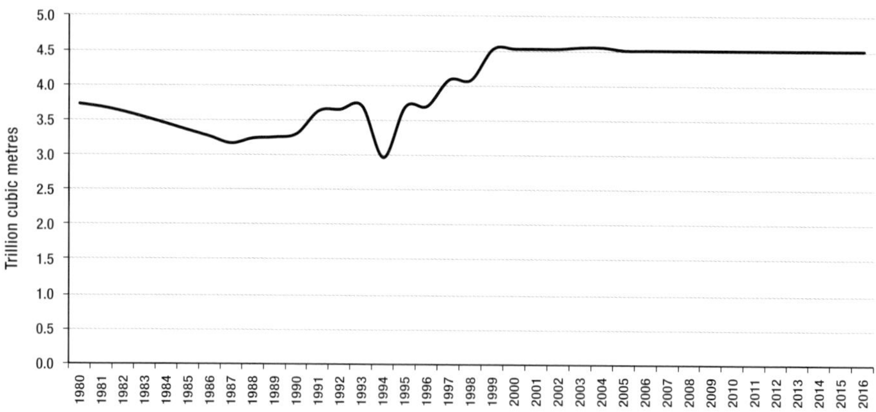

Fig. 1–3. Algerian oil production, 1980–2016
Source: BP Statistical Review of World Energy 2015.

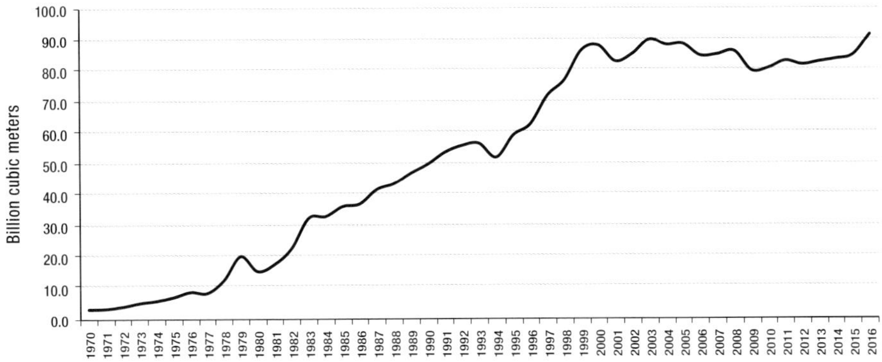

Fig. 1–4. Algerian gas production, 1970–2016
Source: BP Statistical Review of World Energy 2015.

Emergence of Sonatrach

What eventually became the Algerian national oil company (NOC), Sonatrach, was established in 1963, and its role at that time was limited to building a third export pipeline from Hassi Messaoud to the Arzew oil terminal on the Mediterranean.

After independence, Sonatrach played a crucial role in discussions that led to the Franco-Algerian Oil Agreement of 1965. This provided for the Algerian state to participate effectively in the exploration and exploitation of the country's hydrocarbons, while also raising the income tax on oil from 50% to 55%. As a result, Algeria gained a substantial increase in revenues from its O&G resources. In recognition of the part played by representatives of Sonatrach in these negotiations, the company's role was extended by a new Algerian decree in September 1966 to cover all aspects of the Algerian oil industry. On February 24, 1971, a series of ordinances and decrees transferred all nationalized oil and gas assets to Sonatrach, the national company established and owned 100% by the government of Algeria.

Key Players

Sonatrach

As stated on its website,[18] human resources have a key position in Sonatrach's strategy to elaborate on and implement its objectives.

Sonatrach states that it prioritizes HR development, training, and improving qualifications. Nearly 55% of Sonatrach's employees participated in training in 2009, a 7% increase from 2008. At that time, Sonatrach's training budget was nearly US$10 million dollars, an increase of more than 20% from the previous year.

Training objectives include the following:

- Develop managerial competence.
- Prepare takeover for key positions.
- Train specialists in the basic professions (drilling, reservoir, and engineering).
- Embed a health, safety, and environment (HSE) management culture and develop competence in industrial security.
- Professionalize executives in the human resources, finance, juridical, and audit professions.
- Develop specialization and expertise in the fields of finance, oil economy, oil taxation, and project management.
- Generalize the orientation training for all newly hired executives.

The meet these objectives, courses and specializations for employees are offered by three Sonatrach training institutions: the IAP, the CPE, and Naftogaz. These institutions have been consolidated and efforts are being made to bring them up to international standards.

The Company Improvement Center (CPE)

The Centre de Perfectionnement de l'Entreprise (CPE, or Company Improvement Center) is involved in the following fields: management technologies and techniques, languages, training improvement, finance and legal, computer and information systems, and HSE.

Algerian Petroleum Institute (IAP)

The IAP, which is based in Boumerdès, is jointly owned by Sonatrach, 82%; Statoil Hydro, 10%; Naftec, 4%; and Naftal, 4%; it provides operational training to an international standard in the energy and mining fields. These fields are primarily in hydrocarbon upstream, hydrocarbon transportation, hydrocarbon downstream, oil economy, management, HSE, environment engineering, finance, human resources, and company communication.

IAP has three branches based in the following locations:

- Boumerdès, for training of specialized engineers and masters of science in the scientific and technical fields of the oil and gas, mining, and electricity sectors.
- Skikda and Oran, for training focused on specialized technicians and engineers in oil techniques.

Naftogaz Training Institute

The Center for Oil and Gas Techniques Development (Naftogaz) is based in Hassi Messaoud. Its main mission and activities are as follows:

- Training specialized technicians and operational agents
- Upgrading and betterment of the company's executives
- Consultation and applied research

Legislative Hydrocarbon Regime Concerning Local Content

Legal framework for the O&G industry

Exploration for oil in the northern part of the country in the early stages in French colonial Algeria was originally governed by the French Mining Law of 1810, made applicable to Algeria in 1851.[19]

Since that time, the legal architecture for the institutional organization and exploration and production (E&P) aspects of the oil and gas industry evolved through different well-defined phases.

Ordinance of November 1958. The Ordinance of November 1958 (the Saharan Code) originally defined the framework within which foreign companies could carry out activities in Algeria in the field of liquid hydrocarbons E&P. It was not without criticism; many claimed that it was too favorable to foreign international oil companies (IOCs).[20]

Ordinance 71-22 of 1971. With the emergence of Sonatrach as custodian of all nationalized oil and gas assets, Ordinance 71-22 of 1971 redefined the framework within which the activities of foreign companies were to be carried out in the area of exploration and development of liquid hydrocarbons.

At this point, the basic arrangement was a 51%/49% joint venture, in favor of Sonatrach.

Initially, Sonatrach had tried to operate independently, but this proved costly and did not yield the expected results. Due to the nature of the industry, it was virtually impossible for a national oil company to possess the human resource skills, equipment, technology, and finance to develop an effective oil and gas industry. Unfortunately, Sonatrach was trapped in this policy for most of the 1980s, and the return on exploration continued to decline. The result was a tendency for the company's strategic planners to take a pessimistic view about the prospects for hydrocarbon discoveries.[21]

Law 86-14 (1986 Hydrocarbons Law). On August 19, 1986, Law 86-14 reorganized oil activities through the introduction of new legal instruments such as production sharing agreements (PSAs), services agreements, and various forms of gas partnerships. The 1986 Hydrocarbons Law confirmed the monopoly of Sonatrach, the national oil company, over hydrocarbons exploration, exploitation, and transportation activities. Thus IOCs could not carry out hydrocarbon exploration and exploitation activities except through a partnership with Sonatrach, in which Sonatrach automatically held 51%. Foreign workers and contractors could be hired if Algerian entities or personnel were not available.

Law 05-07. With Law 05-07 of April 28, 2005, Algeria embarked on a radical reformation of its hydrocarbons legislation.

The principle of sovereignty over national oilfield resources is stated in Article 17 of the 1996 Algerian Constitution as follows: "Public property belongs to the national collectivity. It includes the underground, the mines and quarries, the natural resources of energy, the mineral, natural and living resources in the various national maritime areas, waters and forests."[22]

This is reiterated in Article 3 of Law 05-07 concerning hydrocarbons, which states that all hydrocarbon resources "whether discovered or not, located in the soil and subsoil of the national territory and maritime spaces under Algeria's sovereignty belong to the national collectivity, of which the state is the product."[23]

Appraising the Algerian local content approach

For the purposes of this chapter, the broad definition of local content provided by Tordo et al. will be used: "Local content can refer to jobs or value-added that is created anywhere in the domestic economy as a result of the actions of an oil and gas company."[24]

Unemployment in Algeria stood at around 25% in 2014.[25] The IMF Country Report stated that to "increase employment, Algeria will need better-functioning labor markets, a workforce equipped with more relevant skills, and more effective labor market programs."[26]

This is despite the significant amount (US$4.7 billion) allocated under the 2010 to 2014 5-Year Plan to address unemployment, as discussed previously.

It is apparent that the Algerian government is unaware or unable to see the key linkages between the oil and gas industry and fuller employment. These linkages are forged not only by direct employment by operators, but by a flourishing service economy of manufacturers, freight companies, engineering concerns, and SMEs linked to the oil and gas industry.

This is borne out by an examination of the most recent major initiative in the regulation of the oil and gas industry, Law 05-07 of April 28, 2005. Law 05-07 was aimed primarily at the liberalization of the Algerian oil and gas industry by radically altering the regime formerly set out in the Hydrocarbons Law of 1986, in order to break the monopoly of the Algerian NOC, Sonatrach.

A detailed appraisal of the Law 05-07[27] reveals that the concept of enhancing local Algerian content in the oil and gas sector was very far from the minds of the Algerian law makers.

In a document that runs to 64 pages and 115 articles, only the following sections appear to relate to the issue of local content. According to KPMG Algeria,

> **Art. 1.** The purpose of this act is to define:
>
> - the legal framework of activities pertaining to the exploration, exploitation, pipeline transportation, refining and processing of hydrocarbons, to the marketing, storage and distribution of oil products, as well as the structures and facilities supporting these activities;
> - the institutional framework supporting the activities mentioned above;
> - the rights and obligations of the persons engaged in one or more of the activities mentioned above.[28]

KPMG Algeria translated Article 4 as follows:

> **Art. 4.** The activities covered by article 1 shall be one of the vehicles for the use and training of national human resources

> and, in this regard, shall benefit from incentives provided for in this act.[29]

When Articles 1 and 4 are examined in the context of local content, their intentions appear extremely vague. There is clearly an awareness of "the use and training of national human resources," but how the incentives, presumably provided from the financial benefits of Algerian E&P, are to be applied for this is not mentioned in detail.

KPMG Algeria translated Article 14 as follows:

> **Art. 14.** The National Agency for the Development of Hydrocarbon Resources (ALNAFT) shall namely be in charge of:...
>
> - helping promote the domestic industry,
> - encouraging research and development activities,
> - collaborating with the minister in charge of hydrocarbons with regard to the sectorial policy and developing regulations governing hydrocarbon activities.[30]

Again, these statements are extremely vague and not particularly helpful. "Helping promote the domestic industry" is indeed a desirable objective of the new policy, but it needs to be articulated much more precisely.

Article 80 was translated as follows:

> **Art. 80.** At the end of an exploration and/or exploitation contract, the transfer of ownership of all the structures necessary to keep conducting the activities shall be made to the benefit of the state. The National Agency for the Development of Hydrocarbon Resources (ALNAFT) shall notify the contracting entity regarding the list of facilities and structures that the state does not want to assume ownership of through a transfer of property at least three (3) years before the end of the exploration and/or exploitation contract.
>
> This transfer of ownership shall be executed free of charge to the state.
>
> At the time of the transfer, the structures to be transferred by the contracting entity must be operational and in good working order.
>
> For any structure that the state does not wish to assume ownership of through a transfer of property, the contracting entity must assume responsibility for all site abandonment and/or

> restoration costs provided for in the contract in accordance with the regulations pertaining to industrial and environmental safety.[31]

This is an interesting approach. It would essentially mean that Algerian managers, technocrats, and engineers would have access to equipment, assets, operating manuals, and protocols that could be used elsewhere in the industry, i.e., initiatives that could lead to a higher level of local content. The downside is, of course, that at cessation of production, these assets would have to be decommissioned by the Algerian state—a very expensive business.

KPMG Algeria offered the following translation of Article 89:

> **Art. 89.** The exploration and/or exploitation activities governed by this act shall be exempted from:
>
> (a) the value added tax (V.A.T.) on the goods and services related to exploration and/or exploitation activities,
>
> (b) the tax on professional activities (TAP),
>
> (c) customs fees, taxes and duties on the importation of capital goods, materials and products intended to be used exclusively for exploration and/or exploitation activities pertaining to hydrocarbon deposits,
>
> (d) any other fee or tax not covered by articles 31, 52, 53 and 67 above and in this chapter, impacting operating results and established for the benefit of the state, territorial collectivities and any public corporation.
>
> The capital goods, services, material and products covered by this article shall be the ones intended exclusively for these activities and appearing on a list established by regulation.[32]

By allowing the tax-free importation of capital goods used in the E&P process, Article 89 ensures that local companies will never develop the skills and technology to manufacture these goods in Algeria. If the policymakers wish to encourage local content in the oil and gas industry, a very high import tariff should be applied to oilfield equipment. This would encourage local manufacturers to meaningfully engage in import substitution.

KPMG Algeria offered the following translation for Article 97:

> **Art. 97.** The activities of hydrocarbon pipeline transportation, of gas liquefaction and of separation of liquefied petroleum gases, shall be exempted from:
>
> - the value added tax (V.A.T.) on the goods and services exclusively related to the activities listed above,
> - customs fees, taxes and duties on the importation of capital goods, material and products intended exclusively for the activities listed above.
>
> The capital goods, services, material and products covered by this article shall be those appearing on a list established by regulation.[33]

Again, this is a short-sighted policy, and its expected outcomes are similar to those resulting from Article 89 as discussed previously. In fact Algeria possesses significant indigenous expertise in gas extraction, pipeline construction, and liquefied natural gas (LNG), developed in the 1960s by Sonatrach. Sonatrach was originally named the Société Nationale de Transport et de Commercialisation des Hydrocarbures (the National Corporation for the Transport and Commercialization of Hydrocarbons). It was initially established to build an export pipeline to the Arzew oil terminal. In addition, Algeria was the first major exporter of LNG, produced from the 260 MMcfd Arzew GL4Z or Camel plant in 1964. This could have been used as a base for Algeria to develop significant exportable global expertise in LNG technology. As discussed previously, however, even though Sonatrach has developed three national training institutes, IAP, CPE, and Naftogaz, the provisions of Article 7 provide no encouragement for a local manufacturing industry to develop.

Positive and Negative Experiences, Issues, and Challenges

The dinosaur which is Sonatrach

By the time of Algerian independence in 1963, it was clear that Algeria possessed within its vast territory very significant hydrocarbon resources. However, the establishment and growth of the vertically integrated giant Sonatrach, which dominates most aspects of the Algerian economy, can be said to be negative for the development of the country.

Starting from the time of President Boumédiène in 1971, Algerian policy makers have repeatedly mandated the "use of hydrocarbons as a tool for industrialisation."[34] In today's world, however, this can be seen as a short-sighted policy. The internationalization of the hydrocarbons industry has in fact opened significant opportunities for Algeria and Sonatrach. If Sonatrach had enhanced its expertise and efficiency, it could have evolved into a significant and progressive international entity. Yet at present, the overall strategy and performance of Sonatrach's activities in the hydrocarbons industry remain remarkably unspectacular and traditional.

As stated by Aissaoui in 2001, "Sonatrach became embroiled in a web of political, economic and administrative relationships which weakened its corporate governance and undermined its incentive structure."[35] This is no less true today as Sonatrach and its senior executives are embroiled in corruption scandals of staggering proportions.[36]

Minister of Energy Chakib Khelil attempted through Law 05-07 of 2005 to liberalize the country's energy industry, making it more attractive to IOCs by limiting Sonatrach's share in domestic upstream ventures. His efforts essentially failed and eventually led to his forced resignation in 2010.

As for the fate of Law 05-07, it was enacted and then overturned a year later by Ordinance No. 2006-10 of July 29, 2006. Algeria continued to pursue a policy of resource nationalism.

The 2014 bidding round

Improved fiscal incentives were introduced in 2013, mainly related to shale gas, small fields, and deposits in underexplored areas, including offshore fields and fields with complex geology and/or those lacking infrastructure.[37] However, in spite of the incentives, the 2014 Algerian licensing round yielded disappointing results, with only 4 of the 31 available blocks awarded.

The outcome represents the lowest success rate of any of the four rounds undertaken following the major fiscal and regulatory changes of Algeria's 2005 Hydrocarbons Act, with just 13% of the blocks awarded.[38]

Clearly the mandatory requirement for Sonatrach to hold 51% in any PSC is seriously undermining IOC interest. Additionally, the corruption scandals that have rocked Sonatrach and its senior management have affected investor confidence.

Security issues

On January 16, 2013, Islamist militants stormed the In Amenas plant in southeast Algeria, located deep in the Sahara Desert and close to the Libyan border and operated by BP, Statoil, and Sonatrach. The attack left 40 people dead.

Production at the site did not resume until 19 months later. Statoil published an 88-page report reviewing the event,[39] concluding that the attack raised serious questions about internal security in Algeria, as well as company security protocols. There is little doubt that the In Amenas attack has had a serious impact on how IOCs view investment in Algeria.

Conclusion

In the heady days of the 1960s and 1970s, the new post-colonial Algerian governments and their successors saw oil and gas as the answer to all of Algeria's economic woes. Currently it is clear, however, that all is not well in the Algerian hydrocarbons industry. Resource nationalism and corruption, compounded by the monolithic and retrogressive policies of Sonatrach, have meant that the country's vast natural oil and gas resources have not translated into economic prosperity. Combined with ongoing chronic unemployment of 25%, particularly among the younger workforce, the situation is potentially dangerous.

Our appraisal of local content in the Algerian oil and gas industry is necessarily brief because of the almost total failure of the Algerian government to address this issue legislatively. It is clear that in a global context, Algeria still has a long journey ahead. It will be a difficult task for Algeria to significantly enhance the use of Algerian national labor, contractors, and equipment in its oil and gas sector. The lessons learned from the initiatives of countries such as Nigeria and Norway in legislating for enhancement of local content have had little effect on Algeria's local content policies for its oil and gas industry.

Notes

1. "Algeria," World Population Review, http://worldpopulationreview.com/countries/algeria-population/.
2. "The Lesson of Algeria," *Economist* (April 19, 2014), http://www.economist.com/news/leaders/21600984-even-if-arab-spring-has-sorely-disappointed-dictators-even-benevolent-ones-are-not.
3. "Bouteflika Buffeted: The Police State Is Not a Happy Place," *Economist* (August 13, 2015), http://www.economist.com/news/middle-east-and-africa/21660988-police-state-not-happy-place-bouteflika-buffeted?zid=304&ah=e5690753dc78ce919090830 42ad12e30.
4. International Monetary Fund, "Algeria 2014: Article IV Consultation—Staff Report; Press Release; and Statement by the Executive Director for Algeria," IMF Country Report No. 14/341 (December 2014), http:/www.imf.org/external/pubs/ft/scr/2014/cr14341.pdf.
5. Ibid.
6. *Algeria 2014: Article IV Consultation.*
7. "The 2010–2014 Public Investment Programme," Embassy of Algeria, London (2015), http://www.algerianembassy.org.uk/index.php/algerian-economy/99-the-2010-2014-public-investment-programme.html.
8. "Portail du Premier Ministère," République Algérienne Démocratique et Populaire (2016), http://www.premier-ministre.gov.dz/index.php?option=com_content&task=view&id=320&Itemid=177.
9. "Upstream Oil and Gas Technicians Training Scheme (UOGTTS)," Wood Group (2016), http://www.woodgroup.com/our-people/students-graduates/uk-apprenticeship-opportunities/Pages/default.aspx.
10. "Corporate Graduate Programme," Statoil (2016), http://www.statoil.com/en/Careers/Students/Graduateprogrammet/Pages/default2.aspx.
11. "MITAS Graduate Programme," Maersk Oil, http://www.maerskoil.com/career/pages/mitas-graduate-programme.aspx..
12. "The 2010–2014 Public Investment Programme," Embassy of Algeria, London.
13. "The Country Seeks to Return to Its Industrial Past," in *The Report: Algeria 2014* (Oxford: Oxford Business Group, 2015), https://www.oxfordbusinessgroup.com/overview/country-seeks-return-its-industrial-past.
14. Ali Aissaoui, *Algeria: The Political Economy of Oil and Gas* (Oxford: Oxford Institute for Energy Studies, 2001).
15. Aissaoui, *Algeria*, 47.
16. Mohamed Sassi, "Compagnie Française des Pétroles and Its Contribution to the Re-establishment of France's Position among the Oil Countries after the Second World War," *Business and Economic History On-Line* (2006), http://www.thebhc.org/sites/default/files/sassi.pdf.
17. Aissaoui, *Algeria*, 66.

18. "Our Human Resources," Sonatrach, http://www.sonatrach.dz/.
19. Aissaoui, *Algeria*, 56.
20. *A Comparative History of National Oil Companies*, ed. Alain Beltran (Brussels: PIE-Peter Lang, 2010).
21. Aissaoui, *Algeria*, 95.
22. "Constitution of the People's Democratic Republic of Algeria, 1989 (amended by the constitutional revision of 1996)," http://confinder.richmond.edu/admin/docs/local_algeria.pdf.
23. *Algeria Hydrocarbon Guide 2007* (Algiers: KPMG Algeria, July 1, 2007), 79, http://www.algeria.kpmg.com/fr/Documents/Guide%20Hydrocarbures%202007en.pdf.
24. Silvana Tordo, Michael Warner, Osmel E. Manzano, and Yahya Anouti, *Local Content Policies in the Oil and Gas Sector* (Washington, DC: International Bank for Reconstruction and Development, World Bank, 2013), 3, doi:10.1596/978-0-8213-9931-6, or http://documents.worldbank.org/curated/en/549241468326687019/pdf/789940REVISED000Box377371B00PUBLIC0.pdf.
25. *Algeria 2014: Article IV Consultation*.
26. Ibid., 1.
27. In the official Gazette of the People's Republic of Algeria, Law 05-07 is published only in French and Arabic. The analysis is based on an unofficial English translation of the law provided by the firm KPMG Algeria SPA in Algeria, *Algeria Hydrocarbon Guide 2007* (Algiers: KPMG Algeria, July 1, 2007), http://www.algeria.kpmg.com/fr/Documents/Guide%20Hydrocarbures%202007en.pdf.
28. Ibid., 79.
29. Ibid.
30. Ibid., 93–94.
31. Ibid., 117.
32. Ibid., 124.
33. Ibid., 127.
34. Aissaoui, *Algeria*, 90.
35. Ibid., 202.
36. Gaia Pianigiani and Stanley Reed, "Eni Scrambles to Contain Damage From Inquiry," *New York Times* (February 8, 2013), http://www.nytimes.com/2013/02/09/business/global/eni-scrambles-to-contain-damage-from-inquiry.html?_r=0.
37. Carole Nakhle, "Algeria's Shale Gas Experiment," Carnegie Middle East Center (April 23, 2015), http://carnegie-mec.org/publications/?fa=59851.
38. Robert Stevens, "Algeria 2014 Bidding Round: Fiscal Changes Fail to Attract Foreign Investment," *GlobalData* (October 30, 2014), http://energy.globaldata.com/resources/expert-insights/oil-and-gas/algeria-2014-bidding-round-fiscal-changes-fail-to-attract-foreign-investment.
39. "The In Amenas Attack," Statoil (2013), http://www.statoil.com/en/NewsAndMedia/News/2013/Downloads/In%20Amenas%20report.pdf.

2

ANGOLA

Patrícia Rosário

Introduction

Angola has been, since the early days of independence, an active sponsor of national human capital[1] and domestic production, with particular focus on the energy sector.

Local content policies, or "Angolanization" policies, have been constantly implemented since the end of the civil war in 2002 as a result of the ongoing dialogue between oil companies and the government. As such, these policies are today seen as a product of collaboration rather than unilateral imposition of compliance by the host nation toward international oil companies.

Governmental measures reflecting the need for domestic capacity development within the energy sector are engraved in most contractual private instruments, such as production sharing agreements, *travaux préparatoires*, parliamentary legislative activity, and in relevant diplomas.

For example, amendments made in August 2015 to the Private Investment Law[2] affected foreign investment in Angola within sectors including electricity and water, hospitality and tourism, transport and logistics, construction, telecommunications and information technology, and media. Foreign investment is now subject to mandatory partnership with Angolan citizens, state-owned companies, or private Angolan companies, in which those Angolan entities should have a capital participation share of no less than 35%. They also must have an effective role in management as provided for in the relevant shareholders' agreements.

These mandatory local content requirements, directed essentially at promoting sustained and inclusive economic growth, coupled with the existing legal framework in force since 2003,[3] are (according to empirical data available) aimed at contributing to the improvement of domestic production and employment. According to the *African Economic Outlook*,

> The government has established an institutional framework for entrepreneurship development and job creation through the Ministry of Employment and Social Security. The ministry's data indicates that professional training capacity has increased significantly and now stands at 159,000 trainees in more than 460 centres. In 2013, the vocational training programme generated 110 000 new jobs, notably in the energy (39,000), transport (29,000), and construction (24,000) sectors. The government currently allocates one-third of the budget to social areas to reduce poverty and reach the most vulnerable people.[4]

Local Content Policy and Baseline

Policy background, including short- and long-term goals

It is a national imperative to diversify the national economy, which has been anchored for the last 30 years solely to natural resources.[5] In order to do so, the Angolan government has been heavily investing in its local communities by encouraging them to participate, notably in the industrial and agriculture sectors.[6] Reduction of bureaucratic procedures and costs and improved access to physical infrastructure and finance have been governmental priorities. Special tax and financing measures have been in place since 2003[7] in order to encourage social and economic development in other relevant sectors of the national economy.

Instituto de Fomento Empresarial (Angola Investment Program) and Instituto Nacional de Apoio às Micro, Pequenas e Médias Empresas (INAPEM, or Institute to Support Micro, Small and Medium Companies) were incorporated in 2012. With an initial budget of US$700 million, they were created to foster local entrepreneurship, to legalize informal sectors, and to reduce poverty and unemployment rates.[8] Two laws were enacted to foster these goals: Law No. 30/11 of September 13, 2011, the Micro, Small and Medium Companies' Law, and Decree-Law No. 43/12 of March 13, 2012, Regulation of Micro, Small and Medium Companies' Law. Provisions include subsidized credit, preferential treatment while contracting with state entities, and fiscal incentives that are available to certified micro, small, and medium companies. In 2014 INAPEM certified 1,730 micro companies, 561 small companies, and 418 medium companies.[9] Other programs to facilitate national goals include the Priority Clusters Creation Program, the Program to Support Large Companies and Their Inclusion in Business Clusters (PAGEC), the Program for Companies' Relocation to Angola, and last, but not least, the Angola Invest Program. Since its inception in 2012, the Angola Invest Program has offered favorable funding conditions to micro, small, and medium companies. The Angola Invest Program alone has already approved 397 projects totaling AKz72,762,000, and budgeted AKz47,609,000 to sponsor 254 projects. According to Minister of Economy Abraão Gourgel, as cited by *Bloomberg*,

> Angola, Africa's largest crude oil producer after Nigeria, plans to use loan guarantees and investment to create thousands of companies and jobs, Minister of the Economy Abrahao Gourgel said. Part of the $435 million plan called Angola Invests is to start or help 9,000 companies and generate 300,000 jobs to cut unemployment from a government-estimated 20 percent, Gourgel said.... "We want to substitute national products for imports, promote exports and support entrepreneurship.... We want to improve the employment rate, diversify the economy and increase national production."[10]

In addition, the EMRC-Rabobank Project Incubator Award/Agribusiness Forum was launched in 2008 to encourage innovation and entrepreneurship in Africa among small and medium companies incorporated in Africa.

In line with the measures mentioned above, Presidential Decree No. 40/12 of March 13, 2012 established the "One Stop Shop for Entrepreneurs," or "Guiché Único da Empresa." This decree regulates all aspects of company formation, including incorporation, registration, and licensing. Recent data published by the Guiché Único da Empresa reveal that since its enactment in 2012, 19,789 companies have been incorporated. This total is comprised of 1,796 single-member companies, 16,701 private limited companies, and 1,292 public limited companies.[11]

According to the director of Guiché Único da Empresa, Pedro Filipe, as reported by *Bloomberg*,

> The government's one-stop shop to set up a local business, called Guiché Único da Empresa, is expanding across the country and reducing bureaucracy of approvals from different

> agencies. Retail and import-export companies are the most popular types.[12]

As a long-term goal, the government of Angola has put education,[13] research, and development at the top of its priorities.[14] According to data provided by the *African Economic Outlook 2014*, Angolan public spending prioritizes investment in education[15] (6%) and health (4%), infrastructure rehabilitation, and stimulus to agriculture production (19.5%).

International oil companies, with the close support of the government, have intensely contributed to the inclusion of local communities in their business plans for the benefit and development of local resources since the early days of independence. By the end of the 1970s, Angolan professionals and students were studying in Italy and Algeria via scholarships to encourage academic excellence and support local skills development.[16] There are now opportunities for students to integrate with advanced teams of industry professionals. Mainstream knowledge is also accessible through the sponsorship of the Angola Partnership Initiative and the Tech Incubator, in partnership with Chevron Angola and BP. The BP Angola Sustainable Development and Community Investment Program focuses on

> education, enterprise development, and Social Inclusion. The company is working closely with the Agostinho Neto University in Luanda to help increase the number of qualified future professionals that graduate each year, so as to meet both the development needs of the oil sector and of the country as a whole. This is achieved by investing in laboratories and staff capabilities to enable the university to serve the community and prepare its students for the market. In addition to this, BP is funding a fully-credited postgraduate programme leading to the award of a Master's of Law Degree (LLM) in Oil & Gas. The programme is a partnership with the Faculty of Law of the Agostinho Neto University. It was launched in April 2007 and produced over 100 graduates from the programme to date. The programme aims to developing a local network of Petroleum law expertise in Angola. The education support programme includes improving access and study conditions of children in various schools across the country as the demand for education far outweighs the supply in Angola's predominantly young population. BP also supports initiatives that incentivize youth in the areas of math & sciences such as the sponsorship of Angolan innovators that initiated in 2013.[17]

The National Program for Development 2013–2017 promotes development of the national private sector and increase in exports. In line with these goals, Sonangol, the national concessionaire, is greatly contributing to the stability and the sustainability of the oil wealth intergenerational transfer through the Strategic Oil Reserve Base Infrastructure and the Sovereign Wealth Fund of Angola.[18]

Overview of key prevailing local conditions

In recent years, Angola has emerged as Africa's second largest oil producer after Nigeria. This achievement was mainly driven by a strong public spending in the industry, notably by developing an adequate legal, regulatory, and business environment, and in particular by creating the best conditions to enhance the participation of local communities. Today the industry generates little direct local employment, with "only about 1% of Angolan workers...in the oil industry."[19] This is particularly evident during development and production phases, where the skills required are very specific and the technologies involved are not easily available in host countries. Nevertheless, Angola is committed to empowering local communities to alter this scenario.

Along with this commitment, Angola is also funding core infrastructure improvements, such as rail transport,[20] power generation,[21] water,[22] and information and communication technology.[23] According to the World Bank,

> Angola already spends approximately $4.3 billion per year on infrastructure, equivalent to 14 percent of its GDP. Spending is calculated to be double the real estimated needs; the rationale of such high investment lies in the intention of the government to pursue an accelerated program of infrastructure reconstruction and hence is spending a massive $2.9 billion a year in that sector alone. As a result, Angola's capital spending on infrastructure is heavily skewed toward transport, which accounts for some 70 percent of the total. In contrast to many of its peers, and reflecting buoyant petroleum revenues, Angola's infrastructure investment is predominantly funded by domestic fiscal resources. China is by far the most significant source of external finance, with only modest private capital inflows and negligible overseas direct investment.[24]

In 2011, the government of Angola established the Fundo Soberano de Angola, the first Angolan Sovereign Wealth Fund. Backed by petroleum revenues, it was launched with an initial capital of US$5 billion. The fund is governed by the following legislation: Executive Decree No. 190/95 of September 8, 1995; Law No. 15/10 of July 14, 2010; Executive Decree No. 28/11 of February 24, 2011; Presidential Decree No. 48/11 of March 9, 2011; Law No. 13/12 of May 2, 2012; Law No. 2/13 of March 7, 2013; Presidential Decree No. 89/13 of June 19, 2013; Presidential Decree No. 92/13 of June 21, 2013; Presidential Decree No. 107/13 of June 28, 2013; and Presidential Decree No. 108/13 of June 28, 2013.

The fund provides investments in the country's social welfare, in social and economic development and capital preservation for its people, in the domestic market, and in several categories of international assets. According to the fund's website,

> As stated in the Fund's investment policy, over the medium to long-term, the FSDEA intends to allocate approximately one third of its initial endowment to interest bearing assets such as sovereign and corporate bonds, mortgage backed, high yield and emerging debt as well as high quality currencies. The other third will be allocated to various investment vehicles and sectors in mainly in emerging and frontier markets. This will also include alternative investments, predominantly in sectors such as agriculture, mining, infrastructure and real estate—particularly hospitality—in Angola and across other African markets to nurture sustainable domestic and regional growth.[25]

Key Players

Host government and its agencies

The Angolan Constitution of 2010 and the Law on Petroleum Activities[26] (PAL) enshrine the fundamental principle of state ownership of petroleum resources.[27] They also confirm and maintain the regime of a sole concessionaire and mandatory association with it for petroleum operations, first established in September 1978 by the Petroleum Law, Law No. 13/78. This legislative matrix envisages not only safeguarding national interests but also endorsing economic and social policy principles in the sector. This primarily concerns the promotion of the development of the employment market, the valorization of mineral resources, the protection of the environment, the rational usage of petroleum resources, and the increase of the country's competitiveness on the international market.

Ministry of Petroleum. The Ministry of Petroleum is the public authority responsible for regulating and supervising the oil sector in Angola. Some of the ministry's notable tasks include formulating the general basis of the national oil policy, proposing the development plan in accordance with the national strategic plan, and ensuring supervision and control over its implementation. The Ministry of Petroleum works with Sonangol as coregulators of the hydrocarbons industry, as discussed further in the "Sonangol" section.

Proposed National Oil Agency. According to Patrícia Rosário and Paolo Esposito, "In order to implement the likelihood of impartiality of the national concessionaire, there are plans underway to create a National Oil Agency in Angola to act as independent regulator and concessionaire separate from the national concessionaire. Nonetheless, and as it will pointed out infra, such intention has not been taken any further."[28]

The Law on Petroleum Activities (PAL). PAL has, inter alia, implemented the following legislation: Law No. 26/12 of August 22, 2012 on transport and storage of crude oil and natural gas; Decree No. 1/09 of January 27, 2009, which approved the regulation on oil exploitation; Executive Decree No. 216/05 of July 10, 2005, which extended the period of the oil research for Block 2/05; Executive Decree No. 133/11 of September 13, 2011, on the subdivision of Maritime Zone of Angola (Cuanza River basin) in six oil blocks; Presidential Decree No. 58/11 of March 30, 2011, which approved the regulation on the oil information system; Executive Decree No. 83/13 of March 19, 2013, which divided the land areas within the Congo Basin in 10 blocks for petroleum exploitation, called "Blocks CON"; Executive Decree No. 89/14 of March 25, 2014, which authorized Alper Oil to give the total of its partnership for Block 9/09 to Sonangol; and Executive Decree No. 88/14 of March 24, 2014, which authorized Alper Oil to give the total of its partnership for Block 21/09 to Sonangol.

National oil company, Sonangol

Sonangol is the Angolan state-owned oil and gas company, national concessionaire, and shareholder in almost every project carried out in Angola. According to Rosário and Esposito:

> In 1976 the Sociedade Nacional de Combustíveis de Angola ("Sonangol") was established as national public company and sole concessionaire for the exploration of oil and gas on the subsoil and continental shelf of Angola, where they operate with foreign companies primarily through joint ventures and

production sharing agreements, as majority shareholder. Today, Sonangol is not only the dominant institution in Angola's petroleum sector: the company made high-profile investment throughout the globe, including Brazil, Iran and Iraq, for ensuring the development of the operational and production capability of its subsidiaries.[29]

As the national concessionaire, i.e., the national oil company, Sonangol plays a fundamental role as coregulator with the Ministry of Petroleum. Other than fulfilling commercial activities (e.g., managing support services for petroleum operations and exports of crude oil), Sonangol's tasks include overseeing and supervising oil operations conducted by foreign oil companies. They also include submitting recommendations to the Ministry of Petroleum about new areas that should be opened up for prospection. Most importantly, it is the institutional body in charge of public procurement and negotiations concerning concession contracts. In sum, it is the national concessionaire's responsibility to regulate the industry and to ensure that the requirements of the Angolan Petroleum Law are adhered to.

Sonangol's corporate management is shared between three key bodies. The board of directors (conselho de administração) is in charge of the corporate strategies of the national concessionaire and also is the main decision-making body. The other two branches of management are the supervisory board ("conselho fiscal") and the governing board ("conselho de direcção"), which is the advisory body to the board of directors of Sonangol.

Today Sonangol is divided into 18 subsidiaries, further divided into the following business segments: upstream, midstream, downstream, noncore (social business and human capital development), and corporate and finance.

Oil and gas companies operating in the country

There are several players operating in the Angolan petroleum industry today: Sonangol, Chevron, Total, Eni, Petrobras, Somoil, Galp Energia, Poliedro Oil, Kotoil, China Sonangol, Ajoco, Naftogaz, Statoil, ACREP, Prodoil, Plus Petrol, Force Petroleum, Cupet, Sonangol P&P, Teikoku, SOCO, Petropars, Ina-Nafta, VAALCO Energy, Falcon Oil, Initial Oil & Gas, Maersk Oil & Gas, Svenska Petroleum, Cobalt, Esso, BP, SSI, Odebrecht E&P, Partex, Geminas, Repsol, Statoil Hydro, ConocoPhillips, WRG, Ecopetrol, and Petronas.

Service providers operating in the country

Major service companies operating in the country include MOSCO, Baker Atlas Wireline, Baker, BJ Services, Expro Gulf, Geoservices, Halliburton, Schlumberger, S-DBS (Security), Kvaerner Oil & Gas, Saipem, Petromar, Seven Seas, SONAMET, Acergy (formerly Stolt Offshore), TechnipFMC, Angoflex, Cameron, Caterpillar, FMC/FKI, Vetco Gray, Weatherford, and several others.

Industry Knowledge

Research and training centers

National Oil Institute. In Angola, several oil companies provide the necessary practical training in support of courses available from universities in Angola. Operating alongside the international oil companies is the National Oil Institute (NOI),[30] a public educational institution located in Sumbe, in the Kwanza Sul Province. The NOI was incorporated in 1983 and operates under the supervision of the Ministry of Petroleum and the Ministry of Education. The NOI provides training courses in oil geology, mining, drilling and production, gas processing, refinery, oil instrumentation, oil installation mechanics, industrial electricity, production, and refinery operation. The NOI also promotes research on oil technology, along with conferences, seminars, study groups, and other initiatives to complement the courses and specializations.[31]

Staffed with 83 teaching professionals, the NOI provides the following medium-level or technical courses: oil drilling and production, mechanics, oil geology, mining, oil instrumentation, refining, and gas processing. Professional courses are also available and provided by the NOI: industrial electricity, mechanical maintenance, Industrial Instrumentation, production operator, refinery operator, electro-hydraulic and electro-pneumatic technician, refrigeration and air conditioning, laboratory analysis—petroleum chemistry, structures locksmith, coppersmith structures, piping locksmith, technical welding, mechatronic technician, CNC machinery operator, CNC machinery programmer, administration techniques, and safety and occupational hygiene.

Moreover, the NOI also provides know-how to workers at all levels in the oil sector and training in the context of the national economy at the request of national and international oil companies and service providers.

Such companies include Sonangol, Chevron, Esso, FMC Technologies, BP, Total, ExxonMobil, BHP, OXY, Schlumberger, Angola LNG, Oceaneering, OPS, and Solar Turbines.

The NOI also has available to its students various laboratories for experiments related to subjects such as electricity, electronics, hydraulics, pneumatics, instrumentation, rotating machinery, chemistry, geology, safety, physics, multimedia, English, and IT. The NOI makes available several classrooms and offers workshops and areas of simulation on topics such as engines, machining, mechanical locks, metal structures and pipes, and welding. Training is also available covering DCS (distributed control systems), and there is a security training center.

NOI has in place several protocols and cooperation agreements with RKK-Norway (Rogaland Training and Education Centre), IAP (Algerian Petroleum Institute), and the government of Mozambique.

In 2013, a training center for the oil industry, sponsored and funded by the Angolan government and Total, was established at the facilities of the NOI. Its purpose is to prepare "young people to enter the labour market and bring vocational training closer to the reality of businesses' needs."

Courses, degrees, and specializations in the oil and gas industry

Several courses, degrees, and specializations are offered in Angola from major local universities. Postgraduate courses in oil and gas law at Agostinho Neto University and the master's degree in economics with a specialization in energy resources at the Catholic University of Angola are two of the most reputable courses of study in Angola.

Under Law No. 13/04 of December 24, 2004 and Decree No. 116/08 of October 14, 2008, Levy for the Training of Angolan Personnel, international oil companies are required to contribute to training programs (the "training levy"). Specifically,

> A levy for the training of Angolan personnel (known as the training tax contribution) is imposed on the national concessionaire's associates, oil and gas exploration and production companies and on service companies that contract with these entities for more than one year. Angolan companies owned or controlled by Angolan nationals (that is, those whose capital is entirely or more than 50% owned by Angolan nationals) are not subject to this levy. Also excluded from this levy are (a) foreign companies supplying materials and equipment,

> (b) services providers and entities engaged in the construction of structures, which perform the entire or most of work outside Angola and (c) entities whose corporate activities are not strictly connected with the oil and gas industry.[32]

The Ministry of Petroleum of Angola ultimately administrates the Angolan Human Resources Development Fund. Decree-Law No. 17/09 of June 26, 2009 stipulates that the Angolan Human Resources Development Fund shall cover the following:

> (a) expenses of the training and specialization of Angolan personnel in the petroleum field;
>
> (b) subsidies to the National Institute of Petroleum and other educational establishments related directly or indirectly with the petroleum industry;
>
> (c) purchase of books, documentation and technical equipment related with training and specialization of Angolan personnel in the petroleum sector;
>
> (d) expenses of visits and traineeships at exploration production refining centers and other petroleum installations;
>
> (e) expenses resulting from participation in seminars or conferences related with the petroleum industry;
>
> (f) appropriations allocated to the Agostinho Neto University the Catholic University and the National Institute of Vocational Training;
>
> (g) appropriations allocated to the execution of development projects in higher education and vocational training according to criteria to be laid down;
>
> (h) any other expenses related with training Angolan personnel in the petroleum field not specified in the above sections.[33]

Despite what is stated above on the availability of higher level, specialized training available in Angola, the following has been stated concerning the general access to, and quality of, education in Angola:

> While secondary schools and higher education institutions exist, the quality of teaching is often low, fees can be high and some courses and degrees (such as vocational training or specialized engineering courses) have restricted entry. A result of this is that enterprises in the private sector often choose to train their staff themselves. This includes a variety of

managerial courses, language training and technical courses relevant for the oil industry. The majority of such training currently takes place abroad—in Portugal, Brazil, China, India, the UK, the United States, Canada and, increasingly, South Africa. Training is also provided by professionals, who are flown in for short periods. Although most government revenue is being generated by mining operations (oil and diamonds), education requirements are not limited to these sectors. There is also a government policy of "Angolanization" aimed at providing greater employment for Angolans. This has created intense competition among businesses to hire skilled Angolans. With Angola's rapidly growing economy requiring ever more quality skills, a number of education and skills providers are looking to stay in Angola for the longer term. These include Portuguese and South African education providers but also less accountable 'cowboy' firms.[34]

Legislative Hydrocarbon Regime Concerning Local Content

Regulatory bodies

The present organization of the Ministry of Petroleum is defined by Presidential Decree No. 103/14 of May 14, 2014. (Earlier decrees on the same subject, no longer in force, were Decree-Law No. 10/96 of October 18, 1996 and Presidential Decree No. 239/12 of December 4, 2012.) Under this decree, the Ministry of Petroleum is in charge of defining the national policy basis on oil and biofuel, licensing, and general supervision of the oil industry. This body is also responsible for the protection of the environment and pollution control.

The National Direction of Petroleum and Biofuel within the Ministry of Petroleum[35] is accountable for defining the national policy basis on oil and biofuel. Its role is also to elaborate the development program on oil and biofuel exploitation according to the National Plan, to submit legislation to the scrutiny of the public bodies, and to carry out research on hydrocarbons. Specifically, research is conducted relating to the production, processing, and trade of crude oil and natural gas, and the storage, transport, and supply of derivatives, including biofuel.

Relevant policies, legislative framework, and key provisions

The terms *local content* and *Angolanization* have no legal definition under Angolan law. Angolan local content has been understood to have "a dual nature, referring to both the 'Angolanization' of human resources and to the procurement of Angolan goods and services." Although there is no legal definition of what local content or Angolanization policy is, the necessity of "local involvement" is clearly stated in the PAL and in several other pieces of legislation. As to the first aspect, the law aims at imposing integration, recruitment, and training of Angolan workers, thus requiring that all oil companies operating within the national boundaries hire Angolans. With reference to the second aspect, all entities carrying out petroleum operations are requested by the law to acquire national materials, equipment, machinery, and consumer goods when they are of approximately the same quality. A further stipulation is that the national goods must be available for sale and delivery in due time and at prices that are not higher than 10% more than that of imported items, inclusive of transportation, insurance costs, and customs charges due. To that end, it is mandatory that Angolan companies be consulted on the same terms as those used for international companies. Moreover, there is a requirement that oil companies must hold competitive tenders to contract the supply of goods and the provision of services. Only in cases of urgency or unavailability of goods or services on the domestic market can oil companies resort to direct hiring from foreign companies. This local preference is further strengthened by the circumstance that any contract entered into in breach of the aforementioned provisions is "deemed to be null and void."

The Angolanization policy is clearly aimed at fostering the country's private sector and technocratic class by means of the oil sector. The "government seeks to use the oil industry to give Angolan business access to capital and to give both business and individuals opportunities to gain valuable experience working on technical and managerially challenging projects alongside capable international partners." Sonangol itself plays an important role in the process, through an enormous web of subsidiaries, joint ventures, and affiliated entities operating in several non–oil-related sectors. It also obligates "international oil consortia to accept minority participation from well-connected Angolan companies and steers contracts during operations to non-Sonangol Angolan companies. The promotion of the 'Angolan-factor' is further revealed by the fiscal incentives and privileges granted to

Angolan investors, as well as by the benefits offered for promoting Angolan entrepreneurship."[36]

The growing legislative proliferation on local content can be interpreted as a reaction from the Angolan government to the increasing participation of foreign players in the economic and social development of the country. However, this does not mean that both host nation and international oil companies are benefiting from it. Local content policies, in order to have the desired effect, should be adapted specifically to the Angolan industry, and this rarely happens. Instead, standard local content policies are introduced in national legislation and regulations. But one thing is certain: royalties and tight local content requirements are no longer sufficient or desirable instruments as the sole means to spread the benefits of the country's wealth. Today, there is a need for real commitment from both the host state and international oil companies. New avenues should therefore be explored through a collaborative approach in order to better align their collective interests.

At present, however, the legal and regulatory framework governing local content in Angola is fundamentally composed of the following pieces of legislation:

- The Law on the Promotion of the Angolan Private Business Community (Law No. 14/03 of July 18, 2003)
- The Local Content Decree, or the General Regulatory Framework for the Contracting of Goods and Services from National, State and/or Private Companies by Companies in the Oil Sector (Decree No. 127/03 of November 25, 2003)
- Petroleum Activities Law (Law No. 10/04 of November 12, 2004)
- The Law on the Petroleum Activities Tax (Law No. 13/04 of December 24, 2004 and Decree No. 116/08 of October 14, 2008)
- The Regulation on Bids for Oil Concessions and Tenders for Goods/Services Supply (Decree No. 48/06 of September 1, 2006)
- The Enterprises Promotion Fund Law (Decree No. 39/08 of June 23, 2008)
- Recruitment and Development of the Workforce in Petroleum Activities (Decree-Law No. 17/09 of June 26, 2009)
- The Foreign Exchange for the Petroleum Sector Law (Law No. 2/12 of January 13, 2012)

The law on the promotion of the Angolan private business community establishes the principles of preferential treatment and promotion of national private companies, initiatives, and national investments. Under this law the government offers special conditions for veterans, displaced persons, disabled demobilized soldiers, and families seriously affected by the war. This law is applicable to public promoters, private individuals, concessionaires, or beneficiaries of incentives for micro, small, medium, and large Angolan companies. "Angolan companies" as defined under the law, are sole traders or forms of companies legally and regularly established or incorporated in Angola that are wholly owned by Angolan citizens or when at least 51% of the share capital is owned by Angolan citizens or Angolan companies.

According to Article 22 of Law No. 14/03, the types of incentives and support offered to the local business community include the following:

(a) fiscal incentives;

(b) financial support;

(c) technical support;

(d) special rights and privileges and guarantees on assets;

(e) to support the creation of professional training centers led by economic or professional associations and their participation in national and international fairs and seminars.[37]

The general regulatory framework for the contracting of goods and services from national, state, and/or private companies by companies in the oil sector aims at establishing the basic rules for contracting with domestic companies for supplies of goods and services in the oil sector. It also guarantees the protection and establishment of a national business community and the use of national goods and services in activities that support or complement oil operations.

As already mentioned, Law No. 13/04 of December 24, 2004 and Decree No. 116/08 of October 14, 2008 stipulate that a "contribution" or levy, calculated as 0.5% of the gross revenue generated from any type of contract, is imposed on national concessionaire's associates, oil and gas exploration and production (E&P) companies, and service companies. Angolan companies owned or controlled (i.e., whose capital is entirely or more than 50% owned) by Angolan nationals are not subject to this levy. Also excluded are (a) foreign companies supplying materials and equipment, (b) services providers and entities engaged in the construction of structures, which perform all or most of the work outside of Angola, and (c) entities having corporate scope not strictly connected with the oil and gas industry.

The Enterprises Promotion Fund aims at sponsoring private entrepreneurship and creating an environment suitable for its development.

Legislation on the recruitment and development of the workforce in petroleum activities sets the amount of the contributions payable by all foreign-registered companies. This is also payable to Angolan-registered companies held by foreign persons or entities that carry out within the national territory activities of petroleum prospecting, exploration, appraisal, and development and production. It also applies to companies engaging in oil refining, production and storage, transport, distribution, and commercial sale of petroleum products. This law includes all foreign-registered companies and Angolan-registered companies held by foreign individuals or entities that provide services to the entities referred to previously on a permanent basis.

Law No. 2/12, Foreign Exchange for the Petroleum Sector, states, among other obligations, that international oil companies are required to use local banks for transactions.

> Under the actual regime, oil & gas companies are now required to transfer into Angola the funds necessary to pay for all petroleum operations and processed through local banks and in local currency. The new FX regime–that will have a massive impact on the international companies operating in the sector concerned, and that will certainly depend on the ability of the Angolan banks to ensure available means and resources consistent with international banking standards–obviously aims at favouring the development of the banking and financial sectors in Angola in the medium and long term. The FX regime will test the maturity of Angolan banking system and assess its capacity to effectively manage complex transactions.[38]

Relevant contracts

Under the currently published Angolan Production Sharing Agreement template, the contracting group, through the operator, and its subcontractors are obligated to contract local enterprises under certain conditions. This obligation exists as long as local services are similar in quality and availability to those available on the international market, and the prices of their services, when subject to the same tax charges, are not more than 10% higher than those charged by foreign contractors for identical services. Moreover, the contractor group should acquire materials, equipment, machinery, and consumable goods of national production, insofar as their quantity, quality, and delivery dates are similar to those of such materials, equipment, machinery, and consumable goods available on the international market. However, such obligation does not apply in circumstances where the local prices for such goods are more than 10% higher than imported goods, before charging customs duties but after the respective costs for transportation and insurance have been included.

Positive and Negative Experiences and Expectations

Local content policies are aimed at generating inclusive economic growth, increasing employment opportunities among local communities, reducing poverty, and eventually easing external assistance. Transition economies, such as Angola, are using foreign investment as a tool to achieve these development goals.[39] Knowledge sharing and technology transfer, along with foreign investment, are at the forefront of the country's mindset. Thus Angola has enacted over the last decades a solid legal framework capable of regulating and adapting to the most cumbersome demands of the industry. Hence, it is expected that the government will continually adapt legislation to changing scenarios to adequately accommodate foreign capital and expertise.

With the aim of generating further benefits to the national economy, it is expected that if the industry requires a more sophisticated technology, perhaps as yet undeveloped, the country will inevitably open up its doors to foreign investors. There would be the possibility of reducing the level of local participation in exchange for the introduction of new specialized skills and resources within the local socioeconomic landscape.

Once local beneficiation and socioeducational growth are achieved, strict local content requirements will likely no longer be needed, and the changed scenario will inevitably require a reshaping of the concept of local content.

Current Issues and Challenges

For some actors and participants in the industry, local content is perhaps the only subject matter that still requires improvement and innovation. The requirement to involve national labor and production should be an ultimate concern of all the players in the energy sector, and achieving a balanced compromise should at all times remain the goal.

On one hand, it is a perfectly understandable that host states sometimes impose local content measures as part of their energy policy. They do so in an attempt to endorse the national development goals of the industry and avoid an incoming stream of foreign element that will further reduce opportunities for local young people.[40] On the other hand, IOCs tend to oppose strict imposition of local content policies and the corresponding costs of complying with them.

How might a desirable balance be reached? Perhaps the desirable balance should be reached through compromise and on a case-by-case analysis, in order to adjust local content requirements to each particular reality. This would be preferable to general and abstract impositions applicable to the energy sector as a whole. For example, in 2010 Sonangol awarded new oil concessions in ultradeep waters in the Kwanza Basin. In 2013 Sonangol publicly announced that it would drill almost half of its offshore wells to test the Atlantic Mirror theory. According to this theory, the geology of both South America and West Africa ultradeep waters are analogous. Therefore, it is expected that presalt oil and gas deposits may be discovered off the Angolan coast, as similar deposits were discovered in the Santos Basin in Brazil.

Upon the discovery of the ultradeep presalt layers, Brazil announced that new, specialized technologies and know-how,[41] as yet unknown and undeveloped, would be required in order to maximize the presalt potential. These presalt fields are located in what is considered by many authors to be the "hottest basin in the world."[42] Faced with the possibility of sharing with Brazil this geologic particularity of presalt oil fields located in ultradeep waters, Angola would benefit in having immediate unconventional technology available. These fields will require massive investments over the next decades for exploration, production, and infrastructure. Thus, in order to attract sophisticated candidates to explore the potential, the author is of the opinion that the government of Angola should lower its local content requirements for this particular project. The development of this geological area should not be hindered by existing local content requirements. Such requirements were conceived primarily to encourage production of the readily available technologies commonly used by IOCs and shared by the host country.

Conclusion

Long-term, effective local content policies have the potential to stimulate inclusive economic and social development and, ultimately, prosperity.

Thus, Angola should endorse the ongoing dialogue with foreign industry players, stakeholders, international institutions, and civil society to include local content policies in the context of a global strategy for the industry. Angola should also carefully assess the benefits and drawbacks of its local content policies currently in force. The government should evaluate whether, for instance, local content measures inhibit sectors from becoming more competitive with foreign providers. It should consider whether there is a decline of investment channeled to research and development and whether Angola is incurring additional costs of specialized (foreign) inputs to the industry. Balanced measures and long-term compromise are keys to achieving the desirable progress of the industry and consequently the productive structure of the economy. Without neglecting national capacity and in order to maximize the potential of the national resources, the government should endorse the employment of mainstream technology and labor. In this way Angola can achieve its goal of attaining positive results in current and new fields, including the newly discovered presalt fields.

Notes

1. According to *African Oil and Gas Law and Regulation*,

   Aiming at promoting Angolan ownership and local involvement, employment and investment in the industry, foreign companies operating in Angola are requested to employ local work-force (with a minimum of at least 70% Angolan national employees), to source locally for goods and services and to submit to the Ministry of Petroleum an 'Angolanization plan' detailing methods and targets to promote local employment. Angolanization or local content initiatives are embedded in several legal statutes such as the Petroleum activities law and the Angolan private entrepreneurship law where preference is given to local products and services suppliers. The priority to local content is e.g. revealed by the Law on the petroleum activities tax, which permits deduction from taxation of several expenditures incurred for or related to the training of Angolan employees, similar provisions highlighting the preference for and governmental intention to boost the local activities can be found in the foreign exchange regime applicable to the oil and gas sector, obliging, amongst others, oil companies to trade on the Angolan banking system to process, for example, salaries.

   Patrícia Rosário and Paolo Esposito, "Angola," in *African Oil and Gas Law and Regulation: A Practical Guide*, eds. Eduardo Pereira and Kim Talus (Globe Business Publishing, 2015), 54.

2. Law No. 14 of August 11, 2015.
3. Order ("Despacho") No. 127/03 of November 25, 2003, "General Regulatory Framework for the Contracting of Goods and Services from National, State and/or Private Companies by Companies in the Oil Sector"; Law No. 14/03 of July 18, 2003, "Law on the Promotion of the Angolan Private Business Community"; Law No. 10/04 of November 12, 2004, "Petroleum Activities Law"; Law No. 13/04 of December 24,

2004, "Law on the Petroleum Activities Tax"; Law No. 11/04 of November 12, 2004, "Customs Regime Applicable to the Petroleum Sector"; Decree No. 20/07 of May 27, 1982, "Mandatory Hiring and Training of Angolans by Foreign Companies Operating in the Angolan Oil Industry"; Law Decree No. 17/09 of June 26, 2009, "Recruitment and Development of the Workforce in Petroleum Activities"; Executive Decree No. 14/10 of February 11, 2010, "Operations of the Fund to Train and Develop Angolan Personnel in the Petroleum Industry"; Decree No. 48/06 of September 1, 2006, "Regulation on Bids for Oil Concessions and Tenders for Goods/Services Supply"; and Law No. 20/11 of May 20, 2011, "Private Investment Law."

4. *African Economic Outlook 2014* (Paris: AfDB, OECD Development Centre, UN Development Programme, 2014), www.un.org.

5. According to the IMF, Angola's exports remain heavily concentrated in a single commodity, crude oil, which in 2012 accounted for close to 97% of total exports. "Angola: Selected Issues Paper" (IMF Country Report No. 14/275, International Monetary Fund, September 2014).

6. Several NGOs, such as ADRA Angola (Acção para o Desenvolvimento Rural e Ambiente), COSPE (Cooperation for the Development of Emerging Countries), Development Workshop Angola, CODESPA, and ADPP, are also present in the area, acting mainly in the agriculture sector.

7. The approval of Law No. 14/03 of July 18, 2003 moved Angola further in this direction.

8. According to the UN Development Programme, the project

   seeks to promote the development of a diverse, robust, micro, small and medium enterprise sector in Angola. This will be achieved by supporting a process to build consensus around a common vision and a national strategy for micro and small business development. As a result of investments contributing to improve the enabling environment, expanding the supply of micro and small business credit, redirecting the offer of vocational training towards the market and introducing pilot models of business development service providers, employment will be generated and incomes raised, making an important contribution to the Government's Interim Poverty Reduction Strategy. It will also contribute to the establishment of a knowledge base on informal, micro and small business sector in Angola. The initial phase of this innovative Public-Private Partnership will be co-funded by UNDP (US$ 1 Million) and ChevronTexaco (US$ 3 Million) in the framework of the Angola Partnership Initiative formally launched by ChevronTexaco's Sustainable Development Company in November 2002. During the implementation of the programme UNDP will mobilize additional donors to support the programme.

   Angola Enterprise Programme—Support to the Development of the Micro Enterprise Sector in Angola, (Project Document ANG/03/011, UN Development Programme, January 2004), 1, accessible from http://www.aboutmicrofinance.com/latin-america-caribbean/africa-4.

9. Globally, Angola stands at 174 in the ranking of 189 economies on the ease of starting a business. *Doing Business 2015: Going Beyond Efficiency* (Washington, DC: World Bank, October 2014).

10. Colin McClelland, "Jobless Targeted in Angola Business Plan, Economy Minister Says," *Bloomberg.com* (May 15, 2013), http://www.bloomberg.com/news/articles/2013-05-15/jobless-targeted-in-angola-business-plan-economy-minister-says.

11. Data available at Guiché Único da Empresa, http://www.gue.minjus-ao.com.

12. Ibid.
13. In Angola, the academic year begins in February and ends in November, and the official primary school entrance age is six. The system is structured so that the primary school cycle lasts six years, lower secondary lasts three years, and upper secondary lasts three years. Angola has a total of 5,912,000 pupils enrolled in primary and secondary education. Of these pupils, about 5,027,000 (85%) are enrolled in primary education. Data available at www.epdc.org.
14. The 2015 state budget allocates 34% of the total expenditures to the social sector, in particular health, education, and social protection measures.
15. According to the United Nations, adult literacy rates reached 73% in 2013. Also in 2013, 636,000 students were enrolled in literacy programs and more than 5 million students were enrolled in primary education, with a gross schooling rate almost three times that of the year 2000. Data available at *African Economic Outlook 2014* (Paris: AfDB, OECD Development Centre, UN Development Programme, 2014), www.un.org.
16. For example, the PROQUADROS Project, conceived by Sonangol and structured by Braspetro, trained 100 specialists between 1982 and 1987. This project was developed under Decree No. 20/82 of April 17, 1982 (The Angolanization Decree) and put in place by ESPA in order to accommodate the Angolanization requirements.
17. Available at www.bp.com.
18. The Sovereign Wealth Fund of Angola was established by Presidential Decree No. 108 of June 7, 2012 and with an initial endowment of US$5 billion. The fund is

    [H]eadquartered in Luanda, Angola, the Fund currently manages a substantial investment portfolio. In accordance with Fundo Soberano de Angola's investment policy and guidelines its portfolio will gradually be diversified across a number of industries and asset classes including global private and public stocks; bonds; foreign currencies; financial derivatives; commodities; treasury bills; and real estate and infrastructure funds, in accordance with its investment policy and guidelines. By pursuing investments that generate long-term and sustainable financial returns, the Fundo Soberano de Angola will play an important role in promoting Angola's social and economic development through the generation of wealth for the people of Angola.

    "Fundo Soberano de Angola," http://www.fundosoberano.ao.
19. *African Economic Outlook 2014.*
20. According to a March 2011 AICD report:

    Angola has three operational railways in place. The railway systems once carried 9.3 million metric tonnes of freight to Angola's ports in Namibe, Luanda, Benguela, and Amboim, before the civil war but freight levels are currently low (World Bank 2005). Similar to its other transport infrastructure, Angola's railways suffered during the 27 years of war: at present, only 30 percent of the total rail network is operating (table 9). The railways are currently under rehabilitation with support from several parties, including Chinese and Indian entities.

    Nataliya Pushak and Vivien Foster, *Angola's Infrastructure: A Continental Perspective*, Africa Infrastructure Country Diagnostic (AICD), Country Report (Washington, DC: International Bank for Reconstruction and Development/World Bank, March 2011), 22, http://siteresources.worldbank.org/ANGOLAEXTN/Resources/AICD-Angola_Country_Report.pdf.
21. According to the AICD report (ibid., 5):

> Angola lacks anything that could be described as a national power grid, let alone regional interconnections. There are a number of isolated power systems, with minimal local transmission links, although a backbone is planned to link the main power assets in the north and south of the country. Angola's national fiber-optic backbone is already much further developed than its power grid, linking up all the major towns on the western side of the country. Hydraulic infrastructure and irrigation is conspicuously absent.

22. According to the AICD report (ibid., 29):

> Angola's urban population is growing at an increasingly fast rate (4.7 percent per year), and the current piped and stand-pipe water supply and management system cannot keep up with the demand for domestic water consumption. This gap is increasingly being filled by private water vendors selling largely untreated water. The supply of water from tankers rose from 10 percent of the population in 2001 to 37 percent in 2007 (figure 17). Indeed, in peri urban areas of Luanda, 70 percent of residents purchased their water from water vendors. This water is extremely expensive, with prices varying from $4/m3 close to the distributional area of the water tank to $20 in more distant areas (Cain, Daly, and Robson 2002), leading to a significant increase in the number of small-scale operators (figure 17). Most of the water comes from the ANGOMENHA filling station and is untreated, posing serious health risks.

23. According to the AICD report (ibid., 32):

> Access to ICT has improved markedly since the end of Angola's civil war in the early 2000s. Mobile penetration rose from less than 1 subscriber per 100 people in 2000 to almost 50 subscribers per 100 people in 2009. Fixed-telephone access has risen marginally, but this is offset to some extent by the development of wireless access.

24. Nataliya Pushak and Vivien Foster, "Angola's Infrastructure: A Continental Perspective," Working Paper 5813 (Washington, DC: World Bank, September 2011), doi: http://dx.doi.org/10.1596/1813-9450-5813.
25. "Fundo Soberano de Angola," http://www.fundosoberano.ao.
26. Law No. 10/04 of November 12, 2004.
27. Decree No. 70/05 of September 26, 2005.
28. Patrícia Rosário and Paolo Esposito, "Angola," in *African Upstream Oil and Gas: A Practical Guide to the Law and Regulation*, eds. Eduardo Pereira and Kim Talus (London: Globe Law and Business, 2015), 75.
29. Ibid., 50.
30. Joint Executive Decree No. 84/83 of September 15, 1983.
31. Additional information is available at "Institute Nacional de Petroleos," http://www.institutodepetroleos.org.
32. Patrícia Rosário and Paolo Esposito, "Angola," in *Upstream Law and Regulation: A Global Guide* (London: Globe Law and Business, April 2013).
33. "Council of Ministers Decree-Law 17/09 of 26 June" (June 12, 2009), http://www.sonangol.co.ao/Style%20Library/Pdf/licitacoes/Law+Decree+17-09-compulsory_en.pdf.
34. Emanuel Gomes and Markus Weimer, "Education in Angola: Partnership Opportunities for the UK," Africa Programme Paper AFP PP 2011/04 (Chatham House, May 2011).

35. Executive Decree No. 186/14 of June 23, 2014, which revoked the earlier decree on the same subject (Executive Decree No. 65/13 of March 15, 2013, which had already revoked Executive Decree No. 145/11 of September 22, 2011).

36. Rosário and Esposito, "Angola," 2015, 64.

37. Angola National Assembly, "National Assembly Law No. 14/03 of 18 July," Section II, Article 22, http://ccsi.columbia.edu/files/2014/03/Local-Content-Angola-petroleum-CCSI-July-2014-.pdf .

38. Rosário and Esposito, "Angola," 2015, 75.

39. Under Article 38 of Angola's Constitution, concerning the right to free economic initiative (available at "Angola's Constitution of 2010," *ConstituteProject.org*, https://www.constituteproject.org/constitution/Angola_2010.pdf?lang=en):

    (1) Private enterprise shall be freely undertaken and exercised with respect for the Constitution and the law.

    (2) Everyone shall have the right to engage in free business and cooperative initiatives, to be exercised under the terms of the law.

    (3) The law shall promote, regulate and protect the economic activities and investments of private, national or foreign individuals and corporate bodies in order to guarantee their contribution to the development of the country, defending the economic and technological emancipation of the Angolan people and the interests of workers.

    Under Article 90 of Angola's Constitution (ibid.):

    The state shall promote social development by:

    (a) Adopting criteria for the redistribution of wealth which prioritises citizens and, in particular, the more vulnerable and needy sectors of society;

    (b) Promoting social justice, as a duty of the state, through a fiscal policy which ensures justice, equity and solidarity in all areas of national life;

    (c) Encouraging, supporting and regulating private sector interventions associated with achieving social rights;

    (d) Removing economic, social and cultural obstacles to genuine equal opportunities for citizens.

40. However, recent data from Silvana Tordo, Michael Warner, Osmel E. Manzano, and Yahya Anouti, *Local Content Policies in the Oil and Gas Sector* (Washington, DC: International Bank for Reconstruction and Development, World Bank, 2013), xiii, doi: 10.1596/978-0-8213-9931-6, or http://documents.worldbank.org/curated/en/549241468326687019/pdf/789940REVISED000Box377371B00PUBLIC0.pdf) suggest that

    LCP design is not solely or principally guided by considerations of economic efficiency and the need to address market failures. On the contrary, it would seem that political imperatives are key drivers, and economic efficiency is often an afterthought. Indeed we were unable to find evidence that countries that had adopted LCPs had assessed the costs and benefits of alternative policy options. It would seem that countries, particularly new producers, with less developed economies and/or large resource bases, tend to adopt more assertive policies by imposing measures that limit investors' choices with respect to employment, purchasing strategies, localization, and the transfer of knowledge. This often translates into quantitative targets for local content, and the establishment of LCP statements and supporting legal and regulatory frameworks.

41. Petrobras plans to invest US$236.7 billion under its 2013–17 business plan. Data are available at www.petrobras.com.
42. "Africa Seen Mirroring Brazil in Atlantic Coast Drilling: Energy," Morgan Stanley, cited by *Bloomberg*, available by subscription at http://www.bloomberg.com/news/articles/2012-01-30/africa-seen-mirroring-brazil-in-atlantic-coast-drilling-energy.

3

AZERBAIJAN

Ummi Jalilova

Introduction

The history of oil, or "black gold," goes back centuries, and its connections with Azerbaijan are ancient. As far back as the 10th century, wells were dug to a depth of 12 meters, producing oil known as "white oil" due to its unique properties. Travelers at the time noted the oil's medicinal qualities, as well as its traditional heating uses.

Situated on the Silk Road, 16th-century Baku became an important transit port. Oil left the city bound for global markets and was transported by camel and mule to every corner of the country. Azerbaijan and the Caspian Sea came to the attention of global powers in the 18th century as a rich source of this black gold, and the country became an important consideration in the foreign policy of many of its neighbors.

In 1803, Haji Kasymbek Mansurbekov from Baku began offshore oil production in Bibi-Heybat Bay from two wells located 18 meters and 30 meters from the shore. These offshore wells were destroyed in 1825 by a violent storm in the Caspian Sea.

Oil production was already well underway in and around Baku in the 19th century, but only from wells dug manually to a very shallow depth. The first mechanically drilled "gusher" in Azerbaijan was struck in Bibi-Heybat, under the direction of Vasiliy Semyonov, head of Transcaucasia.[1] Accounts vary concerning whether this event occurred in 1846,[2] 1847,[3] or 1848.[4] At just 21 meters, this well was not particularly deep. According to one source, "The years of 1847–1848 were characterized by the first production of industrial oil from the dug wells in Bibieybat and later Balakhany fields."[5] The development of Azerbaijan's oil industry had begun.

Not surprisingly, Azerbaijan was also the location for early efforts at oil refining. According to Mike Wysatta in *Reservoir Solutions*, "Baku also achieved firsts in the downstream industry."[6] Most sources agree that the first oil refinery in Azerbaijan was constructed in Baku in 1859.[7] Some sources credit the Canadians for the earliest efforts to refine kerosene. However, its earliest distillation seems to have occurred in Baku. Wysatta states,

> Canadian geologist Abraham Gesner is usually credited as the inventor of kerosene. He first distilled it in 1846. However, in 1834, Nikolay Voskoboynikov, director of the Baku oil fields, invented a special distilling machine that produced kerosene from black and white oil. (He originated the idea to drill, not dig, the 1846 Bibi-Heybat well.)[8]

The development of well drilling technologies led to the discovery of a number of oil wells. As a result, production of oil also increased, along with the development of oil infrastructure and oil refining. Hundreds of companies were founded to engage in oil production, refining, and sales. Azerbaijan became one of the industrial centers of the world.

According to "History of Development of Oil Industry,"

> At that period the state unused lands were commissioned for 24 years for the exploration of oil and the development of new oil fields. The leaseholders had [the] right to export oil they produced and to fix its price. Their net incomes made up 14–15% of the oil sold. The net capital invested into the oil industry accounted for only 4% of the whole investments in the 1870s. The volume of mixed capital with participation of national capital made up nearly 10%. In the late 19th century, 49 (24.8%) out of 167 entrepreneurs operating in the oil sector were Azerbaijanis.... The first stock oil company...was established in Baku in 1874. Swedish Nobel brothers arrived in Baku in 1873 and witnessed the economic growth connected with oil. At that period they owned a number of oil fields, oil refineries, oil tankers in the Caspian Sea, barges, railroads, hotels, etc.[9]

In 1877, Ludvig Nobel built a pipeline to transport oil directly from the production site to the harbor for shipment. In the same year, with design help from Robert, Ludvig's brother, the world's first oil tanker was built in

Sweden. The oil tanker, the *Zoroaster*, was launched in 1878.[10] In 1879, the brothers turned Robert's distillery company, founded in 1876, into Branobel, the world's first oil company with shareholders.[11] Oil refineries followed.

In 1899 to 1901, "in terms of absolute volume of extracted oil, the Baku oil industry was ranked first in the world (producing 11.5 million tons per year, compared to the US, which was producing 9.1 million tons per year)."[12] As a result, the "Nobels and Rothchilds became billionaires. The Rockefellers made a fortune in the Baku fields after the turn of the century."[13]

According to "Interesting Facts from Azerbaijan Oil and Gas History," on "November 13, 1920 the Azerbaijan Polytechnic Institute (the current Azerbaijan State Oil Academy) was established as...Europe and Asia's first higher education facility producing engineers for all branches of the oil industry."[14]

The Caspian's oil played a key role in World War II, with Baku producing 70% of all the Soviet Union's oil. In 1941, "[f]or the first time in the world, a 2000 meter deep inclined well was drilled and operated in Bayil by application of the turbine method invented by Aga Nematulla."[15] In Soviet times Azerbaijan's scientists made a large number of oil and gas discoveries, including the world-famous Oil Rocks, which gave a strong boost to oil production in Azerbaijan. In 1949 "the world's first oil and gas related scientific research and design institute, 'Denizneftqazlayihe' was opened in Baku to initiate designing of hydro-technical equipment for high sea conditions."[16]

Oil production in Azerbaijan exceeded 1 billion tons in March 1971. Later, the "Deep Water Jackets plant (the current Baku Deep Water Jackets plant named after Heydar Aliyev) was constructed and launched in Baku" in 1984.[17]

Independence marked the start of a golden age for Azerbaijan's economy. When the country faced a difficult situation following the breakup of the Soviet Union, President Heydar Aliyev of Azerbaijan took a number of momentous steps. In September 1994, the "Contract of the Century" was signed. This agreement between the State Oil Company of Azerbaijan Republic (SOCAR) and a consortium of leading international oil companies was signed for development of the offshore Azeri–Chirag–Gunashli fields. With the launch of this ambitious project, Azerbaijan became a symbol of global cooperation for countries and companies around the world. The contract was not only the first of its kind in the Caspian region but also was a clear example of successful cooperation between Azerbaijan and leading international investors, who brought with them the world's most advanced oil production technologies. Its signing set the course for large-scale development of the Caspian's resources and marked a new era in the petroleum history of Azerbaijan.

In the very same year, Azerbaijan's first exhibition event dedicated to the oil and gas industry, Caspian Oil & Gas, was organized on the initiative of Heydar Aliyev. It continues to be a key event in the oil and gas industry, making important contributions to the region's fuel and energy sector.

On June 4, 1996, "[BP], Statoil, LUKAgip, Elf Aquitaine (France) (the current TOTAL), OIEC (Iran) (the current NICO), TPAO and SOCAR signed in Baku a Joint Development Agreement for the Shahdeniz gas condensate field.... First gas was produced from the Shahdeniz field in autumn [2006].... The Baku-Tbilisi-Erzurum South Caucasian Gas Pipeline was put to service."[18]

Azerbaijan is still carrying out large oil and gas projects. Production at the Shah Deniz field, the largest gas field discovered in the world in the past 20 years, has begun. A new Absheron gas field was discovered in 2011. Production plans have been developed and approved for the Umid, Zafar-Mashal, Nakhchivan, and Shafag-Asiman offshore fields. Azerbaijan is also involved in the Baku-Novorossiysk, Baku–Tbilisi–Ceyhan, and Baku–Tbilisi–Erzurum pipelines, as well as the TANAP pipeline (gas), for which an agreement was signed in 2012. Work has also begun on the major Oil & Gas Processing and Petrochemical Complex (OGPC) project not far from Baku.

The country's government is implementing an oil strategy centered on the interests of the country, with revenues from hydrocarbon sales and transportation being plowed into the nonhydrocarbon sector to make Azerbaijan's economy stable and sustainable. Visitors to Baku can easily see how much has been achieved in the last 20 years, and changes are planned for another 20 years. As time goes on, a brand new chapter is set to be written in the history of black gold in Azerbaijan.

According to "Oil and Gas in Azerbaijan,"

> Currently, Azerbaijan's main export routes are the Baku–Supsa pipeline ("western route"), which mainly carries AIOC's "early oil" from ACG to Georgia's Black Sea coast (Batumi) and the Baku–Tbilisi–Ceyhan pipeline ("BTC"), main outlet for the Azeri–Chirag–Gunashli (ACG) field in the Caspian. Oil products such as lubricants also are exported by rail in tank wagons to Georgia's Black Sea ports.[19]

New, extended PSAs on development of Azeri-Chirag-Guneshli block of oil fields was held in Baku on September 14, 2017.

The existing PSA signed in September 20, 1994, in Baku by and between the government of Azerbaijan and a consortium of 11 foreign oil companies was due to expire in 2024. By the reinstated PSA dated September 14, 2017, the period has been extended for an additional 25 years until the end of 2049. BP is the current operator of the field and will remain the operator under the extended ACG PSA. As part of the new PSA, SOCAR will increase its equity share from 11.65% to 25% and the participating interest of all partners will be revised.

Following completion of the contract, the new ACG PSA participating interests will become as follows: BP: 30.37%; AzACG (SOCAR): 5.00%; Chevron: 9.57%; INPEX: 9.31%; Statoil: 7.27%; ExxonMobil: 6.79%t; TP: 5.73%; ITOCHU 3.65%; and ONGC Videsh Limited (OVL): 2.31%.[20]

Currently, gas export projects are more popular and promising than oil. Thus expansion of the South Caucasus pipeline (SCP), which exports Shah Deniz gas from Azerbaijan to Georgia and Turkey, is expected.

Local Content Policy and Baseline

Since Azerbaijan gained its independence and the country entered into the so-called Contract of the Century, local human resources and industrial capacity have expanded. This is due in part to state programs to fund education abroad. The big oil companies such as SOCAR and BP have also played a role by providing financial support to their employees for studies at Western universities. The Employment Strategy of the Republic of Azerbaijan has been influential in encouraging improvements as well.

The government's goal is to continue developing the present human and industrial capital relating not only to the petroleum industry, but also to other nonhydrocarbon sectors of the economy.

The State Program on Implementation of the Employment Strategy of the Republic of Azerbaijan (2011–2015) included steps to increase competitiveness of locally produced goods in the international industry market. Both the program and the amendments made to the Law on Public Procurement in 2016 gave preference to local goods (services or supplies) at public procurement tenders.

The volume of oil and gas reserves is determined based on estimates of expected oil and gas reserves from the Azeri–Chirag–Gunashli, Shah Deniz, and other fields under agreements on production sharing among

SOCAR and international companies. It also includes other fields exploited by SOCAR and joint ventures established with its participation.

According to the Decree of the President No. 128, 2004:

> Oil and gas prices shall be forecast on the basis of a pessimistic assumption (on a price lower than the one in the baseline assumption) (baseline assumption: the oil and gas price forecast for the respective period).
>
> The sources of oil and gas revenues are as follows:
>
> (a) Azerbaijan's share of profit oil under production sharing agreements and other sources of income of the State Oil Fund of Azerbaijan Republic as envisaged by law;
>
> (b) profit tax paid by investors to the state budget under production sharing agreements;
>
> (c) the tax payments of the State Oil Company of the Azerbaijan Republic relating to oil and gas production and its other liabilities to the state budget;
>
> (d) tax payments made to the state budget by joint ventures established by the State Oil Company.[21]

Also according to the Decree of the President No. 128, 2004:

> The principles for long-term use of oil and gas revenues are as follows:
>
> - when forecasting the amount of long-term expenditures from oil and gas revenues, the "constant real expenditures" principle shall be used as a basis and annual limits shall be set for these expenditures that are to be made within the period covered by the strategy;
> - when incomes from oil and gas revenues peak, at least 25 percent of them shall be saved;
> - the regulations adopted for spending oil and gas revenues shall remain unchanged during the effective period of the long-term strategy on management of oil and gas revenues and the expenditure limits projected on the basis of the constant real expenditures principle shall be observed;
> - the volume of medium-term expenditures shall be determined based on the non-oil deficit (the difference between revenues and expenditures of the consolidated budget of the country, excluding the oil sector) and taking

account of the long-term expenditure limit. Sharp year to year fluctuations in expenditures are undesirable and the non-oil deficit may not be abruptly changed;

- investment expenditures shall be made in the framework of the medium-term State Investment Program that is drafted annually.[22]

The government believed that the large initial investments that would be necessary, along with growth of expenses at the beginning, would create conditions for development of nonhydrocarbon sectors of the economy. Development of other sectors of the national economy could then gradually ease the country's dependence on the income from oil and gas. It was expected that development of the nonhydrocarbon sector during the long-term period could reduce the need for external loans.

According to the Decree of the President No. 128:

> Macroeconomic stability shall be ensured and priority expenditures from oil revenues shall be coordinated based on the medium term policy.
>
> The size of the medium term real expenditures may be higher or lower than the established long-term expenditure limit. Although this will accelerate development of the non-oil sector and lead to increased investments in different periods, annual real expenditures shall be made taking account of the established non-oil deficit limit.
>
> The annual size of real expenditures and the non-oil deficit shall be established consistent with the long-term expenditure limit. When the size of the non-oil deficit is adjusted, concrete justification for such adjustments shall be provided.[23]

Further, according to Decree No. 128:

> The strategy of use of oil and gas revenues includes the following objectives, while aiming at retaining macroeconomic stability:
>
> - developing the non-oil sector, regions, SMEs;
> - large-scale development of infrastructure;
> - fulfillment of poverty reduction measures and the solution of other social problems;
> - stimulating the improvement of the intellectual, material, and technical base of the economy;

- development of "human capital" (training[,] including in foreign countries[,] highly qualified specialists and improving the professional capacity of employees);
- consolidating the defense capabilities of the country;
- executing projects relating to reconstruction activities in liberated territories and the return of internally displaced persons to their native lands.[24]

Industry Knowledge

Since its independence, economic growth and social development in the country have been closely connected with development of the oil sector. This has been true both in terms of the industry producing revenues and of the industry creating jobs and giving assignments to local manufacturers and service providers. The government has launched major public sector investment programs through financing of utilities, roads, and highways. The number of individuals living at or below the poverty level has dropped remarkably via increases in wages and social transfers. However, the nonhydrocarbon private sector has not reached the desired level of development, and less attention has been given to expenditure management.

According to Aitor Ciarreta and Shahriyar Nasirov,

> Current and potential energy pipelines are major economic, political and strategic components of Azerbaijan's foreign and transport policies. Most Azeri oil exports pass through the Baku–Tbilisi–Ceyhan (BTC) pipeline, with an estimated cost of $3.9 billion, which was designed to deliver up to one million barrels per day of crude oil, crossing 1,055 miles from Sangachal Terminal near Baku in Azerbaijan, via Georgia, to Ceyhan Terminal in Turkey. From there oil is shipped by tankers to European markets.... This giant project has further prospects: BTC can also expand export and transit opportunities for other countries such as Kazakhstan, Turkmenistan, and Russia....The rest of Azeri oil is carried through the Baku–Supsa and Baku–Novorossiysk pipelines.[25]

Ciarreta and Nasirov also note:

> The exploitation and development of the Shah Deniz gas field and other new gas fields, considered to be the largest natural gas discoveries in the last 20 years, have led Azerbaijan to

> export its natural gas reserves to Turkey and onward to Europe via the South Caucasus Pipeline (SCP). The SCP is the first pipeline to transmit Caspian oil to Europe, bypassing Russia. It is 680 km in length and has an annual capacity of 6.6 BCM crossing Azerbaijan and Georgia to end in Turkey.[26]

Agriculture is a meaningful component of Azerbaijan's nonhydrocarbon economy and has significant potential for boosting export revenues for the country. While the sector only accounts for 7% of GDP, it is still a key source of jobs and is a priority in the context of food security.[27]

Tourism has been developing rapidly since 2015. The number of tourists constantly grows, especially after the Ministry of Foreign Affairs simplified the visa process (i.e., introducing online visa applications, visas upon arrival, and so on).

The Baku-Tbilisi-Kars railway (BTK) was opened on October 30, 2017, in Baku. The railway's peak capacity will be 17 million tons of cargo per year. At an initial stage, this figure will be one million passengers and at least 5 million tons of cargo. Baku-Tbilisi-Kars railway will accelerate the access of central Asian countries (Turkmenistan, Kazakhstan, Uzbekistan, Tajikistan, Kyrgyzstan and Afghanistan) to the European market. Along with Azerbaijan, Georgia, and Turkey are also expected to receive significant revenues from transit transportation.

The healthcare system in the country is better than in the past, but it still leaves room for improvement. Reforms and government efforts have made healthcare services available to all people, regardless of income or whether the person is insured. However, the quality of services offered by most health clinics remains average or below average.

The education system of the country also has been undergoing reforms. Universities have switched to Western curricula and teaching methodologies. Joint degree programs with foreign universities, guest lecturers invited from different countries, skill-building short-term and long-term professional programs, and seminars and conferences have led to improvements in skill-based employment and training.

Azerbaijan faces big challenges on the road to development, including social and political challenges, slower-than-desired progress in improving institutional capacity, implementation of the existing portfolio, global economic uncertainties, and regional security issues. With better calibration to client demands and capacity, stronger emphasis on analytical services, faster implementation of the existing portfolio, and a cross-cutting filter of governance, Azerbaijan can mitigate these risks.[28]

Key Players

Azerbaijan and its agencies

The Ministry of Industry and Energy. According its website, the Ministry of Industry and Energy is the central executive body carrying out state policy and regulation in the industry and [in the] fuel and energy complex of the country. The Regulations of the Ministry were approved by presidential decree of the President of the Republic of Azerbaijan dated May 15, 2006.

> Activity of the ministry includes manufacturing of competitive industrial goods...[and] the efficient use of energy resources.[29]

The Ministry of Petroleum is also tasked with protecting Azerbaijan's interests related to the hydrocarbons industry, including field development, applicable programs, and appropriate development concepts.

According to the Law on Energy, the Ministry of Energy must supervise, and is vested with authority to issue special permits for, the exploration, exploitation, production, processing, storage, transportation, distribution, and use of energy materials and products, including oil and natural gas. Furthermore,

> The ministry prepares forecast[s] of separate energy production and consumption performance...[and coordinates] activities of other bodies of executive power in the relevant field, as well as coordinates the activities of state-owned enterprises operating in the relevant field, or entities which have a controlling part of the shares at state-ownership.
>
> The Ministry of Industry and Energy participates in the preparation of international agreements on cooperation in the area of industry and energy, ensures implementation of international agreements, joined by Republic of Azerbaijan within its power, and coordinates the work done in this area.[30]

State Oil Fund of the Azerbaijan Republic (SOFAZ). SOFAZ was created in 1999 to guarantee the saving and effective use of oil revenues. The fund is a legal entity with an independent administrative structure and reports directly to the president of Azerbaijan. Its main goal is to ensure collection and proper management of revenue flows from oil and gas production sharing agreements (PSAs).

According to revenue rules, all revenue flows from PSAs go directly to SOFAZ, with the exception of taxes paid by foreign oil companies and SOCAR, which go directly to the state budget. SOFAZ's expenditure policy does not assume any limit on expenditures from the oil fund. However, the articles of association of the fund state that withdrawals from the fund in one year cannot exceed inflows obtained in that year. The operations of the fund are under control of the president. Azerbaijan's Parliament has the power to approve or disapprove transfers from the fund to the state budget. Transfers from SOFAZ are determined through a discretionary process within the budget framework. The funds are actually spent mostly on covering the budget deficit, though there can also be special expenditures determined by presidential decree.

The State Oil Company of the Azerbaijan Republic (SOCAR). SOCAR was established by presidential decree on September 13, 1992. SOCAR is a 100% state-owned entity.

Heidi Kjaernet noted in the *Caucasus Political Digest*:

> Baku has not shown signs of moving towards resource nationalism, but Azerbaijan's national oil company SOCAR plays an important role in the country's petroleum sector. In addition to being partner to the international oil companies present in Azerbaijan, the company is an actor in policy formation. Close ties between the government and the company ensure that SOCAR, in addition to being one of the biggest taxpayers in Azerbaijan, also carries out political and social tasks for the government. At the same time, commercial tasks are intertwined with petroleum policy and regulation, and Azerbaijan's national petroleum sector seems to be moving in a more opaque direction.[31]

According to "State Oil Company of Azerbaijan: History of SOCAR,"

> [SOCAR] was established...[from the entities] "Azneft" State Concern and "Azerneftkimya" Production Unit in order to carry out use of oil resources of Azerbaijan Republic in [support of]...state policy, [to] improve management structure of [the] oil industry, [and] ensure development of [the] fuel-energy complex...[through] increasing the effectiveness of...oil producing and oil processing enterprises.[32]

Additionally, SOCAR is

> involved in exploring oil and gas fields, producing, processing, and transporting oil, gas, and gas condensate, marketing petroleum and petrochemical products in domestic and international markets, and supplying natural gas to industry and the public in Azerbaijan. Three production divisions, one oil refinery, and one gas processing plant, a deep water platform fabrication yard, two trusts, one institution, and 23 subdivisions are operating as corporate entities under SOCAR.
>
> Joint ventures (including ventures in Georgia and Turkey), consortia, and operating companies established with SOCAR's participation, are doing business in different parts of the petroleum industry. SOCAR has representative offices in Georgia, Turkey, Romania, Austria, Switzerland, Kazakhstan, Great Britain, Iran, Germany and Ukraine and trading companies in Switzerland, Singapore, Vietnam, Nigeria, and other countries.[33]

According to the "State Oil Company of Azerbaijan: Duties and Object of Activity of SOCAR," SOCAR seeks to assure the

> effective activity and development of [the] oil and gas industry as a sole complex, meeting the demands of consumers...[for] energy resources in a reliable manner.
>
> SOCAR...[implements] exploration, preparation, [and] exploitation of oil and wells within [the] territory of Azerbaijan Republic, including offshore and onshore areas, [and the] transportation, processing and sale of oil, gas, condensate, and [their] products..., meeting the consumer demands...[for] energy resources in a stable and constant manner, [completing] other works and services, and...[generating] revenue.
>
> To fulfill its major duties, [the] Company carries out [the] following types of activity in a manner set by legislation:
>
> - elaborate long-term forecasts on development of the sphere [oil and gas industry], [and develop] purposeful and complex scientific technical, economic and social programs;
> - increase effectiveness of production, carry out policy of saving energy and material resources, create effective

production-technological and economic ties among field enterprises;

- ensure mutual ties between [the] Company and state agencies with respective authorities in determining the future development of the sphere;
- develop production and social base of the sphere, modernize, enlarge, [and] reconstruct the enterprises of [the] Company…[to ensure use of] new technique[s] and technologies [and to] implement investment activity…[in compliance with] legislation;
- implement geological-exploration works…using modern building technologies and technical means, increase effectiveness of exploitation of wells, accelerate construction of oil and gas wells, as well…[as improve the] quality and speed of preparation of wells to exploitation by observing the [appropriate] legislation…;
- improve [domestic] methods and forms, [and] apply scientific achievements in production.[34]

SOCAR is also tasked with fostering domestic and foreign banking involvement to achieve various governmental program objectives. Furthermore, SOCAR is expected to create and support a personnel policy that prepares the domestic workforce for employment both domestically and abroad in the oil industry.

In addition, SOCAR is tasked with translating oil revenues to social improvements within Azerbaijan, as well as creating cooperative relationships with domestic and international business partners. SOCAR also oversees geological and geophysical work, as well as directing the development of new fields. Furthermore, SOCAR reviews annual plans of oil and gas companies and determines their annual budget funds. It arranges for the refining and sale of the produced hydrocarbons, and arranges and pays for the necessary transportation of the produced hydrocarbons. SOCAR also is tasked with restoration of the area after hydrocarbon production and for other environmental protection concerns. Finally, SOCAR is responsible for conducting foreign trade in accordance with Azerbaijan's current laws or treaties.

SOCAR contributes to the education of local specialists through scholarships (the Foreign Scholarship Program) to undertake bachelor's and master's degree programs at highly reputable universities abroad. SOCAR also provides stipends to students at local universities with the goal of educating professionals in fields of interest to SOCAR and hiring from the pool of well-educated candidates. Other opportunities offered by SOCAR include summer schools at SOCAR and summer internships.

SOCAR employs more than 70,000 people, primarily citizens of Azerbaijan.[35] SOCAR has many decades of exposure to international business and hiring. However, SOCAR can and does exert control over the number of foreign employees in the oil and gas sector. Without SOCAR's approval, the State Migration Service is reluctant to issue a work permit for a foreign employee.

Oil and gas companies operating in the country

According to "Azerbaijan: Oil Policy":

> On September 20, 1994 the first agreement was signed on the "Development and production sharing of Azeri, Chirag, and Guneshli (deepwater) fields in the Azerbaijan sector of the Caspian." The agreement, later called "Contract of the Century" paved the way for foreign companies for signing new contracts in Azerbaijan. 27 contracts ha[ve] been signed between SOCAR and foreign oil companies on the exploration, production of hydrocarbons and production sharing.[36]

The companies that were parties to the Contract of the Century formed a steering committee, an advisory council, and the Azerbaijan International Operating Company (AIOC), which is currently managed by BP. Azerbaijan International Operating Company is comprised of BP (UK), Chevron (ABS), SOCAR (Azerbaijan), Inpex (Japan), Statoil (Norway), ExxonMobil (US), TPAO (Turkey), Devon Energy (United States), Itochu (Japan), and Hess (United States).

Besides major oil companies, there are local companies that act as subcontractors. The oil companies' reputation and strength have impacted the subcontractors' businesses considerably. For instance, being a subcontractor of BP was and is reputable and profitable. Until a couple of years ago, the majority of the subcontractors were foreign legal entities with branches and representative offices in Azerbaijan.

Since late 2014 and early 2015, the intention of PSA parties to strengthen the local content has been accomplished on a practical level. This is, perhaps, partially due to the local currency devaluation and drop in oil price. Oil companies are making less profit than before 2015; consequently, they need to cut expenses. When selecting subcontractors through tenders, the cost now matters more than ever, and at this point the local entities are in a better position to be competitive. Of course, total elimination of foreign content (employees or subcontractors) in the oil industry will not be possible any time soon. Steps are being taken, however, to slowly replace foreign employees and businesses in the oil industry with local employees, companies, and infrastructure. Infrastructure investments in Azerbaijan are the large investments and associated activities indirectly impacting the communities or country. Such investments could include new hotels, refurbishment of old hotels, opening of new restaurants, and establishment of new airline connections between Baku and Europe. These activities generate more local employment, more disposable income within Azerbaijan, and more tax revenue for the government of Azerbaijan.

An example of the partnership between the state company and its international partners can be seen in the following news release:

> UK-Azerbaijan Oil & Gas Partnership Facilitation Program [was] launched today in Baku [on October 29, 2015]. UK Trade and Investment, the Ministry of Economy and Industry of Azerbaijan, SOCAR (the State Oil Company of Azerbaijan) and BP with an active support of the British and Azerbaijani Embassies confirmed their cooperation with the launch of the programme which will assist UK and Azerbaijan oil and gas supply and manufacturing companies to form mutually beneficial and sustainable partnerships. A delegation of 26 UK companies was introduced to over 30 companies from Azerbaijan, as a first phase of a program that will continue over the next 3 years. The program will initially focus on increasing local manufacturing capacity of components...with the aim of improving efficiency and reducing the costs of delivering Azerbaijan's oil and gas production.[37]

Legislative Hydrocarbon Regime Concerning Local Content

Azerbaijan's legislation provides for three non-PSA mechanisms for the oil industry:

- Joint ventures with local companies

- Concessions agreements
- Energy contracts

Legislation in Azerbaijan

The non-PSA regime is regulated by the Law on Energy. The law allocates the land for use and requires that appropriate permits be obtained from the Ministry of Energy. The law also regulates the tender or negotiation of energy contracts, requires independent environmental expert opinions, and stipulates that energy contracts must be registered with the Ministry of Energy. The PSA regime concerns delegation of authority from the president to SOCAR to negotiate and then later sign the contract reviewed by government bodies, with ratification of the signed contract by the Parliament.

Among issues related to the non-PSA regime is the Law on Energy's lack of clarity concerning whether it applies to crude oil or oil from bituminous minerals. Among the gaps is an absence of a mechanism for allocation of land for use. The Law on Energy refers to the Law on Subsoil Use and Land Code; however, those laws lack clear mechanisms for implementation. Other problems relate to licensing and permitting.

The law does not make clear when an entity needs to apply for the permit, i.e., before entering into an energy contract or after (Articles 10 and 13 of the Law on Energy). Also, according to the Law on Energy, the term of energy contracts is limited to the term of the permit. However, according to the Law "On licenses and permits," the term of licenses and permits is unlimited unless cancelled. In addition, there is no specific procedure for bidding in the tender under the Law on Energy.

In addition to these gaps and uncertainties, one of the legislative framework problems is that there is no law on hydrocarbon resources. Although the draft law on oil and gas was prepared and submitted to Azerbaijan's Parliament (Milli Majlis) in 2009, it has not been adopted yet. Also, the Law on Energy and Law on Subsoil Use and Land Code contain contradictory provisions.

Laws passed by Azerbaijan's Parliament. Laws passed by the Parliament include the following documents (not an exhaustive list):

- The Law on Application of Special Economic Regime to Export Oil and Gas Activity, February 2, 2009
- Law of the Azerbaijan Republic on Acceptance, Approval and Issuing Permits for the Implementation of the Agreement on Rehabilitation, Exploration, Development and Production Sharing for the Block,

including Zykh and Hovsan Oil Fields in the Azerbaijan Republic between the State Oil Company of Azerbaijan Republic and Russneft Absheron Investments Limited and SOCAR Affiliated Oil Company, dated March 6, 2007

- The Law of the Republic of Azerbaijan about Adoption and Approval, Giving Permission for the Implementation of the Agreement between State Oil Company, Rafi Oil Efzei and Affiliated Oil Company SOCAR about the Restoration, Development and Production Sharing on the Block, Which Includes the Oil Field Surakhani in the Republic of Azerbaijan, dated September 20, 2005
- Law of the Republic of Azerbaijan on Approval of the Framework Agreement on Organizational Principles for the Establishment of Interstate Oil and Gas Transportation Systems, October 24, 2000
- The Law on Energy, dated February 6, 1999
- The Law on Subsoil Use, dated April 30, 1998

Non-PSA regulations also include the Law on Protection of Foreign Investments (1992) and the Law on Use of Energy Resources (1998).

Presidential decrees. A number of decrees, including those promoting local content, have been passed since Azerbaijan has gained independence, including the following:

- Presidential Decree on the Application of the Law of Azerbaijan Republic, On Application of Special Economic Regime to Export Oil and Gas Activities, April 14, 2009
- On Some Issues Relating to the Payment of Wages of Employees of the State Oil Fund of the Azerbaijan Republic, October 7, 2008
- Presidential Decree on Approval of the Rules of Preparation and Execution of the Annual Program of Revenues and Expenditures (Budget) of the State Oil Fund of the Azerbaijan Republic, September 12, 2001

Cabinet of Ministers' decrees. The Cabinet of Ministers has also issued decrees relating to oil and gas resources, including the following:

- Resolution No. 174, On the Approval of Certain Legal Documents Relating to the Application of Special Economic Regime to Export Oil and Gas Activities, November 6, 2009
- Resolution No. 1, On the Change of the Level of Existing Excise Taxes on Petroleum Products, January 6, 2007

- Resolution No. 167, "Security Zones" Establishing Rules on Exploitation of Oil Export Pipelines, November 2, 2004
- Resolution No. 267, On the Transportation of Early Oil That Will Be Produced as a Result of the Joint Development of "Azeri," "Chirag" and the Deep Part of the Field "Gunashli" in the Azerbaijani Sector of the Caspian Sea, December 30, 1995
- About the Software and Responsibilities of the Government of the Azerbaijan Republic under the Agreement dated on September 20, 1994, On the Joint Development and Production Sharing for the Azeri and Chirag Fields and the Deep Water Portion of the Gunashli Field in the Azerbaijan Sector of the Caspian Sea, Resolution No. 385, November 11, 1994

International treaties. Among international treaties entered into and conventions ratified by the Republic of Azerbaijan, production sharing agreements attract attention due to their provisions for giving preference to local employees, strengthening and developing the infrastructure, and training personnel. One such agreement was the Agreement on the Joint Development and Production Sharing for the Azeri and Chirag Fields and the Deep Water Portion of the Gunashli Field in the Azerbaijan Sector of the Caspian Sea, September 20, 1994, as referred to previously. Other agreements include the following:

- A Protocol between BP and SOCAR on Cooperation in the Area of Nationalization of Staff Involved in BP-Operated Projects, November 14, 2013. Under this agreement BP commits to continue to develop national professionals so that they can take over the maximum number of jobs within BP. BP also undertakes implementation of a five-year nationalization plan for the years 2014 to 2018, including progress of nationalization targets, a revised recruitment strategy, and various learning and development initiatives. BP and SOCAR will continue to work together to arrange recruitment campaigns and educational and training programs to support the development of national oil and gas industry professionals.
- A Memorandum of Understanding with SOCAR on Cooperation in the Area of Training of National Workforce at Gobustan Regional Training Center, dated November 14, 2013. Under this agreement BP commits to support vocational and technician training for 100 representatives of the local communities residing in the neighborhood of the Sangachal terminal: Sangachal, Umid, Azimkend, and Gobustan settlements. BP will cover the tuition

and other expenses related to the training of these students at the Gobustan Regional Training Center.

- A Memorandum of Understanding with Baku Higher Oil School on Cooperation in the Area of Education, November 14, 2013. Under this agreement BP commits to engage the students of Baku Higher Oil School in its various programs, initiatives, and development activities, as well as deliver business lectures at this educational institution.[38]

Production Sharing Agreements

Under a PSA, a contractor and its subcontractors, and the joint operating company and its subcontractors, are free to employ such personnel as, in the opinion of contractor, are required for the purposes of carrying out petroleum operations. Nevertheless, PSAs in Azerbaijan provide for local content obligations to give preference, when consistent with efficient operations, to employing citizens of the Azerbaijan Republic in the performance of petroleum operations to the extent reasonably practicable. This remains applicable as long as such citizens have the required knowledge, qualifications, and experience. According to the PSAs, such citizens shall be eligible for training.

Among other obligations of contractors pursuant to a PSA is an obligation for the joint operating company (JOC) to periodically provide SOCAR with projections for expected employment numbers and job specifications for Azerbaijani citizens. This requirement also applies to subcontractors, who also must comply with periodic reporting concerning employment of Azerbaijani citizens. In turn, within two weeks of receiving the list, SOCAR must provide the JOC and any subcontractors with a list of candidates recommended by SOCAR. Persons from the list provided by SOCAR shall enjoy a priority right to employment by the JOC and the subcontractors if they meet the necessary requirements. In the event that vacant positions remain after the JOC or its subcontractors select persons from SOCAR's recommended list, the JOC and its subcontractors are entitled to fill these vacant positions with such Azerbaijani citizens as it chooses. In the event that the candidates selected independently by the JOC and its subcontractors include SOCAR employees, such persons shall be hired by the JOC or subcontractors after consultation with SOCAR.[39]

Overall target personnel levels of Azerbaijani citizen employees pertaining to Petroleum Operations shall be as follows according to Azeri-Chirag-Deepwater Gunashli PSA:

Azerbaijani Citizens	
Prior to commencement of full field development	
Professionals	30%–50%
Nonprofessionals	70%
Upon commencement of first oil from full field development	
Professionals	70%
Nonprofessionals	85%
Five years from commencement of first oil from full field development	
Professionals	90%
Non-professionals	95%

Source: https://www.bp.com/content/dam/bp-country/en_az/pdf/legalagreements/PSAs/ACG_PSA.pdf.

The Shah Deniz and Shafag-Asiman PSAs have similar or the same requirements.

PSAs also require that contractors give preference to Azerbaijani suppliers (production, economic, and other entities, regardless of ownership, legally operating in the Azerbaijan Republic) with respect to the purchase of imported materials and equipment. Azerbaijani supplies must be given preference when they are competitive in price, quality, and availability with commodities or services available from other sources.

Positive and Negative Experiences, Issues, and Challenges

The legal framework for the regulation of oil and gas contracts is based on the Subsoil Law of February 13, 1998 and the Energy Law of November 24, 1998. Although both of these acts provide a general framework for exploiting energy resources, in many instances their provisions clash. Furthermore, it is unclear which law has priority when there are contradictions or overlapping provisions. Another concern is that most existing PSAs do not fully comply with the provisions of these acts.

According to Ciarreta and Nasirov,

> In order to improve the preparation and implementation of state policy on oil and gas production, the president of Azerbaijan signed a decree on 15 May 2006 for the founding of the Ministry of Industry and Energy (MIE). SOCAR is thus charged with conducting commercial functions while the MIE is responsible for noncommercial functions such as preparing, negotiating and implementing PSAs and other types of contract on behalf of the government. However, in reality, the MIE has been accorded only nominal responsibility for concluding PSAs. SOCAR has its own share in all contracts and is thus one of the contractors. At the same time, SOCAR represents the government in these contracts. As a contractor, SOCAR is interested in increasing its share in the contract. But this might reduce the revenues of the government. This situation has come about because of the lack of a law on oil and because SOCAR continues to have considerable power and influence over the oil and gas sector.[40]

Apart from issues related to the regulations of the oil industry, the national oil sector has developed significantly since Azerbaijan signed the Contract of the Century. However, despite development of the oil sector, the nonhydrocarbon sector remains relatively undeveloped. Nondevelopment of other sectors of industry causes obstacles to the oil sector on one hand, and overdependence of the state economy on the oil sector on the other hand. Devaluation of the local currency, staff redundancies in hydrocarbon sectors and other sectors, and liquidation of some subcontractors of the AOIC since the drop in oil prices have proven that this overdependence may lead to a serious negative impact to the country's economy.

Access to finance, tax rates, practice of the "informal sector," and tax management remain the biggest obstacles for entrepreneurs. Difficulties in obtaining credits from financial institutions and high interest rates make such investments unattractive for entrepreneurs, especially for local ones. Lack of funds to invest in equipment and technologies makes local entrepreneurs less competitive than international suppliers.

Despite anticorruption measures taken by government and noisy investigations of corruption and bribery, embezzlement, misappropriation, and bribery still remain issues that entrepreneurs face.

Conclusion

As seen in the above discussion of history, achievements, and challenges in the oil industry, Azerbaijan has come a long way toward developing its national oil and gas industry. Some attention has been given to developing local content, yet more is needed. This will be increasingly true given the drop in petroleum prices and the subsequent devaluation of local currency in 2015. The oil and gas sector will require considerable cost cutting. This will not be possible if there is still some level of dependence on foreign experts and imported goods. Currently, there is active implementation of three cooperation agreements signed between BP and SOCAR in support of the former's plans for development of Azerbaijani nationals involved in its activities as the operator of the Azeri–Chirag–Gunashli (ACG), Shah Deniz, Baku–Tbilisi–Ceyhan (BTC), and South Caucasus Pipeline (SCP) projects.

Notes

1. "Interesting Facts from Azerbaijan Oil and Gas History," *SOCAR Plus Extra*, 5, http://www.socarplus.az/pub/book/spextra_autumn_2012/.
2. Ibid.; and Mir Yusif Mir-Babayev, "Azerbaijan's Oil History: A Chronology Leading up to the Soviet Era," *Azerbaijan International* 10, no. 2 (Summer 2002), 34–40.
3. "History of Development of Oil Industry," http://www.azerbaijan.az/_Economy/_OilStrategy/oilStrategy_02_e.html.
4. Sabit Bagirov, "Azerbaijani Oil: Glimpses of a Long History," http://sam.gov.tr/wp-content/uploads/2012/01/1.-AZERBAIJANI-OIL-GLIMPSES-OF-A-LONG-HISTORY.pdf.
5. "History of Development of Oil Industry."
6. Mike Wysatta, ed., "166-Year-Old Oil Well Largely Unknown to West Until 2002," *Reservoir Solutions* 15, no. 3 (Sept.–Nov. 2012), 7, https://www.ryderscott.com/wp-content/uploads/news-2012-sep.pdf.
7. Bagirov, "Azerbaijani Oil."
8. Wysatta, "166-Year-Old Oil Well Largely Unknown to West Until 2002," 7.
9. "History of Development of Oil Industry."
10. "Zoroaster: The World's First Tanker Ship," Maritime Insight News Network (July 22, 2016), http://www.marineinsight.com/maritime-history/zoroaster-the-worlds-first-tanker-ship/.
11. Brita Åsbrink, "Brothers Nobel: Branobel Celebrates Its 30th Anniversary," http://www.branobelhistory.com/themes/the-branobel-company/Branobel-celebrates-its-30th-anniversary/; and "Branobel," Wikipedia, https://en.wikipedia.org/wiki/Branobel.

12. "Interesting Facts from Azerbaijan Oil and Gas History."
13. Wysatta, "166-Year-Old Oil Well Largely Unknown to West Until 2002," 6.
14. "Interesting Facts," 11.
15. Ibid., 15.
16. Ibid., 17.
17. Ibid., 23.
18. Ibid., 24–26.
19. "Oil and Gas in Azerbaijan: Oil Production," Azerb.com, http://www.azerb.com/az-oil.html?i=1.
20. "The Azerbaijan Government and Co-venturers Sign Amended and Restated Azeri-Chirag-Deepwater Gunashli PSA" (press release, September 14, 2017), https://www.bp.com/en_az/caspian/press/pressreleases/acg_psa_amended.html.
21. "Long-Term Strategy on the Management of Oil and Gas Revenues: Approved by the Decree of the President of the Republic of Azerbaijan No. 128 dated September 27, 2004," Article 1, http://www.oilfund.az/uploads/5-eng-long-term.pdf.
22. Ibid., Article 2.
23. Ibid., Article 3.
24. Ibid., Article 4.
25. Aitor Ciarreta and Shahriyar Nasirov, "Analysis of Azerbaijan Oil and Gas Sector" (working paper, USAEE, 2011), 13–14, https://www.usaee.org/usaee2011/submissions/OnlineProceedings/Ciarreta_Nasirov-Article1.pdf.
26. Ibid., 14.
27. "Azerbaijan Overview," World Bank, http://www.worldbank.org/en/country/azerbaijan/overview.
28. Ibid.
29. "Ministry of Industry and Energy of Republic of Azerbaijan," http://www.azerbaijans.com/content_530_en.html.
30. Ibid.
31. Heidi Kjaernet, "The State Oil Company SOCAR: A Microcosm of Azerbaijani Development?" *Caucasus Analytical Digest*, No. 16 (April 26, 2010), 5, http://www.css.ethz.ch/content/dam/ethz/special-interest/gess/cis/center-for-securities-studies/pdfs/CAD-16-5-8.pdf.
32. "State Oil Company of the Republic of Azerbaijan: History of SOCAR," http://www.azerbaijan.az/portal/StatePower/Committee/committeeConcern_01_e.html.
33. "About SOCAR: The State Oil Company of the Republic of Azerbaijan," http://socar.az/socar/en/company/about-socar/discover-socar.
34. "State Oil Company of the Republic of Azerbaijan: Duties and Object of Activity of SOCAR," http://www.azerbaijan.az/portal/StatePower/Committee/committeeConcern_01_e.html.
35. "State Oil Company of the Republic of Azerbaijan, SOCAR," LinkedIn, https://www.linkedin.com/company/state-oil-company-of-the-republic-of-azerbaijan-socar-.
36. "Azerbaijan: Oil Policy," http://www.azerbaijans.com/content_777_en.html.

37. "UK-Azerbaijan Oil & Gas Partnership Facilitation Programme Launched Today in Baku 29.10.2015" (October 29, 2015), https://www.facebook.com/notes/british-embassy-azerbaijan-britaniyan%C4%B1n-az%C9%99rbaycandak%C4%B1-s%C9%99firliyi/uk-azerbaijan-oil-gas-partnership-facilitation-programme-launched-today-in-baku-/1014394611936881/.
38. "BP Signs Agreements in Support of Its Staff Nationalization Plans for Azerbaijan" (press release, November 14, 2013), http://www.bp.com/en_az/caspian/press/pressreleases/BP-signs-agreement-for-nationalization.html.
39. "Agreement on the Joint Development and Production Sharing for the Azeri and Chirag Fields and the Deep Water Portion of the Gunashli Field in the Azerbaijan Sector of the Caspian Sea among the State Oil Company of the Azerbaijan Republic and Amoco Caspian Sea Petroleum Limited, BP Exploration (Caspian Sea) Limited, Delta Nimir Khazar Limited, Den Norske Stats Oljeselskap AS, Lukoil Joint Stock Company, McDermott Azerbaijan, Inc., Pennzoil Caspian Corporation, Ramco Hazar Energy Limited, Turkiye Petrolleri AO, Unocal Khazar Limited," Article 6, http://www.bp.com/content/dam/bp-country/en_az/pdf/legalagreements/PSAs/ACG_PSA.pdf.
40. Ciarreta and Nasirov, "Analysis of Azerbaijan Oil and Gas Sector," 23.

4

BRAZIL

Daniel Szyfman and Daniel Lírio Simões

Introduction

Choosing, structuring, implementing, and monitoring public policies is a topic widely studied and discussed from many different angles, including economic, sociological, business, and legal viewpoints. When one adds the seasoning of political and ideological ideas, the result is a recipe for a long and rich debate.

One common feature, however, in most of the public policies implemented worldwide is the need for financial resources. Save for a few (not unimportant) initiatives that are based on voluntary efforts, implementation of policies will most likely consume funds.[1]

Invariably, countries with oil and gas industries have implemented programs to capture part of the inherent flow of investments. It is not uncommon for some of these funds to be used to support public policies to foster positive results in the local economy and community. These policies are called local content policies (LCPs).

In this article, the authors examine the case of Brazil: how the LCP was implemented, its current challenges, and perspectives for the near future. Important background information about local policies on a worldwide basis is presented first, along with the Brazilian E&P regime.

In Brazil, to date the most important tool of the Brazilian LCPs is a clause in the licensing contract requiring E&P companies to hire goods and services locally. The obligation is set in a percentage applied to the overall expenditures; until the latest-implemented concession bid round, bidders who offered higher local content thresholds increased their chance of winning the relevant block. That rule is now changing, and the percentages will be set forth by the government at the outset as a condition of the relevant licensing contract.

Compliance with this clause, the legal challenges of regulating the local content obligation, and the gap in time between the bidding date and the expenditure have shown, in practice, that this obligation causes a great deal of concern for investors and Petrobras (the Brazilian NOC).

Over the last couple of years, the market has seen a global meltdown in the industry caused by plummeting oil prices. In the case of Brazil, this is combined with the country's own hardship: an unprecedented economic downturn,[2] political turmoil resulting in the second presidential impeachment in less than three decades,[3] and a compliance-related scandal that has led Petrobras to its most challenging moment ever.

After 25 years, changes in the LCPs are being implemented by governmental authorities. These changes are aimed at reducing the negative impact of the local content obligation in attracting investments to the Brazilian E&P market. The full picture is not known yet, in particular because the new federal administration is still being formed.

Local Content Policies: A Global Overview

Extractive industries have many common features, such as large capital expenditures, exposure to commodity-price fluctuations, and a deep concern with social responsibility and interaction with local communities. In addition to that, they generate an economic impact in the area in which they operate, as the projects consume large amounts of investment.

Policy makers are usually faced with the challenge of demonstrating to constituents and citizens that the side effects of the extractive industries (such as environmental hazards, as mitigated as they may be) can be offset with gains for the local and regional community.

One method of obtaining this positive outcome is through designing and implementing LCPs, which are defined by a World Bank study as "government interventions that look to increase, in the long term, the share of employment or of sales to a sector that are locally supplied at each stage of the value chain."[4]

The same World Bank study, an important guide for this article, indicates the main features an LCP should take into account:

- It cannot be limited to increasing local content on an immediate basis, but rather should focus on a long-term increase of enhancing local capabilities. These include education and training programs, along with relationships with other industries that can also benefit from the service chain.
- It will vary throughout time, accompanying the E&P cycle (from exploration to production). For instance, if a cluster is in the exploration phase, service providers more linked to that stage should have incentives to produce locally. However, the policy should anticipate when the cluster will divert to production and plan to maintain incentives.
- It should also consider the direct provision of social welfare contributions, such as the building and operation of local schools and hospitals.

In this section, we explore examples of LCPs around the world and discuss methods to measure their efficiency and hot issues relating to the LCPs.

International experience[5]

United Kingdom. The United Kingdom was a lead country in the E&P industry with the major North Sea cluster, shared with Norway. The 1970s were particularly important given the petroleum crisis and the perception that the UK-based companies were not directly benefiting from the growth of oil and gas activities. In this context, the Offshore Supplies Office (OSO) was created, with the aim of being

> responsible to the Secretary of State for ensuring the maximum possible involvement of UK manufacturers, consultants, contractors and service companies in the provision of supplies and services to the offshore hydrocarbon industry. This includes the creation of new industrial capacity to meet existing and emerging needs and measures to ensure that such new capacity is as fully and continuously utilised as possible.[6]

A series of local content measures were then implemented, including discretionary licensing (as opposed to auctions), audits of purchases made by oil companies, and provision of financial assistance to domestic suppliers. According to a study by the UN Conference on Trade and Development,

> The successful implementation of these measures increased the UK local content in the North Sea oil sector from 30% in 1973 to 82% by 1986, with nearly 100% local content in postdevelopment operations. Much like their Norwegian counterparts, local content policies have been successful in turning UK firms into competitive suppliers of petroleum services at the international level.[7]

The first step was the requirement for E&P companies to submit information on their procurement efforts. Every contract in excess of £100,000 had to be reported, including information on bidders and the winner. Another feature provided local companies with financial aid. Still, research conducted by Aberdeen University in 1980 showed that many of the legal entities established in the United Kingdom were in fact subsidiaries of foreign groups.

The study also noted that the "focus of local content requirements was on creating a high value added chain, rather than mere local incorporation, low levels of transformation or local ownership." They did not opt for the application of fines.[8]

As long as the United Kingdom legally remains a member of the EU, its policies are subject to the EU regulations on subsidies at the country member level.

Norway. Norway enjoyed the same enthusiasm as the United Kingdom deriving from the large discoveries in the North Sea. In the late 1960s, the policy mandated that the state hold a stake in the blocks through Statoil or Norsk Hydro. The Norwegian LCPs were formalized by means of the Royal Decree of 1972, and the Goods and Services Office was established. It was created to (1) support the local supply industry through JVs and encourage research and development and transfer of technology; (2) review tendering procedures to ensure that local companies were given a fair chance to participate; and (3) establish minimum local content requirements and monitor their implementation.[9] LCPs in Norway have been driven mainly by a focus on research and development and the transfer of technology.

Local suppliers also enjoyed a statutory preferential regime,[10] so long as they could be competitive from price, delivery, and quality standpoints. When Norway adhered to the European Union, its policy was also affected by the EU regulations.[11] Decisive moves allowed the policy to foster internationalization of the Norwegian service providers. With the benefit of the experience and capacity developed over previous years, the companies were very successful in this process, and global players are currently active in the market.[12]

Malaysia. Malaysia is one of the leading LNG players. While oil is an important part of the economy, it is not the main activity in the Malaysian economy. Private companies are present, along with Petrobras.

The LCP objectives in Malaysia were to strengthen value-added activities across the oil industry and to create a hub for service companies.

Petronas created learning centers to facilitate training local manpower, which has worked well for both Petronas and private companies. In addition, tax incentives related to research and development were created. In fact, the production sharing contract requires an annual contribution equivalent to 0.5% of the sum of cost oil and contractor's share of profit oil. The tax incentives can be enjoyed so long as companies spend at least 1% of their gross sales on domestic research and development, and at least 15% of their manpower has a scientific or technical background.

Implementing and Measuring Effectiveness of LCPs

A state that is willing to implement an LCP should first do its homework. There are several examples of LCPs implemented around the globe, along with a sizable body of literature that investigates their successes and failures. Even in relation to the successful programs, the state should mind the social, economic, and legal system differences in relation to the benchmark program. As with many other public policies, there is no "one-size-fits-all" solution.

The next step would be to establish the objectives for the LCP to be successful. Countries with a higher level of accountability and public scrutiny are more used to running this exercise before spending public resources on public policies. Policy makers who fail to carry out this step may become the subjects of criticism at a later stage. It is equally important to frequently evaluate the program and to publish the findings, so the state and others may assess whether the objectives are being met and if improvements can be made.[13]

Part of the objective should be defining "local." As a matter of fact, the results of the program could drastically change if, for instance, the state considered origin of capital when deciding whether a company's output is local. For some jurisdictions, the fact that the company is established in its territory would suffice; for others, it would be local only if at least 50% of the capital is owned by local-based investors.[14]

The organizational structure of the public bodies involved in the LCP must also be clearly set, establishing lines of responsibility and cooperation. The LCP is usually one public policy among a set of others. Many public bodies tend to be involved in the E&P industry (oil and gas regulator, environmental agency, labor authorities, etc.), and thus the effectiveness of an LCP is dependent on efficient functioning of its regulators.

Issues and Concerns for Establishing an Effective Local Content Program

A common conclusion of researchers is that the LCP should not be a one-off isolated policy. Rather it should be part of a wider range of programs aimed at capturing the goodwill generated by the oil industry and funneling efforts and funds into training the local workforce and fostering local businesses.[15]

An efficient LCP should also aim to increase the capacity of the local labor force to absorb more complex tasks. This is typically a role of the state, despite the support (financial or otherwise) it might get from oil companies as part of the LCP. Because the oil industry is technology intensive, the oil companies will need the support of a well-trained labor force to perform E&P activities. A long-standing LCP should then aim to increase the percentage of locals who are prepared to fill such positions.

One of the tools at the disposal of policy makers is the use of sanctions against companies that breach the LCP obligations. Evidently, sanctions may be effective as a deterrence mechanism and as a means to level the playing field between E&P competitors. However, the response must be one that does not repel investors from coming in the first place. In the same tone, complexity in complying with the LCP's obligations should also be examined carefully so as to avoid a material burden on the E&P company. Otherwise, the compliance cost could overcome the social benefit of the LCP.[16]

E&P Regime and the Local Content Obligation in Brazil

The most relevant tool of the LCP for the oil and gas industry in Brazil is an obligation imposed on the E&P companies to purchase goods and services "made in Brazil," as set forth in the relevant licensing contract (the local content obligation). Therefore, understanding and exploring the relevant aspects and nuances of such policies requires a brief overview of the E&P legal regime, which is the subject matter of this section.

E&P in Brazil—from monopoly to the current framework

As with many other jurisdictions, petroleum in Brazil is viewed as strategic and to some extent a matter of national sovereignty. The Federal Constitution (Section 20, V, IX, and paragraph 1),[17] sets forth that petroleum and natural gas reserves located within the Brazilian territory (including the continental shelf, territorial sea, and exclusive economic areas) are an asset of the union.

Before 1995, where the Constitution was ruled by its original 1988 wording in Section 177, first paragraph, the regime for oil- and gas-related activities was monopolistic: exploration, production, refining, import, export, and transportation of hydrocarbons[18] were to be conducted by Petrobras, a mixed-capital public company controlled by the national government.

In 1995, among other reforms aimed at opening the infrastructure market for the private sector, Amendment No. 05/1995 was approved. Together with the Oil & Gas Law (Federal Law 9,478/1997), these new statutes allowed private entities to explore, produce, and own[19] hydrocarbons under the regulation of the federal government.[20]

This reform established the framework of authorities in charge of oil and gas matters in Brazil, which is centered on the National Petroleum Agency (ANP) and the National Energy Policy Board (CNPE).[21]

The CNPE is an advisory chamber to the president, with the following goals: (1) fostering rational use of the nation's energy resources; (2) ensuring the functioning of the national fuels inventory system; (3) reviewing energy matrices for different regions of Brazil; and (4) establishing guidelines for specific energy programs. The ANP sets out legal requirements and technical standards, enacts ordinances, approves and reviews plans, inspects activities (from both paperwork and on-the-ground standpoints), and is the executive regulator for oil and gas matters.[22]

The ANP is an autonomous regulatory agency (*autarquia especial*) run by a board of directors and the head of the agency. The "autonomy" is based on three main features:

- Directors are appointed by the president for fixed terms that do not coincide with the president's own term, and they cannot be dismissed at the president's sole discretion.
- Its decisions are not subject to review by the relevant ministry (Mines and Energy).
- It enjoys budgetary autonomy.[23]

In terms of granting E&P rights, there are currently three different regimes in force in Brazil: the concession regime, created by the Petroleum Law in 1997; the production sharing regime, applicable for the granting of E&P rights in the presalt and strategic areas from December 22, 2010 onward; and the onerous assignment of E&P rights, a regime applicable to Petrobras only.

Starting in 1999 with the first concession bid round,[24] and until the 10th concession round in 2008, the ANP managed to carry out the bid rounds on an annual basis. This stream was broken with the astonishing discovery of large volumes of hydrocarbons under a thick layer of salt, which unleashed a series of political reactions that led to a material change in the oil and gas framework in Brazil.

Mr. Lula da Silva, at that time the incumbent president, led a political movement that resulted in (i) the CNPE passing resolution 6/2007, which removed blocks in the presalt area from the upcoming ninth concession round just days before the auction; and (ii) the federal government submitting bills of law to Congress suggesting changes to the E&P legal framework. The argument used by the government at that time was that the presalt was a lottery ticket with near-zero risk; hence the rules should be changed so as to achieve a better risk-return balance between the state and the concessionaire.[25]

Mr. da Silva's administration submitted a set of four bills of law to the National Congress. Three federal laws that were passed in this context are essential to understanding how Brazil ended up with three different regimes for E&P activities, as follows:[26]

- *Federal Law 12,351/2010 (PSC Law)*. This law established production sharing contracts as the applicable regime for presalt and "strategic" areas.[27]
- *Federal Law 12,276/2010 (Onerous Assignment Law)*. This law authorized the federal government to grant directly to Petrobras (i.e., without competition) the right to explore and produce hydrocarbons in a determined presalt area, limited to the maximum volume of 5 billion barrels of oil equivalent (boe).[28]
- *Federal Law 12,304/2010*. This law authorized the creation of a new wholly state-owned entity, Pre-Sal Petróleo SA (PPSA), that would represent the interests of the federal government in the consortia that would become a party to the PSCs and the federal union's rights under unitization arrangements where the tract outside the border is located within the boundaries of the presalt and strategic area.

An important feature of the 2010 reform was that vested rights under the concession contracts were *essentially* unaltered by the PSC Law. The main impact on existing concession rights was the changes in the unitization rules (consolidated later on with the approval of ANP Resolution No. 25/2013). These rules were changed to regulate situations in which the deposit associated with a concession block extends to a neighboring tract that is governed by a different regime. This could be a PSC contract, an onerous assignment contract, or an area not yet granted, where at least part is located in the presalt and strategic area.

Local content obligation

The main legal and regulatory framework of the local content obligation is set forth in the relevant licensing agreement (concession, PSC, or onerous assignment contract).[29] Because nearly all existing E&P contracts are concession contracts, this article will be based on such regimes, except as otherwise noted.

Both the tender protocol and the concession contract drafts have continued to evolve since the first concession round. In the following subsections, the authors analyze the evolution of the local content obligation through time, taking as main sources latest concession contracts that were executed (in the context of the 13th concession bid round) and the draft tender protocol and concession contract for the 14th concession bid round.[30]

Local content obligation in the concession contract—current regime. The local content obligation comprises the following subset of obligations imposed on the concessionaire:

- They must comply with the committed percentages of goods and services hired locally.
- They must include local[31] suppliers in the process of awarding contracts to service providers.

The second obligation is less controversial than the percentage of local goods and services. The first, percentage of the local content obligation, is then subdivided by the later bid rounds as follows.

The 13th concession bid round:

- There is a global percentage for the exploration phase, and a second global percentage for the development/production phase.
- Within each phase, there is a set of items (e.g., drilling unit, support vessel) and a set of subitems per each item (e.g., wellhead, naval support).

The 14th concession bid round:

- The LC percentages are different depending on (1) the blocks being onshore or offshore and (2) the phase of the concession contract (either the exploration phase or the development/production phase).
- Specifically, during the development/production phase, the local content obligation is split into three different macro groups: (1) wells construction, (2) delivery and offload systems, and (3) stationary production unit.[32] The idea behind this is to facilitate compliance with the obligation by giving more freedom to the concessionaire, as it can maneuver within one group to facilitate compliance with the clause.

The ANP is in charge of inspecting and auditing compliance with local content. For contracts entered into on or after the seventh concession round, compliance is conducted via the accredited agents or *certificadoras*, companies that are authorized by the ANP to audit local content obligations. Essentially, the *certificadoras* gather data in relation to the contract between the E&P company and the relevant provider of goods or services. They issue certificates of compliance, which are then delivered to the ANP[33] as evidence that the obligation was met.

Under certain extraordinary circumstances, up to the 13th concession bid round, the concessionaire could require a waiver from the ANP in complying with a certain item or subitem. If the waiver is granted, the relief is on the respective item and subitem only, and thus the relevant global percentage (whether for the exploration or development/production phase) remains unchanged. The language on waivers has been excluded from the 14th concession bid round draft contract. (For more on waivers, please refer to the "Waiver" section later in this chapter.)

If the local content obligation is not met, the concessionaire is subject to a penalty calculated with a predetermined formula[34] set forth in the concession contract.[35] The formula is applicable over the noncomplied monetary value of the items or subitems of relevant goods or services, also considering the noncomplied percentages (NCPs). One or the other of the following is applied.

The 13th concession bid round:

- If the breach is equal to or higher than 65% of the original committed percentage, then the ANP applies a mathematical formula, which is [Fine = 8 × (NCP – 1) / 7] of the *monetary value*.
- If the noncompliance is lower than 65% of the original committed percentage, then 60% of the *monetary value* is assessed.[36]

The 14th concession bid round:

- If the breach is equal to or higher than 65% of the original committed percentage, then the ANP applies a mathematical formula, which is [Fine = NCP – 20%]
- If the noncompliance is lower than 65% of the original committed percentage, then 40% of the *monetary value* is used.

Therefore, as an example, the financial impacts for a potential concessionaire that awards a concession contract under the 14th round and has a minimum work commitment to invest 200MM in the exploration phase, being 50% of LC and thus 100MM, would be the following:

- Breach of 20% (20MM) of the original committed LC% (lower than 65%) = 40% of NCP.
  - 20MM × 0.4 = 8MM
- Breach of 90% (90MM) of the original committed LC% (equal or higher than 65%) = 90% – 25% = 65% of NCP.
  - 90MM × 0.65 = 58.5MM

Local content obligation in the concession contract—historic background. Bidding round zero did not provide any local content requirement whatsoever. Petrobras was (and still is) free to invest in its assets of bidding round zero at any local content percentage until 2025, when round zero concession contracts will expire.[37]

In the contracts of the first round (1999) and beyond, concessionaires assumed a minimum local content commitment in respect to acquisition of local goods and services. Thus competitors could freely offer different values of goods and services to be acquired from Brazilian suppliers in order to perform the exploration and development activities. The offered local content percentages were computed for the purposes of pointing the offers during the bid proposals phase. This regime was effective until the fourth concession round.

In the fifth and sixth concession rounds, the local content clause was modified to require minimum local content percentages. The bidding floor for local content percentages varied according to location category (i.e., onshore blocks, shallow water blocks, or deepwater blocks) for both exploration and development phases.

In the seventh concession round, the ANP reformed the system of compliance with the local content obligation. The main change was the creation of the local content certification system, under which private institutions accredited by the ANP are in charge of reviewing the local content documents and issuing certificates, which are later reviewed by the ANP.[38] Also in such rounds, a ceiling to the local content percentages was inserted.

Figure 4–1 shows the evolution of the average local content committed by winning bidders.

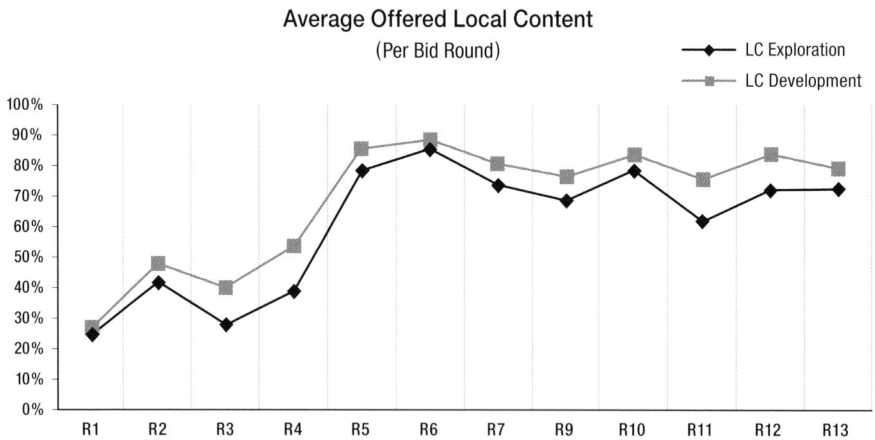

Fig. 4–1. Average offered local content
Source: ANP.

Setting the local content percentage target—current framework. Having explained the importance of local content percentages and how they apply, this section aims to explore how the percentages applicable to local content obligations at all levels (global, items, and subitems) are set. This depends primarily on the relevant E&P regime.

Concession regime. Up to the 13th concession bid round, the decision on the winner of a block in a bid round relies on the combination of three criteria (table 4–1), balanced by the winning formulas (shown in table 4–2).

Table 4–1. Three criteria for bid round selection

Element	Percentage of Global Formulas
Signing Bonus Upfront cash payment	40%
PEM Minimum employment program	40%
Local Content Obligation	20%

Source: PSC's tender protocol, "Edital de Licitação para a Outorga do Contrato de Partilha de Produção Disposições Aplicáveis às Atividades de: Exploração.

Table 4–2. Winning formulas

$$\text{GRADE 1} = 40 \times \left[\frac{\text{Offered Bonus Signature (R\$)}}{\text{Highest Offered Bonus Signature (R\$)}}\right]$$

$$\text{GRADE 2} = 5 \times \left[\frac{\text{CL (\%) Offered in Exploratory Phase}}{\text{Highest CL (\%) Offered in Exploratory Phase}}\right]$$

$$\text{GRADE 3} = 15 \times \left[\frac{\text{CL (\%) Offered in Development Phase}}{\text{Highest CL (\%) Offered in Development Phase}}\right]$$

$$\text{GRADE 4} = 40 \times \left[\frac{\text{PEM Offered in UTs}}{\text{Highest PEM Offered in UTs}}\right]$$

$$\text{FINAL GRADE} = \text{GRADE 1} + \text{GRADE 2} = \text{GRADE 3} + \text{GRADE 4}$$

Evidently, the higher the local content obligation percentages submitted, the more points the bidder gets with the formulas, thus increasing the chance of winning the block. Of course the flipside of offering higher percentages is the possibility (and cost) of complying with the obligation once the concession contract is signed, and the risk of being penalized.

We note that the bidders were required to fill out a form in which they would set out not only the proposed global percentages for the exploration and development phases, but also the breakdown of items and subitems per system.[39] This created an obvious difficulty in foreseeing the future CAPEX, since the bidder does not even know if hydrocarbons will be discovered. For more on this matter, please refer to the "Timing and 'futurology'" section later in this chapter.

A major change of direction happened for the 14th concession round. As part of the ongoing reforms, the local content obligation is no longer part of the bidding criteria,[40] but rather a given input in the concession contract.

PSC and onerous assignment regimes. In both PSC and onerous assignment regimes, the percentages are set by the ANP. In the former case, it is disclosed in the tender protocol; in the latter case, it is set forth in the specific onerous assignment contract.[41]

Please see table 4–3 for the minimum percentages of local content under the PSC bid rounds.

Table 4–3. Minimum percentages of local content under the PSC bid rounds

<table>
<tr><th rowspan="2">ROUND</th><th rowspan="2">Field</th><th>Exploration Phase</th><th>Development Stage (modules with first oil until 2021)</th><th>Development Stage (modules with first oil until 2022)</th></tr>
<tr><th>Minimum (%)</th><th>Minimum (%)</th><th>Minimum (%)</th></tr>
<tr><td>1st Production Sharing Bid Round</td><td>Libra</td><td>37</td><td>55</td><td>59</td></tr>
<tr><td rowspan="4">2nd Production Sharing Bid Round</td><td>Carcará</td><td>35</td><td colspan="2">30</td></tr>
<tr><td>Sapinhoá</td><td>35</td><td colspan="2">30</td></tr>
<tr><td>Gato do Mato</td><td>38</td><td colspan="2">60</td></tr>
<tr><td>Tartaruga Verde</td><td>55</td><td colspan="2">65</td></tr>
</table>

The role of the service provider

As one can see, the local content obligation is imposed, prima facie, on the concessionaire.

Concessionaires sometimes try to pass this obligation through to the chain of service providers. The use of the pass-through is widespread, but the extent of it (all or just part of the local content penalty), along with other terms and conditions, will vary. Such pass-through mechanisms are defined in the contract between the concessionaire and the service provider, on a private and unregulated basis. The concessionaire cannot appeal to the ANP concerning noncompliance by the service provider.

In practice, a service provider interested in selling goods and services to E&P companies in Brazil must be aware that an important part of its success is its capacity to deliver the highest possible percentage of local content. Many of the leading global service providers have established a base of operations and a local subsidiary and are able to sell local-content goods and services.[42]

Critical overview of the results of the LCP in Brazil

The Brazilian LCP was criticized by the Brazilian General Accounting Office (TCU) in the context of an audit requested by ANP to evaluate the systematics and effectiveness of the Brazilian LCP. Their assessment was based on extensive research, and the findings were very conclusive and strong in condemning its effectiveness. Below are four excerpts (based on a free translation of the original in Portuguese) that speaks volume about their conclusions:

- "There is no mistake in stating that the LCP imposes costs to the Federal Government by means of Petrobras. Besides, the LCP makes the manufacturing of goods more expensive and delays the processes, and therefore, delays the tax revenues due to the Government, which encompasses royalties and special participation. The current subsidy policy in place cannot ensure real benefits for the country, and also, there is a risk of the LCP being responsible for deliberately channeling resources from the operators (especially Petrobras) to certain segments protected by protectionist rules."[43]
- "The evidence of such inefficiency is the TCU's own report published in respect of the construction of P-66 platform: the delivery term for a platform (FPSO) produced in Brazil is 4 times higher (44 months) than a similar one produced in a shipyard in South Korea (CLOV – 8 months), for instance. For information purposes, P-66, produced in Brazil, took 44 months to be delivered and has a hull length of 288 m. The 'Prelude' FPSO, manufactured in South Korea took 12 months to be delivered and has a hull length of 488 m. Therefore, a vessel with a hull that measures almost double that of another produced in Brazil was delivered in a period of time almost 4 times shorter."[44]
- "LCP has certainly provoked additional costs that are relevant for the oil and gas industry without a corresponding benefit for the LCP goals. Besides making the projects more expensive, the LCP has not contributed to the enhancement of competition in the sector,

and has also created an environment favorable to the assessment of large fines, the sole consequence of which is a lack of interest by investors."[45]

- "The conclusion is that LCP has generated additional costs to the oil and gas sector without proper benefits created by the same LCP."[46]

According to ANP's financial statements and consolidated reports,[47] the agency has so far (by the end of 2016) applied an aggregate amount of R$663 million or US $202 million[48] in local content penalties. These high figures were used by those who criticized the policy as being focused on collecting monies rather than fostering development of the local industry and promoting development. These findings came out in the same period when the new administration was taking office and certainly contributed in forming a view that the LCP needed reforms.

Current Issues and Challenges

The local content obligation draws a lot of attention from both the government (including the ANP) and industry. Discussions over the past years have often become politicized.[49]

Apart from that, there are other issues that are the subject of major discussion among industry agents. This section aims to give a high-level overview of some of these issues.

Timing and "futurology"

One common complaint by E&P companies, particularly the ones that participate in bid rounds for the first time, is that bidders are faced with a challenge: filling out the local content form. As previously mentioned, local content is one of the three criteria that will govern the result of the bids.

Until the 13th concession bid round contracts, such criteria were measured by the sum of the global local content percentages (one for exploration and one for development/production). Each global local content percentage represented a result of the particular form, in which bidders were required to input two sets of information: the proposed percentage of each subitem and item and the ratio (weight) each item should have in the formula. The form—a spreadsheet until the 13th concession bid round contracts—applied such formulas to determine a weighted result of the global local content percentages.

To give an example, the mechanism meant that bidders should commit, at the bid process, to a local content percentage for the purchase of seabed manifolds (the floor was 70% in 13th concession round), umbilicals (the floor was 55% in 13th concession round), and the stationary production unit's anchoring system (the floor was 80% in 13th concession round). This is at a time during which the bidder does not even know if hydrocarbons will be discovered. If the effort is successful, it may not be initially apparent whether or not it is commercially feasible. Even if it is commercially feasible, the characteristics of the discovered hydrocarbons (oil and/or gas, volume, quality, etc.) are unknown. Moreover, since development could take another 7–10 years, even the available surrounding infrastructure for gas evacuation, for instance, could be dramatically changed between the bid date and the time when gas is produced.

As part of the reforms, the 14th concession bid round concession contract simplified the breakdown, creating three macrogroups. As a result, if a given item proves to be more difficult to achieve the percentage of local content, concessionaires can try to achieve increased percentages for other items within the same macrogroup.

Waiver

Perhaps the most contentious matter when it comes to local content in Brazil today, the waiver is a mechanism in the licensing contracts under which concessionaries can apply to the ANP seeking relief from the local content penalty. It is an exception to the general rule that noncompliance with the local content obligation triggers the local content penalty.

Over the past five years, the industry has held many discussions with the ANP over the interpretation of the rules applicable to the waiver. As a matter of fact, the rules have evolved, given that they were set forth in the concession contracts (rather than being regulated via ANP ordinance), and the relevant clauses have changed over time. Hence, the ANP has reviewed its position in light of the changes that were made to the regime set forth by the 7th round concession contract template, particularly the later modifications with the 11th and 12th concession bid rounds.

The draft concession contract for the 14th bid round did not contain the waiver language of the previous rounds. If this is maintained for the final version,[50] and save for regulation that could come after the concession contract is executed, this means that concessionaires of the 14th concession bid round contracts would not have the right to seek waivers.

Meeting the exceptionality criteria. In general terms, the relief applies when the local provider's proposed terms (price and date of delivery) were less favorable than those available from international sources.

The requirements have been marginally softened from the 7th through 10th concession rounds to the 11th through 13th concession rounds, as seen in table 4–4.

Table 4–4. Availability and price requirements, 7th–10th and 11th–13th bidding rounds

Bid Round	7th – 10th	11th – 13th
Availability	20.2 (f): "if the concessionaire receives proposals from Brazilian suppliers for the purchase of goods or rendering of services with delivery limits **much above what is practiced by the market** or that might harm the proposed investment schedule approved by ANP"	20.7 (a): "There is **no Brazilian supplier** for the purchased product or contracted service" 20.7 (b): "All proposals received from Brazilian suppliers **present an excessive delivery limit** in relation to non-Brazilian counterparts"
Price	20.2 (e): "if the concessionaire receives proposals from Brazilian suppliers for the purchase of goods or rendering of services with **prices excessively above the Brazilian market, when compared to the international market**"	20.7 (c): "All proposals received from Brazilian suppliers present an **excessive price in relation to non-Brazilian counterparts**"

In any case, it is evident that the requirements, however different, are not objective. How does one determine what qualifies as an "excessive price"?

The threshold of the 11th through 13th concession contracts could be interpreted as simply requiring that the price of the local provider not be higher than the one offered by a foreign supplier. On the other hand, the ANP could hold that the term "excessive" requires a material deviation.

A similar discussion exists when it comes to the terms for delivery of the good. This element adds a layer of complexity: while the price could be objectively assessed at a private bid where the underlying contract sets a fixed price (thus allowing the E&P company to compare the bids on an equal basis), delivery is more complicated. In fact, a delay is perceived only months or years later, at a time when the bid would have already been selected. Evidently, remedies such as delay liquidated damages could be an incentive for bidders to commit to a feasible timeline, but those usually have limited effectiveness.[51]

In addition, in a solely private and not-subject-to-regulation tender, E&P companies could value track record as a key point within the decision-making process.[52] In the Brazilian local content regulatory framework, if a newly established shipyard submits a bid that is competitive in price and delivery terms, and another bidder has a similar (marginally more expensive) proposal by a much more experienced shipyard, the difference in track record is not an express safe harbor in the concession contract. Hence the E&P company would most likely have difficulties in obtaining a waiver.

Timing: When must the waiver be requested? The timing of when a concessionaire must request a waiver from the ANP has evolved throughout the bidding rounds.

Between the 7th through 10th concession rounds, the E&P companies had to submit the waiver "before contracting with any foreign supplier of goods and/or services."[53] On the other hand, there was nothing in the concession contract that would indicate the deadline that the ANP would have to confirm or deny the waiver.

This absence of a time estimate proved challenging for E&P companies. As a matter of fact, when one considers the inherent complexity of the procurement process of the E&P companies, it is clear that the concessionaire had to make a difficult decision. It could simply take the risk, i.e., hire a party that would not meet the local content obligation, and then wait (and hope) for the ANP's waiver. Alternately, the concessionaire could take into account in the procurement process that, if a bidder with insufficient local content capacity (overall or for a certain item) wins the bid, the proposal's validity would have to bind the vendor while the ANP considered the waiver request. This way, if the ANP rejected the waiver, the E&P company would have the chance to cancel the bid or the contract, as applicable. Obviously, the longer the ANP takes, the more commercial difficulties arise.[54]

The 11th concession round concession regime changed the framework. Under such contracts (and also the 12th and 13th concession round contracts), the waiver should be submitted to the ANP "during the performance of the phase in which the concessionaire intends to have its waiver."[55]

Removing the nuance that the waiver had to be the subject of a prior waiver was certainly an improvement. Moreover, ANP's in-house legal attorneys have issued an opinion holding that this change is not only applicable to the newer concessions contracts (which have this consideration written in), but also applies retroactively to prior concession contracts. Their position is based on the fact that this is an improved technique of inspecting compliance with the local content obligation, and such improvements of a regulatory nature should be considered a "regulatory evolution."[56]

Despite this change, some players still manifest difficulties in managing their exposure to local content penalties. If they submit a waiver request in the early stages of a phase, the ANP would most likely not be able to respond to it until other contracts are signed and the ANP has a wider vision of such item (usually one item is comprised of many contracts).

Effects of the waiver. Having passed the high bar of the waiver test, the next question is, What are the effects of a waiver granted by the ANP? The wording of the concession contracts deals with this matter but does not clearly and objectively set out the result of the waiver.[57]

ANP's in-house attorney office analyzed such matters and issued an opinion over four possible outcomes of the waiver.[58] Two of them were ruled on in the opinion.[59]

One interpretation that was admitted by the office was to apply a deemed-local-content percentage: for the waived item, the formula will consider that the complied local content was the minimum (floor) percentage set in the bid documents. That means, for instance, that if the minimum was 55% and the concessionaire had bid 75%, 55% will be the complied local content. Conversely, the concessionaire would have to offset this 20% gap in another subitem of the relevant item in order not to be penalized in the item or global local content threshold.[60]

The other possibility accepted by the attorney's office was that the waiver should (i) consider the relevant item/subitem as complied with but (ii) not affect the global local content percentage, which shall remain unchanged.[61]

The 11th round concession contract language was clearer in saying that the waiver applies to the "percentage with which the concessionaire is committed in relation to recruitment of a specific good or service."[62] Following this change, the attorney's office updated the prior opinion and considered that the waiver should apply for the offered local content also in the prior rounds, once again by applying the concept of regulatory evolution.[63]

Double dip effects on penalties

As explained, ever since the 5th concession round, the E&P companies' obligation to comply with a minimum percentage of local content has been applicable on a per subitem, per item, and per global local content basis.

Both at early stages (5th and 6th concession rounds) and in the latest versions of the concession contracts (11th to 13th concession rounds), provisions are included to prevent the double dip effect. Thus the E&P company cannot be harmed twice; for example, by being penalized both at the item and global levels as a result of one contract having a lower local content percentage than required. However, the concession contracts of rounds 7 through 10 were silent regarding this matter.

In this context, ANP's in-house attorneys issued an opinion holding that even in such contracts where there was no express provision, applying a double dip penalty would be illegal.[64]

A recent case study is the Libra case. Petrobras filed a waiver request with ANP seeking permission to hire a foreign company to build a floating production storage and offloading (FPSO) unit in the presalt Libra field. Petrobras's request was based on the fact that the LC rules made the price offered by the winning local competitor 40% higher than the estimated budget. In addition, Petrobras alleged that the bidders/proponents replied to the respective request for proposals, stating the risks caused by difficulties of local procurement for the construction of the FPSO.

Given the relevance of the matter and the repercussions amidst the reforms, ANP decided to run a public hearing with the purpose of gathering support to review regulations.

Legislative Hydrocarbon Regime Concerning Local Content

Given the current political and economic context described in the introduction, two changes were implemented, both aimed at easing the burden of the local content obligation.

PEDEFOR

Ms. Dilma Rousseff, then acting president, passed a presidential decree instituting the PEDEFOR.[65] According to the decree, the federal program aims to foster the competitiveness of the local base of suppliers and encourage technological innovation. The program is still in the early stages of implementation.

It is managed within the federal administration by an advisory committee, which has issued two resolutions to date: (1) Resolution 3/2016, suggesting to CNPE LC rules for the 2nd PSC bid round; and (2) Resolution 1/2017, suggesting to CNPE LC rules for both the 14th concession bid round and also the 3rd PSC bid round.

The industry has reacted positively to one of the PEDEFOR features, which is the possibility of a concessionaire to be relieved of LC percentage commitments by purchasing goods in Brazil that are meant to be used in overseas operations.[66] Another main incentive is to allow for a surplus of local content to be considered in the local content obligation formula if the concessionaire and/or the supplier meet certain requirements associated with investments that match the program objectives. The decree says that in such cases a bonus will be granted to the concessionaire, although there is no provision that explains how this bonus will be calculated.[67]

ANP Resolution 20/2016

The ANP has issued Resolution 20/2016 with the objective of changing the method of calculation of the global local content percentage when the concessionaire is able to exceed the local content percentage for a given item.

The resolution created a formula whereby an exceeded achieved local content percentage may result in a changed (reduced) global local content percentage,[68] which is a breakthrough in terms of the relief that concessionaires enjoy from a waiver. If well implemented, concessionaires will have greater incentive to pursue a higher-than-committed local content for a given item.

Conclusion

Local content policies have been widely used by countries that have a meaningful oil and gas industry. The idea of capturing the goodwill of at least part of the massive inbound and outbound flow of cash and other kinds of investments that are part of oil and gas projects to foster transformation of the local economy is obviously attractive to any policy maker.

Norway is widely regarded as a successful example of a country with an LCP that generated material benefits to its local economy, creating a network of service providers that not only were able to meet the local demand, but also allowed the country to become an exporter of high-technology oil services worldwide.

Other countries did not enjoy the same result. To name a few, the programs of Australia, Malaysia, and some South American countries have had their share of difficulties.

For any country that is considering implementing or reviewing an existing LCP, our view is that policy makers should be constantly assessing and reassessing the results. Are the goals that were announced (or should have been announced) at the outset being met? Periodic in-depth reviews should be carried out with the participation of industry players and society in general. The vast examples of past experience cannot be ignored, and countries should be aware of their own local framework (legal, political, and economic) to pick and choose the right tools, and to adapt those that require improvement.

A full reassessment of Brazil's LCP is underway, which is welcoming. A meaningful debate with the main actors (such as ANP, the Ministry of Mines and Energy, Petrobras, industry players, service providers, experts, and society) has taken place, and the first measures are being implemented. The federal administration should also consider which actions have to be implemented by other branches of government (such as the Ministry of Education and Labor) in order to ready the local workforce.

The scenario we live in can be summarized as countries with reserves competing for the scarce available exploration investment resources of the IOCs, NOCs, and independents. These E&P companies are used to commitments to develop local capacity, but excessive burdens will be considered in their strategic decisions of whether they should invest in such a country.

In the case of Brazil, the industry has passed a clear message to the government that it does not consider itself to be responsible for creating and improving a local service capacity. While they accept the obligation to hire locally, they expect that other parts of the state machine will be able to make the local suppliers more efficient, so that ultimately local suppliers do not need to rely on the local content obligation to be chosen by E&P companies.

There is certainly a lot of work to be done, and the information that has been gathered over the past decades is invaluable. If government and industry find a way to work together, learning from international experience, setting objectives that are achievable and clearly segregating the role of each agent, an improved local content policy is definitely achievable.

Notes

1. The fundamental contribution of NGOs, for instance, is based on some sort of funding, whether from donations or public sources.
2. Brazil's 2015 GDP was reduced by 3.5%, the worst result in 25 years.
3. President Michel Temer took office following the impeachment of Ms. Dilma Roussef in August 31, 2016, in a proceeding that began in May 6, 2016, in the House of

Representatives and was confirmed by the final vote in the Senate on August 31, 2016. At the time this article was written, Mr. Temer was suffering immense political pressure as a result of allegations put forward by a whistle-blower that he was involved in corrupt acts, and the Supreme Court had authorized an inquiry on such allegations.

4. Alberto Melo and Andrés Rodríguez-Clare, "Productive Development Policies and Supporting Institutions in Latin America and the Caribbean" (Inter-American Development Bank, Felipe Herrera Library Publisher, 2006), 5.

5. This subsection has benefited from the lessons set out in the following Institute for Applied Economic Research (IPEA) study: Carolos Eduardo Ramos Xavier Jr., "Local Content Policies in the Petroleum Sector: The Brazilian Case and the International Experience," in *Text for Discussions No. 1775* (Rio de Janeiro: IPEA, October 2012), http://repositorio.ipea.gov.br/bitstream/11058/1072/1/TD_1775.pdf; and Silvana Tordo, Michael Warner, Osmel E. Manzano, and Yahya Anouti, *Local Content Policies in the Oil and Gas Sector* (Washington, DC: International Bank for Reconstruction and Development, World Bank, 2013), doi: 10.1596/978-0-8213-9931-6, or http://documents.worldbank.org/curated/en/549241468326687019/pdf/789940REVISED000Box377371B00PUBLIC0.pdf.

6. "Memorandum of Understanding between Department of Energy and UK Offshore Operators Association Ltd., Which Created the OSO," Appendix 1, Paragraph 2 (1973).

7. Sacha Silva, Bonapas Onguglo, and Alexey Vikhlyaev, *Local Content Requirements and the Green Economy*, United Nations Conference on Trade and Development (UNCTAD) Secretariat, UNCTAD/DITC/TED/2013/7 (Geneva: United Nations Publications, November 2014), 9, http://unctad.org/en/PublicationsLibrary/ditcted2013d7_en.pdf.

8. Ibid.

9. Ulrich Klueh, Gonzalo Pastor, Alonso Segura, and Walter Zarate, "Inter-Sectoral and Local Content in Extractive Industries and Beyond" (IMF Working Paper No. 07/213, IMF, September 2007), 42, https://www.imf.org/external/pubs/cat/longres.aspx?sk=21123.0.

10. In accordance with the Norwegian Petroleum Law from 1965.

11. In fact, Article 54 of 1996 Petroleum Act still requires Norwegian suppliers to have equal bidding opportunities and be notified in advance. Article 23 of same statute requires the operator to perform at least 50% of the R&D obligations in Norway.

12. Besides Statoil and Norsk, other well-established groups are Aker, Det Norske Oljeselskap, Hydro Oil & Gas, Norse Energy, NOV, Petoro, Petroleum Geo-Services (PGS Group), Prosafe, and Teekay Petrojarl, among others.

13. Brazil is experiencing this. Officers of the new interim federal administration are questioning the benefits of the PSI program, which has been run by BNDES. Political views aside, one harsh criticism is that there were no clear statistics to allow one to assess whether the subsidies achieved the objectives of the program.

14. In Brazil, the option was to consider the company's place of incorporation. Even if the ultimate parent company is a wholly foreign group, the local subsidiary would be eligible to provide local content goods and services. In fact, since Constitutional Amendment No. 6, 1995, the general rule in the Brazilian legal system has been not to distinguish the origin of capital, with few exceptions in the Constitution, such as the airline and nuclear industries, and E&P activities in the border regions.

15. Tordo, et al., *Local Content Policies in the Oil and Gas Sector*, xiii; and Melo and Rodríguez-Clare, "Productive Development Policies and Supporting Institutions," 5.

16. Tordo, et al., *Local Content Policies in the Oil and Gas Sector*, xvi. The same study brings additional contribution to the matter, as follows (pp. 137–142):

    Complex LCPs may make it difficult for policy makers to monitor their implementation, assess their impact, and enforce their application. Target groups may have difficulty in applying LCPs, which may result in misapplication or lack of application, and intended beneficiaries may lack the capacity to take full advantage of implementation. Especially for small businesses, assimilating and complying with complex administrative and/or technical rules can create an unreasonable burden. Complexity may translate into additional administrative costs for all parties involved, and inefficient or harmful outcomes. Therefore, in designing LCPs, regulators should be mindful of introducing measures for which the cost of compliance exceeds the social benefit. As the experience of our case study countries show, the use of standardized definitions of local content and standard reporting templates improve the transparency and the comparability of information provided by different companies. These measures can also help to reduce the regulator's administrative costs.... Compliance with local content regulations would necessarily involve additional costs for investors, especially when the regulatory requirements are complex, the measurement and reporting procedures are unclear, and modifications to the investors' procurement and/or reporting systems are necessary. Additional costs should be carefully considered since they may put small and medium investors at a disadvantage. Since these costs are normally recoverable for the purpose of calculation of petroleum sharing and/or corporate tax, regulators should have an incentive to design regulatory requirements that minimize compliance costs.

17. Brazil's Constitution sets forth in Section 20 the assets that constitute property of the union (the federal government). Among those, Item V includes the natural resources from the continental shelf and from the exclusive economic zone, and Item IX addresses the mineral resources, including subsoil resources, that encompass the petroleum and natural gas reservoirs, whether onshore or offshore. Moreover, the first paragraph of Section 20 establishes the guarantee of participation of the federal states and the municipalities in the results of the exploitation of such resources.

18. The original wording of the first paragraph of Section 177 is, "The monopoly provided in this Section includes the risks and profits resulting from the activities thereof. The Union cannot assign or grant any type of participation, in kind or in value, in the exploration of the petroleum and natural gas deposits, except for the provision of Section 20, First Paragraph."

19. The licensee can book reserves and own legal title of the hydrocarbons upon production. The government has title to all petroleum in the ground, but title passes to the concessionaire at the production metering point, where production is measured for tax and government take purposes.

20. The same statutes regulate midstream and downstream activities, except for the local distribution of piped gas, which falls under the jurisdiction of each federation state.

21. The Conselho Nacional de Política Energética is composed of nine federal ministers and three representatives of society. It ordinarily meets every six months and also occasionally on an ad hoc basis.

22. In terms of hierarchy, a simple summary would be that while the ANP does not respond to CNPE or the MME, it carries out the policies and decisions at an executive level that were taken by such entities. For instance, in terms of the bid

rounds, the CNPE defines and decides the blocks for future concession or production sharing regime bid rounds (Section 2, Item VIII of the Petroleum Law). The ANP performs the geotechnical studies and implements the bid (Section 8, Item II of the Petroleum Law).

23. In the past years, the ANP and regulatory agencies have experienced a decrease in autonomy, with budgetary constraints, long periods with vacant board seats, and more interference by the central government.

24. Round zero was not a public bid process but rather the entering into of concession contracts between the ANP and Petrobras to govern the then-existing E&P rights held by Petrobras, which were granted before the Petroleum Law was passed. In August 6, 1998, the ANP executed 397 concession contracts with Petrobras in relation to fields already in development of production phases.

25. As some pointed out at the time, the federal government could arguably have achieved the same economic result by changing the government takes structure then in existence. For example, by simply raising the percentage of one of them (e.g., special participation), the concessionaire would have to forego a higher stake of production earnings as government take, without disrupting a framework that was considered a success, having been proven effective for more than a decade with the participation of many foreign investors and the strengthening of Petrobras.

26. The reforms also resulted in the creation of the Social Fund, fully owned and controlled by the federal government to manage part of the income earned under E&P activities of the presalt and strategic areas (Federal Law 12,351/2010), and also a material change to how royalties and special participation are shared within the federation bodies (i.e., federal union, state, federal district, and the municipalities).

27. The PSC Law also made amendments to the Petroleum Law, such as altering the powers of the CNPE and the ANP.

28. Following the execution of the onerous assignment contract granting to Petrobras the right to explore and produce up to 5 billion boe, Petrobras found out that the limits of the hydrocarbons deposits extrapolate the boundaries of the relevant block. In June 24, 2014, the CNPE authorized the execution of PSC contracts for the production of the "exceeding deposits" against payment of R$2 billion as a signing bonus and compensation for the barrels that supersede 5 billion boe (Resolution No. 1/2014). The payment of this compensation is now being renegotiated between Petrobras and the federal union.

29. In addition to the concession contract, other important statues are ANP Resolutions No. 37/2007, 38/2007, 39/2007, and 19/2013, which constitute the regulatory framework for the local content certification system (the system whereby the E&P companies must hire an accredited entity to evidence compliance with the local content obligation). The particularities of the local content certification system will be explained further in this chapter.

30. The 13th concession round occurred in October 2015. At the time this chapter was written, ANP had just released the 14th concession round draft tender protocol and concession contract. The round is expected to occur in fourth-quarter 2017, and the final versions of both such documents can change when compared to the draft version.

31. The expression "local" in these bullets is used in a generic way. There are several rules that determine whether a service or good is local. For the purposes of this article, it suffices to say (i) that it is possible for suppliers of goods and service providers headquartered abroad to furnish local content goods and services; and (ii) a good can

be partially local, meaning it would have a percentage of local content out of its total cost.

32. Section 20.1.4 of the draft of the 14th concession round contract.

33. One aspect that has been criticized by the industry is the possibility admitted in administrative regulation that the certificate may be revoked by the *certificadora*. According to Section 50 of ANP Resolution 19/2013, a certificate may be cancelled if the *certificadora* incurred formal errors when measuring the percentages, or if an item or subitem cannot be identified within the certificates, or even if during an audit conducted by ANP, the willful misconduct of the *certificadora* may be evidenced in regard to the listing (or absence in listing) of an item or subitem.

34. The concession contract has an overarching provision that sets out that noncompliance with obligations set forth in the contract may subject the concessionaire to sanctions, which are imposed by statute. This would be the case, for instance, if the concessionaire does not include Brazilian suppliers in an award for contracting a good or service (where exceptional waiving conditions set out in the contract are not met).

35. Penalties are admitted under Brazilian law differently from certain common law jurisdictions. It is not clear in the contract whether the payment of the penalty would liquidate the obligation or if the concessionaire could be subject to other penalties. Although there are excerpts that indicate that an additional claim for damages could be made, we are not aware of any precedent in that sense.

36. Please see an example with the different financial impacts for the noncompliance of local content percentages based on the ANP's tender protocol information for the 13th round. For the purposes of this example, we assumed a default by the E&P company during the exploration phase of a deepwater block:

13th Round tender Protocol Deep Water Blocks (Exploration Phase) **Min. 37% – Max. 55%**	**Offered percentage of Local Content** (Committed)	**Accomplished Local Content** (Realized)	**Contractual Penalty NCP% < 65% = 60% NCP% = or > 65% = 8.NCP – 1/7**	**Financial Impact**
Total Value: 200M	50% 100MM	80% 80MM	NCP = 20%, 20MM Penalty = 60% or (0.6) 0.6 × 20MM = 12MM	**12MM**
Total Value: 200M	50% 100MM	10% 10MM	NCP = 90%, 90MM Penalty = (8 × 90%) – 1/7 = 0.88 0.88 × 90MM = 79.2MM	**79.2MM**

37. The CNPE authorized the ANP to extend the contracts term of bidding round zero (Resolution No. 2/2016). Petrobras or the relevant current concessionaire that assumed Petrobras' rights in such round zero assets (by farm-in operations) shall undertake an additional investment commitment on production initiatives. The ANP has already authorized four extensions, Marlim and Voador, Araças, and Ubarana for an additional 27-year period.

38. The certification system is composed of four ANP resolutions. ANP Resolution 19/2013 (which revoked and replaced resolution 36/2007), which defines the criteria and procedures to perform the certification activities; ANP Resolution 37/2007, which defines the procedures to register and accredit certifying entities; ANP Resolution

38/2007, which provides the audit procedure for the local content certification of accredited entities; and finally ANP Resolution 39/2007, which sets forth the methodology for the investment reports that must be submitted to ANP quarterly to evidence the fulfillment of the local content commitment.

39. The ANP divides the local content proposal in systems. Each of the systems is divided into items and subitems. The exploration phase has, for instance, seven systems (e.g., logistics support, drilling and completion, assembling and integration of modules, etc.). Within the logistics support system, for instance, there are four different items (e.g., acquisition of data, interpretation and processing of G&G data, drilling units, and Christmas trees).

40. The following formula has been adopted for the 14th bid round: 80% for the signing bonus and 20% for the minimum employment program (PEM).

41. Please see below the minimum percentages of local content under the PSC (note that P* = depth in meters):

Location of the Block	Exploration Phase	Development Stage (Modules with first oil until 2021)	Development Stage (Modules with first oil until 2022)
	Minimum (%)	Minimum (%)	Minimum (%)
Deep/Ultra Deep Water P* > 400m	37	55	59

Source: PSC's tender protocol, "Edital de Licitação para a Outorga do Contrato de Partilha de Produção Disposições Aplicáveis às Atividades de: Exploração e Produção de Petróleo e Gás Natural Primeira Licitação de Partilha de Produção," Agência Nacional do Petróleo, Gás Natural e Biocombustíveis (ANP), 41, http://www.brasil-rounds.gov.br/arquivos/Edital_p1/Edital_autorizado_030913.pdf.

42. The place of incorporation or origin of funds of the ultimate parent company of the local subsidiary of the service provider is irrelevant for the purposes of establishing whether or not the good or service provided qualifies as local content.

43. TCU decision no. TC 030.511/2015-1 Group I, Class V, Plenary (3072–2016), paragraph 256.

44. Ibid., paragraph 219.

45. Ibid., paragraph 478.

46. Ibid., paragraph 254.

47. Available at http://www.anp.gov.br/wwwanp/fiscalizacao/fiscalizacao-de-conteudo-local.

48. Approximate numbers, applying an exchange rate of US $1.00 = R$ 3.28.

49. In his first successful campaign for the presidency, Mr. Lula da Silva raised the flag that Petrobras was procuring production units abroad, claiming that as president he would take actions to increase local content, ultimately generating jobs and adding value to the procurement chain in Brazil.

50. This feature was a suggestion set forth in section 1, item IV of the PEDEFOR Resolution No. 1/2017 dated March 28, 2017.

51. For instance, it could be subject to a liability cap; there may be difficulties with enforcement depending on the magnitude of the liquidated damages vis-à-vis the financial situation of the service provider, etc.
52. This is true for several quality reasons, but also because hiring from a more experienced player should mean a greater chance of the good being delivered on time and on price.
53. Section 20.2 of ANP's 10th concession round agreement.
54. Usually the service companies need some ability to foresee when mobilization will begin. If the E&P company has an option to hire for a long window, the service company may suffer an opportunity cost that may end up being factored in the price, thus causing inefficiency in the process.
55. Sections 20.7.2 and 20.12 of ANP's 12th and 13th round concession contracts, respectively.
56. Excerpt of ANP Attorney's Office Legal Opinion: *PROGE 109/2014/PF-ANP/PGF/AGU*, paragraph 13, 3.
57. Failure to comply with the local content obligation results in a penalty that is calculated by applying a formula, as discussed in the text. Therefore, if the waiver has been granted, the question then is how that decision is considered in the calculation of the penalty.
58. This opinion was issued in March 2012 and was based on the 7th to 10th concession contracts.
59. The interpretations were: (i) to exclude altogether the investment in the relevant subitem, which would result in a change of the invested amount and the weight (proportion) set for the item; and (ii) to consider the relevant item with 100% local content (ANP Attorney's Office Legal Opinion: *PROGE 50/2012/PF-ANP/PGF/AGU*).
60. This opinion seems to be a deviation from the concept of the waiver, given that the concessionaire will be punished for an event that is completely beyond its control (unavailability of local providers). In addition, the use of the floor percentage does not seem to be adequate, as it was a mere indicative element in the bid, not binding the parties. If it has legal value, E&P companies that meet the minimum set forth in the bid but do not meet the threshold submitted in the bid (i.e., meet the 55% but not the 75% in our example) could then allege that failure to achieve the additional 20% should not result in penalties, as the 55% seems to be the effect that ANP was seeking when disclosing the bid documents.
61. Even this line of thinking needs clarification, as it is not clear whether the waiver should be requested in relation to all contracts signed for a particular item or on a per-contract basis.
62. Section 20.7 of 11th concession round agreement.
63. ANP Attorney's Office Legal Opinion: *PROGE 109/2014/PF-ANP/PGF/AGU*, paragraphs 6–10 and 12–14.
64. Curiously, the two attorneys subscribing the opinion had different views as to how this effect would be achieved: if by means of an offset of a penalty already applied to a category, or if in the calculation of the penalties, the investment that was already the subject of a penalty being calculated in the context of another category. The latter was the system chosen in the 13th round concession agreement.
65. Decree No. 8,637 from January 15, 2016.

66. Section 2, item II, letter "d" of Decree No. 8,637 from January 15, 2016. Such a mechanism still needs further development.
67. The effectiveness of this program requires further regulation from the ANP.
68. Please see the formula below:

$$CGN = \left[\sum_{i=1}^{n} Pl_i \times PMLI_i\right] \times \frac{CGO}{PGML} \qquad \sum_{i=1}^{n} Pl_i \ i{=}1$$

Where

CGN	is the neutralized global local content commitment,
CGO	is the offered global local content commitment,
PGML	is the minimum global local content percentage to offer at the bid,
PMLI$_i$	is the minimum local content percentages to items and subitems (*i*) to offer at the bid, and
PI$_i$	is the weight effectively performed in relation to the items and subitems (*i*).

5

CANADA

Nigel Bankes

Introduction[1]

Canada is a federal state, and the provinces and territories play a dominant role in the ownership of oil and gas resources and in making laws for the development of those resources.[2] However, the federal government does have some significant involvement in both the offshore and northern Canada. While the Supreme Court of Canada has consistently ruled that resources located in marine areas outside the provinces fall under federal control,[3] the federal government has preferred to negotiate political accords with coastal provinces and to establish cooperative arrangements with these jurisdictions for the development of offshore resources.[4] These arrangements are reflected in so-called mirror legislation, which to date has established separate joint management boards for the offshore areas pertaining to Nova Scotia and Newfoundland and Labrador.[5]

The legal and political picture in northern Canada is characterized by the progressive devolution of responsibilities for onshore oil and gas activities from the federal government to the three territories. Thus, Yukon obtained control of its oil and gas resources in 1998, and the Northwest Territories assumed control of its onshore oil and gas resources in 2014. The oil and gas resources in Nunavut are still controlled by the federal government, as are the offshore resources throughout Canada's Arctic.

Given this geopolitical structure, there is the potential for considerable variation in the development of local content rules in the oil and gas sector. Accordingly, this chapter discusses the local content requirements of four different jurisdictions: Alberta, Yukon, the federal requirements in northern Canada (i.e., Nunavut and all Arctic offshore), and the Newfoundland and Labrador offshore. Alberta is selected because it is the most important oil- and gas-producing jurisdiction in Canada; Yukon because it recently obtained control over its oil and gas resources; Nunavut and the Arctic offshore because they remain under federal control; and Newfoundland and Labrador as the jurisdiction that has been the most aggressive in the development and application of local content requirements.

In examining these jurisdictions, we would expect to see considerable diversity in local content requirements, principally due to the maturity of the oil and gas sector. In general we would expect that the more mature the sector, the less need there is for prescriptive local content requirements. Another important variable in the development of local content requirements is the significance of the Aboriginal population in the jurisdiction and the existence of modern land claim agreements that address local employment and benefits issues.[6]

Nevertheless, one factor may lead to greater uniformity over time: the role of Canada's international investment law commitments, particularly those found in the North American Free Trade Agreement (NAFTA).[7] While this agreement grandparented some existing "domestic performance requirements," Article 1106 of NAFTA will prevent the introduction of any new measures that accord preferences to domestically sourced goods and services or which require an operator to achieve a prescribed domestic content.[8]

The next section provides some background on each of the four jurisdictions, and then the following four sections examine the law and practice with respect to local content requirements in each of the four jurisdictions. By far the longest section deals with Newfoundland. Of the jurisdictions covered here, Newfoundland has the most experience with the implementation of local content requirements.

Background: The maturity of the oil and gas industry in each subjurisdiction

Alberta is a mature oil and gas jurisdiction, with production going back to the first decades of the 20th century. Conventional reserves are declining; however, there is considerable potential for growth in the oil sands sector, even though current oil prices (ranging between US$50 per barrel in the first quarter of 2015 and as low as US$30 in the first quarter of 2016) will certainly dampen investment.[9] All of the major international oil companies and service companies are active in Alberta, along with domestic companies. Alberta-based companies are also active globally. There is a sophisticated service structure to support the industry, and the two major universities in the province, the University of Calgary and the University of Alberta, have strong research and teaching programs in geology and engineering.

Yukon is a very young oil and gas jurisdiction. There was some limited natural gas production in southeast Yukon until 2012, but only 76 wells have been drilled in that territory.[10] Nunavut and the Arctic offshore generally present a similar picture. There has been some exploratory drilling and a small amount of production from the Bent Horn field in the 1980s.[11] Known reserves in the shallower waters of the Beaufort Sea and in the Arctic islands of Nunavut remain undeveloped because of high costs and the absence of necessary transportation infrastructure. There are no oil and gas companies based in Yukon or Nunavut, no specialized service sector to support the industry, and no university programs in engineering and geology.

Newfoundland falls somewhere between Alberta and the two Arctic jurisdictions just discussed. While it has no onshore production, the Hibernia field was discovered in 1979 and continues to produce, along with two other fields, Terra Nova and White Rose. Several other projects are in the development or advanced exploration phase, including the Hebron Project.[12] Many major international oil companies are active in this offshore area. Memorial University, Newfoundland's main university, offers programs in earth sciences, engineering (including ocean and naval architectural engineering), as well as a variety of programs related to marine resources.[13]

While Canada and its subnational jurisdictions have some experience with (state) national and provincial oil companies and direct equity investment in the oil and gas sector,[14] the main national oil company, Petro-Canada, was privatized beginning in 1991. Provincially owned oil and gas corporations have also been privatized.[15] Nova Scotia did take an active role in developments in its offshore area through a provincial Crown corporation Nova Scotia Resources Limited (NSRL), with only limited success and some investment failures. Nova Scotia subsequently divested itself of NSRL assets between 1995 and 2003.[16] Newfoundland has taken a direct equity investment in some of the more recent projects that have proceeded to production in its offshore area.[17]

Local Content Policy and Baseline

Local content requirements: Alberta

Alberta has no legally prescribed local content requirements for the oil and gas sector (including the oil sands) beyond the purely formal requirement of a registered presence in order to hold a crown petroleum and natural gas lease or license.[18] Thus, local content commitments are not a condition precedent to acquiring Crown oil and gas rights or developing them. The main provincial regulator is the Alberta Energy Regulator (AER), formerly the Energy Resources Conservation Board (ERCB). The AER may consider local and provincial benefits as part of assessing a major project, such as a new oil sands mine, but only as part of a general cost-benefit analysis rather than to assess whether the proponent has complied with any particular local content requirements.[19]

While there are no legally prescribed local content requirements in Alberta, oil and gas companies, and in particular major oil sands operators, do negotiate impact and benefit agreements with First Nation and Métis communities who may be affected by their projects. These agreements are generally confidential and are negotiated as part of the implementation of corporate social responsibility policies and as part of obtaining a social license to operate rather than a legal license.[20] Such agreements undoubtedly facilitate operators in navigating regulatory approval processes and also may be referred to and relied upon by the government as satisfaction of its constitutional obligation to consult and accommodate First Nations and Métis communities whose rights may be affected by project approval decisions.[21]

Local content requirements: Yukon

The Yukon Oil and Gas Act[22] largely follows the model of Alberta's legislation, but it does contain certain local content provisions designed to provide benefits for members of Yukon First Nations and their businesses. Thus, Section 68 of the Yukon Oil and Gas Act provides that no person may carry out any oil and gas activity within the traditional territory of a First Nation in the absence of a "benefits agreement." A benefits agreement must provide undertakings by the licensee:

> to provide opportunities for the Yukon First Nation, its citizens, residents of communities affected by the oil and gas activity, and other Yukon residents

i. to obtain employment and training, and
ii. to supply goods and services for the licensee and the licensee's contractors.

The agreements should provide "benefits that are commensurate with the nature, scale, duration, and cost of the work to be undertaken by the licensee" but should "not place an excessive burden on the licensee." Since the act was passed, there has been very little exploration in Yukon, and no significant discoveries. Hence there is very little practice to draw on in order to evaluate whether the provisions are being implemented and how effective they are.

Local content requirements: Nunavut and the federal offshore area

As noted above, the federal government of Canada has exclusive jurisdiction over oil and gas developments in Nunavut and in the Arctic offshore area. The main legal instruments are the Canada Petroleum Resources Act[23] (CPRA) and the Canada Oil and Gas Operations Act[24] (COGOA). The CPRA provides the leasing regime and the COGOA the regulatory regime. Both statutes address local content requirements, the latter in somewhat more detail than the former. Section 21 of the CPRA simply provides that "no work or activity on any frontier lands that are subject to an interest shall be commenced until the Minister has approved, or waived the requirement of approval of, a benefits plan in respect of the work or activity [which requires an approval under the COGOA]."

This provision has a very broad ambit since it applies to all activities and thus covers both exploration activities and development and production activities.

By contrast, the more prescriptive COGOA provisions on local content are triggered when a proponent presents a development plan that is effectively a concrete proposal addressing how it proposes to develop and produce a pool or field that it has discovered.[25] Section 5.2(2) stipulates that the National Energy Board (the primary federal energy regulator for these lands) shall not approve a development plan until the minister of Aboriginal Affairs and Northern Development Canada (AANDC) has approved a benefits plan or waived the requirement for such a plan. Section 5.2(1) defines a benefits plan as

> a plan for the employment of Canadians and for providing Canadian manufacturers, consultants, contractors and service

> companies with a full and fair opportunity to participate on a competitive basis in the supply of goods and services used in any proposed work or activity referred to in the benefits plan.[26]

This basic definition, however, is evidently qualified by Section 5.2(3), which indicates the following:

> The Minister may require that any benefits plan...include provisions to ensure that disadvantaged individuals or groups have access to training and employment opportunities and to enable such individuals or groups or corporations owned or cooperatives operated by them to participate in the supply of goods and services used in any proposed work or activity referred to in the benefits plan.[27]

The Department of Aboriginal Affairs and Northern Development Canada has elaborated on these requirements in a document entitled Benefits Plan Guidelines for the North.[28] The guidelines apply to the benefits plan provisions of both the CPRA and COGOA. Section 4 of the guidelines establishes the principles and objectives for benefits plans. The guidelines emphasize the importance of communication between an operator and northern stakeholders and northern Aboriginal organizations. The guidelines do not prescribe any particular targets, but they do insist that a "Benefits Plan should ensure that Northern Aboriginal residents and Northern Aboriginal businesses, as well as other Northern residents and Northern businesses, are provided an opportunity to participate in and benefit directly from oil and gas work or activities on Northern frontier lands."[29] Plans should include a "Northern preference" that reflects a preference "in which first consideration for employment and business opportunities should be extended to qualified Northern Aboriginal and other Northern residents and businesses."[30] More specifically, the plan must respect a preference for local northern Aboriginal residents and businesses, followed more generally by northern residents and businesses in any of the three territories, followed by other Canadian residents and businesses. The plan must include a description of both the operator's training and employment strategies, including forecasts as to planned hiring and the same for procurement and business opportunities.[31]

The guidelines are very general in nature, but the introduction to the document indicates that "once a Benefits Plan is approved by the Minister, an operator is obligated to endeavor to fulfill the commitments therein."[32] Approval takes the form of a decision report that may impose further conditions on the proponent.[33] Much as in the case of Yukon, it is difficult to evaluate actual practice in Nunavut and the Arctic offshore areas since there has been so little activity in recent years.

Local content requirements: Newfoundland and Labrador

The Canada-Newfoundland and Labrador Atlantic Accord Implementation Act[34] (hereafter Accord Act) effectively combines in one statute the basic contents of both the CPRA and the COGOA, with some important additional elements. One of these elements is the Canada-Newfoundland and Labrador Offshore Petroleum Board (CNLOPB), which is charged with implementing the legislation. The Accord Act is much more specific than the CPRA and the COGOA when it comes to local content requirements. The board's guidance documents and actual practice are certainly much more detailed, prescriptive, and demanding than their federal counterparts in relation to Nunavut and the Arctic offshore discussed in the previous section. In addition, the decision-making structure of the Accord Act makes it possible for the province to play a very active role. As a result, current practice requires a proponent to negotiate a benefits contract with the province as well as securing the board's approval of a benefits plan. There is therefore effectively a twin track process to the negotiations of local content requirements. One track, that expressly prescribed by the Accord Act, is under the supervision of the board; the second track involves direct negotiations between the proponent and the province. It is not expressly provided for in the Accord Act but rather depends on the proponent's need to secure the province's approval for its overall development plan. Paradoxically, the latter is much more significant than the former.

The account begins with a discussion of the express provisions of the Accord Act dealing with benefit *plans* and then deals with the legal basis for provincial demands for benefit *contracts*. The subsequent section examines the contract and benefit plan provisions for one recent project, the Hebron Project, as a way of illustrating how these schemes fit together.

The chapter then turns to examine the efforts of a third party to enforce the terms of a benefits plan. This is in turn is followed by a discussion of the CNLOPB's efforts to elaborate on the Accord Act through a set of guidelines. The final section discusses the board's contested efforts to establish and enforce a set of guidelines designed to achieve what it considered to be an appropriate level of research and development investment by project proponents during the production phase of a project. These efforts resulted in both domestic administrative law litigation and an investment treaty arbitration.

The Accord Act. The Accord Act is designed to implement the terms of the agreement between Canada and Newfoundland for the management of offshore petroleum resources and revenue sharing with respect to those resources (the Atlantic Accord).[35] Two elements of the Accord Act are crucial here. The first and most obvious is Section 45, which deals with benefits plans. The second element is the provisions of the Accord Act that afford the province an effective veto over the terms of development and which therefore provide the anchor and leverage for the benefit contracts.

Section 45 of the Accord Act follows the CPRA and the COGOA in providing that the board should not approve any work or activity or any development plan until it has approved a Canada-Newfoundland and Labrador benefits plan. The definition of such a plan follows that quoted above in COGOA, although the Accord Act goes on to state that

> (3) A Canada-Newfoundland and Labrador benefits plan shall contain provisions intended to ensure that
>
> (a) before carrying out any work or activity in the offshore area, the corporation or other body submitting the plan shall establish in the Province an office where appropriate levels of decision-making are to take place;
>
> (b) consistent with the *Canadian Charter of Rights and Freedoms*, individuals resident in the Province shall be given first consideration for training and employment in the work program for which the plan was submitted and any collective agreement entered into by the corporation or other body submitting the plan and an organization of employees respecting terms and conditions of employment in the offshore area shall contain provisions consistent with this paragraph;
>
> (c) expenditures shall be made for research and development to be carried out in the Province and for education and training to be provided in the Province; and
>
> (d) first consideration shall be given to services provided from within the Province and to goods manufactured in the Province, where those services and goods are competitive in terms of fair market price, quality and delivery.
>
> (4) The Board may require that any Canada–Newfoundland and Labrador benefits plan include provisions to ensure that disadvantaged individuals or groups have access to training

> and employment opportunities and to enable those individuals or groups or corporations owned or cooperatives operated by them to participate in the supply of goods and services used in any proposed work or activity referred to in the benefits plan.

The board must consult with both federal and provincial governments as part of reviewing a plan.

At the development stage of a project, a proponent's benefits plan will be subject to a public review. For the purposes of that review, the proponent must file a preliminary benefits plan.[36] The review panel report is then considered by the board, which must decide whether to accept the plan as submitted, with or without conditions. A recent example is provided by the Hebron Project. The project proposal, including a lengthy benefits plan,[37] was subject to a review by a hearing commissioner.[38] Next the board had its own staff assess both the plan and the subsequent commissioner's report,[39] in which it issued its decision report on the benefits plan and the development plan. The board then approved the plan subject to terms and conditions.[40] In the case of the benefits plan, there were three conditions. One required the proponent to update its human resource requirements for the construction phase and to take the necessary steps to develop a plan to deal with possible labor shortages. A second required the proponent to develop more detailed and descriptive information on the contracting opportunities for the construction of the production facility, a stand-alone concrete gravity base structure (GBS), and the topsides. A third condition required the proponent to develop a project-wide policy to ensure that bid packages will be "scaled to enable the participation of Canadian companies with first consideration given to the capacity of local industry."[41]

While the legislation contemplates that the approval of a benefits plan is the responsibility of the board, it is equally clear that the provincial government also plays a crucial role because of the development plan provisions of the Accord Act. The Accord Act designates certain decisions that must be made as "fundamental decisions." A fundamental decision requires the approval of both the federal and provincial ministers.[42] One such decision is the approval of a development plan.[43] The development plan must detail the scope and timing of the proposed development, the production rate and modes of recovery, and the production system, including possible alternatives. While a development plan (as already noted) cannot be approved without the prior approval of a benefits plan, the proponent will also engage in direct discussions with the province as to the economic terms of the development. These terms will include the fiscal arrangements for the project, which will be recorded in an agreement. The terms will also include (at least with respect to recent projects such as the Hibernia South Expansion, the White Rose Expansion, and Hebron) a separate benefits agreement. This agreement takes the form of a contract between the project joint venturers and the province. Thus, local benefits issues will be dealt with in two documents: a benefits contract between the proponents and the province, and a benefits plan approved by the board.

The Accord Act does not itself require a benefits contract, but it is clear that proponents are motivated to enter into such contracts in order to secure provincial approval of the overall development plan. The benefits contract will typically already be in place by the time the proponent applies to the board for approval of the development plan.[44] The Hebron Benefits Plan describes the relationship between the plan and the contract as follows:[45]

> The Benefits Plan also takes into account the Hebron Benefits Agreement.... The Agreement requires that certain expenditure and activities occur in the Province, and specifies plans, processes and mechanisms for delivering these benefits. This Plan will ensure that the Project meets the commitments in the Agreement.

Denstedt and Thrasher describe the actual practice as follows:[46]

> In practice, developing an acceptable benefits plan (*sic*) requires involvement from project proponents, provincial authorities, federal authorities and the...Board. Because of the highly political nature of a benefits plan, the typical experience involves an initial negotiation between the provincial government and the project proponents. Once an agreement or understanding on appropriate benefits has been reached, this will form the basis of the benefits plan submitted by the proponents to the...Board.

The Hebron Project Benefits Agreement and the Hebron Benefits Plan. The Hebron Benefits Agreement is a 35-page contract between the province and the working interest parties.[47] The agreement is signed at the highest level on the provincial side by both the premier and by the minister of Natural Resources, and it contains a lengthy annex dealing with dispute resolution procedures. The core of the agreement is Section 5, which contains prescriptive values for a number of elements of the project, including the following:

- Proponent's project team and office
- FEED engineering (front-end engineering and design)
- Detailed engineering
- Choice of production method
- Fabrication
- Procurement and contracting
- Research and development (R&D)

For example, under the heading of project team and office, the proponents agree to establish a project office in the province and agree that there will be not less than 1 million person hours of project team activities in the province before production start-up. On the subject of production method, the proponents agree that concrete gravity base structure (GBS) is the selected production platform for the project and that the GBS will be constructed in the province. Under "fabrication and other works," the proponents recognize that the construction of certain components will take place in the province (e.g., flare boom, helideck, subsea drilling template). Finally, under the heading of R&D, the proponents agree to invest $120 million over the life of the project in consultation with the province.

By contrast, the benefits plan is not a negotiated contract. Instead the benefits plan represents a proposal from the project proponents as to how they will conduct their business so as to meet their contractual commitments as well as the requirements of the Accord Act. The plan is written in much less prescriptive terms. Following an introduction to provide context with respect to both the project and the subject of benefits, the plan describes the joint venturers' benefits approach, underlying principles, and policy guidelines and procedures. This section of the plan, for example, describes how the proponent will approach engineering procurement, how it will create a benefits culture, and how it will engage in supplier development, including details as to its communication plans. The proponents indicate how they will identify supply gaps and new opportunities, and the plan describes how the proponents will recruit and select qualified candidates. An important part of the plan is the proponents' capacity assessment for two different phases of the project (construction and operations) and for the different project activities.

In sum, whereas the benefits agreement with the province contains firm targets and commitments, the benefits plan is more concerned with describing in concrete and detailed terms how those targets will be realized and the information and policy parameters that will inform that effort.

Enforcing the terms of the Terra Nova Benefits Plan. It is apparent that a benefits plan may need to be framed in general or aspirational terms, leaving the details to be worked out in the future. This may lead to conflict. A case in point is the benefits plan developed for the Terra Nova Project, which ultimately led to litigation when the board agreed to accept "best efforts" compliance with one of the conditions that had been included.[48] In that case the board had approved the proponent's benefits plan subject to a number of conditions. One condition that became contentious was that "as soon as practicable after Project Sanction, the Proponent relocate engineering and procurement activities for the Project to Newfoundland."[49] This condition was imposed because of the review panel's concerns that jobs would be lost if this work were not undertaken in the province. The board noted as follows:[50]

> The Proponent has chosen to establish an alliance with a group of contractors and key suppliers to execute project-phase work up to first oil production. Work to be undertaken by the Terra Nova alliance includes engineering, procurement, construction and modification, installation, commissioning and pre-drilling. In its report, the Panel stated its opinion that the alliance approach provided an improved contracting approach to ensure that the provisions of the Accord Acts were met. In its Benefits Plan, the Proponent committed to use its "best efforts within the competitive bidding process to cause the project management and engineering work for the production platform, associated subsea facilities, mooring and loading systems and production risers for the development to take place in Newfoundland."

The proponent subsequently took the view that there would be a significant incremental cost to moving these activities to Newfoundland and that such a move would also cause delay. Instead, the proponent proposed taking some Newfoundland-based staff to the UK for further training. The board ultimately, and reluctantly, accepted this proposal as satisfying the condition. While the board was therefore not in a position to complain about noncompliance, the city of St. John's, believing that it and its residents would be major beneficiaries of any relocation, took the view that the board ought to have enforced the condition. The city commenced an action to force the board to do so. The court ultimately held that the city was not in a position to maintain the litigation and that the board did not owe the city a duty to enforce such a condition, but the litigation does illustrate a couple of propositions. First, it confirms that benefits plans will inevitably have to be adjusted as projects evolve and as implementation challenges become apparent. This does not mean that benefits plans are unenforceable, but it does mean that regulators will have to be practical in assessing whether or not it is realistic to demand performance of commitments made at the time of project approval. In making that assessment, the regulator may need to be cognizant of trade-offs between strict observation of commitments and delays in bringing the project onstream.[51] Second, while benefits plans are designed to benefit residents and resident corporations, the principal party responsible for enforcing the terms of such a plan is the regulator rather than the third-party beneficiaries. The regulator has some discretion in deciding to accept substandard performance,[52] but it should be emphasised that the board's authority is limited to the plan. The board has no authority with respect to benefits agreements or contracts between a proponent and the province.

The CNLOPB's guidelines. The board has elaborated on the statutory requirements of the Accord Act in a set of consolidated guidelines.[53] The consolidated guidelines expand upon each of the elements of the statutory section, i.e., office in the province, full and fair opportunity and competitive basis, first consideration (employment), first consideration (procurement), R&D and education and training, and disadvantaged individuals or groups.

The consolidated guidelines contain additional and more specific guidance in a set of appendices. In particular, appendix I provides guidance on exploration benefits, and appendix II, adopted in October 2004, provides guidance for R&D expenditures. These latter guidelines offer an important statement as to the philosophy behind these specific guidelines, although the statement is evidently applicable more generally to the consolidated guidelines:[54]

> Research & Development represent one avenue whereby the exploration for, and the development and production of the petroleum resources in the Newfoundland Offshore area can make a contribution to the sustainable development of the Province. This was the vision or intent of the legislators at the time when they inserted the requirement for Research & Development and Education & Training "in the Province" into the Atlantic Accord legislation. The petroleum resource is finite and exhaustible, and it is the intent of this provision of the legislation that its exploitation create a lasting economic legacy for the people of the Province. This is best achieved by building on the intellectual capital and human resources

> of the Province. Achievement of this legislative intent is a key reason why some parameters or guidance are required in respect of the requirement in the Act that there be expenditures in the Province for R&D.

The guidance on exploration benefits (appendix I) is directed at operators engaged in exploration activities, including geophysical and geotechnical activities as well as drilling.[55] As might be expected, the guidance is more general than the guidance for a benefits plan to accompany a development plan. Nevertheless, it still requires an operator to address a number of factors, including: an assessment of supply capabilities for the province and for Canada, plans for communicating its requirements in a timely way to suppliers and manufacturers, the preparation of bid specifications and bid packages in such a manner so as to ensure that they do not unfairly prevent provincial and other Canadians from participating, and mechanisms to ensure that any commitments made by the operator are passed on to its contractors.[56] A key part of preparing a benefits plan in the exploration stage is the presentation of a work plan listing such things as proposed geophysical or other surveys (length, number, duration), seismic processing, number of wells, drilling units, seismic vessels, and support vessels. It will also list total expenditures, a forecast of direct employment and crew rosters, and a list of contracts and procurements over $100,000.

The guidelines offer a more detailed statement of expectations of a benefits plan designed to accompany a development plan. At this stage the operator "should provide an analysis of the potential for Canadian companies, and in particular Newfoundland and Labrador companies, to participate in the engineering, supply, fabrication, construction, operation and support activities of the project."[57] A similar analysis should be prepared of the labor needs of the project. Recognizing that project construction has the potential to offer significant opportunities, the guidelines distinguish between the construction and production phases of a development.

The guidelines also recognize that there will be a transition phase between the time when an operator makes a discovery and the time when it proceeds to file a development plan. The board clearly indicates that the more rigorous expectations that apply during the development phase should also apply in this transitional phase. Thus the guidelines warn the operator as follows:[58]

> Any pre-development procurement and contracting activity, which is in support of preparing a development application, will be monitored in accordance with the provisions of the approved Exploration Benefits Plan.

> An operator's activity with respect to any long lead procurement or contracting for the project under consideration, prior to the approval of the project Benefits Plan, will be at the operator's own risk, if it does not conform to the Benefits Plan which is subsequently approved. However, this risk may be mitigated to the extent such activity is conducted in accordance with these guidelines, and to the extent there have been diligent consultations.

Monitoring and reporting are a key part of implementing the plans.

The CNOPB's R&D rules. As noted above, Section 45(3)(d) of the Accord Act expressly addresses research and development and education and training (compendiously, R&D). The R&D guidelines adopted by the board in 2004 have proven to be very contentious and indeed have been the subject of both domestic administrative law litigation[59] and an investment treaty arbitration commenced by Mobil and Murphy Oil in respect of both the Hibernia and Terra Nova projects.[60] The board first issued guidelines in 1986 and revised them successively in 1986, 1987, 1988, and 2004. The early versions of the guidelines focused on the exploration phase rather than the development phase of project activities. The 2004 guidelines introduced the idea of an industry-wide benchmark against which to measure R&D investments. The board made this change to the guidelines because of concerns that R&D investments were falling as the projects moved into the production phase.[61] The next section discusses the evolution of the guidelines and the ensuing litigation.

Benefit plans were established for both Hibernia (1985) and Terra Nova (1997). These plans were heavy on principles rather than detailed commitments. The 2004 R&D guidelines are considerably more prescriptive and require operators to make certain levels of expenditure on R&D matters for each of the exploration, development, and production phases of the project. The required expenditure is determined by reference to Statistics Canada data for average R&D expenditures in the upstream oil and gas industry. In default thereof, operators are required to contribute any shortfall to a board-administered R&D fund. The guidelines are made enforceable by making them a condition of the operator's production operations authorization (POA), which is renewed annually.[62] When the board sought to apply these new guidelines not only to new projects but also to the existing Hibernia and Terra Nova projects, the investors in these projects sought to contest the guidelines in both the domestic courts and ultimately before an international investment tribunal.

In the domestic judicial review application, the operators sought to question not only the application of the guidelines to these two preexisting projects but also the board's authority to issue the guidelines at all. The operators were unsuccessful both at first instance and in the court of appeal.[63] The majority of the court of appeal concluded that it was reasonable for the board to apply the new guidelines to existing projects. The court noted that Section 45 of the Accord Act requires that a benefits plan "contain provisions intended to ensure" R&D expenditures within the province, and the operator knew that its POA was subject to annual renewal depending on its performance[64] and knew that its performance would be monitored.[65] The court also concluded that the guidelines themselves were reasonable and that it was reasonable for the board to base the level of required expenditures on the standard in the industry.[66] Finally, the court concluded that it was reasonable for the board to establish the fund as an alternative means of securing compliance with the R&D objectives[67] and reasonable as well for it to tie compliance to the POA.[68]

Justice Rowe in dissent saw the matter very differently and emphasized the "colossal" sums of money involved[69] and the need for stability in the offshore investment climate.[70] In language that reflects the idea of the obsolescing bargain, Justice Rowe concluded:[71]

> [I]t is to me inconceivable that the Parliament of Canada intended to confer on the Board authority of a nature that would allow the Board to approve a project for development and, then, when billions of dollars had been invested to put the project into production, to say to the operators that they will no longer be permitted to operate the project unless they spend phenomenally large sums on matters un-related to the project because the Board has decided that it (and not a government) wishes to create for the province a "legacy", as the Board conceives this from time to time.

The operators were more successful in their NAFTA arbitration. The majority of the tribunal concluded that the 2004 R&D guidelines breached the domestic performance provisions of NAFTA (Article 1106) and were not grandparented by the exceptions provisions (Article 1108 and Annex I) of that agreement.[72] As a result, the operators/investors were entitled to damages. The tribunal issued a final award on February 25, 2015, whereby the majority awarded Mobil damages of C$13.893 million and Murphy Oil C$3.401 million.[73] Both companies have since filed notice of intent to launch additional arbitrations against Canada on the basis that the guidelines continue to breach Canada's obligations under NAFTA.[74]

Conclusions

At the outset I observed that we should expect to see considerable diversity in the local content requirements of the subjurisdictions under consideration here due to the differences in the maturity of the oil and gas sector in each subjurisdiction. The description and analysis here confirm aspects of this expectation. Thus, Alberta, the senior producing jurisdiction in Canada, does not maintain any prescriptive local content rules. On the other hand, it is Newfoundland, rather than the younger jurisdictions of Yukon and the federal lands of Nunavut and the Arctic offshore, that has the most prescriptive local content rules.

Newfoundland and the joint CNLOPB have also had the most experience in the design and implementation of local content rules. The board has experienced considerable pushback from operators in the implementation of new requirements. The board has also come under pressure from third-party interests who stand to benefit from local content requirements not to relax the terms of those requirements.

The experience with local benefits arrangements on the Newfoundland coast emphasizes that our understanding of these arrangements requires that we consider both law and practice. Thus, while the Accord Act refers to benefit plans approved by the board, the actual practice emphasizes the central importance of the contractual benefits arrangements between the province and the project proponents. The practice also draws attention to the importance of understanding the leverage that a government might be able to use to secure meaningful benefits commitments. In the case of Newfoundland, this leverage comes from the province's authority to approve a development plan rather than from the express authority for the approval of benefits plans (which is part of the board's mandate). The practice in Alberta illustrates the same points, although perhaps much more softly in the sense that while there is no formal legal requirement for a proponent to make local content or local benefit commitments, the need to secure and maintain a social license to operate and to facilitate regulatory approvals may motivate project proponents to enter into impact and benefit agreements with local communities, especially indigenous communities.

Notes

1. I would like to thank the late Jim Thistle of the McInnes Cooper law firm, St. John's, Newfoundland for his assistance. Jim gave me important feedback on a draft of this chapter in June 2015; Jim died of ALS in March 2016. I dedicate this chapter to his memory.
2. Peter Hogg, *Constitutional Law of Canada*, 5th edition (Toronto: Carswell, 2007).
3. *Reference Re: Offshore Minerals Rights*, [1967] SCR 792; *Re: Strait of Georgia*, [1984] 1 SCR 388; *Re: Newfoundland Continental Shelf*, [1984] 1 SCR 86.
4. For the background see Constance D. Hunt, *The Offshore Petroleum Regimes of Canada and Australia* (Calgary: Canadian Institute of Resources Law, 1989); and Shawn Denstedt and Jack Thrasher, "The Accord Acts Twenty Years Later," *Dalhousie Law Journal*, 30 (2007), 287.
5. For the federal legislation, see Canada-Newfoundland and Labrador Atlantic Accord Implementation Act, SC 1987, ch. 3 (hereafter Accord Act); and Canada-Nova Scotia Offshore Petroleum Resources Accord Implementation Act, SC 1988, ch. 28. These two statutes are essentially identical. For a discussion of local benefit rules in the context of the Nova Scotia offshore, see Peter Clancy, *Offshore Petroleum Politics: Regulation and Risk in the Scotian Basin* (Vancouver: UBC Press, 2011), especially chapter 7, "The Politics of Backward Linkage: Industrial Benefits."
6. For a listing of agreements and for links to the texts, see the website for Indigenous and Northern Affairs Canada, https://www.aadnc-aandc.gc.ca/eng/1100100030583/1100100030584. I do not examine those requirements here as they are properly the subject of a separate paper. These agreements generally require the negotiation of project-related impact and benefit agreements between a proponent and an affected community.
7. North American Free Trade Agreement, December 17, 1992, in force January 1, 1994, 32 ILM 289, Article 1106. Canada's model investment agreement (2004) also contains a clause forbidding the adoption of new domestic performance requirements; see Article 7 at http://www.italaw.com/documents/Canadian2004-FIPA-model-en.pdf.
8. Canada found out to its cost that NAFTA may also preclude significant revisions to existing measures: *Mobil Investments Canada Inc. and Murphy Oil Corporation v Canada*, ICSID Case No. ARB(AF)/07/4, "Decision on Liability and Principles of Quantum," May 22, 2012; redacted version released in the fall of 2012. The tribunal issued its final award on February 20, 2015. Links are available on the ICSID website, https://icsid.worldbank.org. For comment on the "Award on Liability," see Nigel Bankes, "From Regulatory Chill to Regulatory Concussion," May 6, 2013, http://ablawg.ca/wp-content/uploads/2013/05/Blog_NB_Mobil_Investments_May20131.pdf. Canada's effort to overturn the arbitral award failed: *Attorney General of Canada v Mobil et al.*, 2016 ONSC 790.
9. *Alberta's Energy Industry: An Overview* (Alberta Energy Resources Conservation Board, June 2012), http://www.energy.alberta.ca/Org/pdfs/Alberta_Energy_Overview.pdf.
10. "History of Oil and Gas Development in Yukon," Yukon Government, Energy, Mines and Resources, http://www.emr.gov.yk.ca/oilandgas/oil_gas_history.html.
11. For Bent Horn (and more generally), see Northern Oil and Gas Directorate, *Petroleum Exploration in Northern Canada: A Guide to Oil and Gas Exploration and Potential,*

1995, http://www.aadnc-aandc.gc.ca/DAM/DAM-INTER-HQ/STAGING/texte-text/nog_pubs_penc_1324579829739_eng.pdf.

12. Canada-Newfoundland and Labrador Offshore Petroleum Board, http://www.cnlopb.ca/offshore/.
13. Memorial University of Newfoundland, http://www.mun.ca/.
14. The high point of state involvement in the upstream sector in Canada was the early 1980s, which saw the adoption at the federal level of the National Energy Program under a liberal administration led by Pierre Trudeau. The main elements of this program were largely dismantled by the late 1980s under the federal conservative administration of Brian Mulroney. The Mulroney administration cemented the return to market values by negotiating the Free Trade Agreement with the United States in 1987 (the predecessor agreement to NAFTA).
15. Privatized provincial Crown energy corporations include Alberta Energy Company and SaskOil. See generally, Anthony Boardman and Aidan Vining, "A Review and Assessment of Privatization in Canada," 5, no. 4 (School of Public Policy Research Papers, University of Calgary, 2012).
16. Peter Clancy, "The Provincial State and the Entrepreneurial Impulse: NSRL," in *Offshore Petroleum Politics: Regulation and Risk in the Scotian Basin* (Vancouver: UBC Press, 2011).
17. These interests are held by Nalcor Energy, a provincial Crown corporation. Nalcor holds a 4.9% interest in Hebron, a 5% interest in the White Rose Expansion Project, and a 10% interest in the Hibernia Southern Extension Project; see http://www.nalcorenergy.com/development-and-production.asp. Some background is provided in Jack Thrasher and Simon Baines, "Developments in Newfoundland and Labrador Offshore Royalties: From Hibernia to Hebron and Back," *Dalhousie Law Journal* 37 (2014), 33.
18. Mines and Minerals Act, RSA 2000, Ch. M-17, Section 23.
19. See, for example, *Report of the Joint Review Panel: Joslyn North Mine Project*, ERCB Decision 2011-005, CEAA Reference No. 08-05-37519 (January 2011), 109, https://www.aer.ca/documents/decisions/2011/2011-ABERCB-005.pdf. In one older decision the Supreme Court of Canada struck down an attempt by the ERCB to impose an affirmative action program to create employment and other opportunities for First Nations in connection with one of the first oil sands mines in Alberta; see *Athabasca Tribal Council v Amoco Canada Petroleum Co. Ltd. et al*, [1981] 1 SCR 69.
20. On the social license to operate, generally see Jason Prno and D. Scott Slocombe, "Exploring the Origins of 'Social Licence to Operate' in the Mining Sector: Perspectives from Governance and Sustainability Theories," *Resources Policy*, 37, no. 3 (September 2012), 346; and Neil Gunningham, Robert Kagan, and Dorothy Thornton, "Social Licence and Environmental Protection: Why Businesses Go beyond Compliance," 29 *Law and Social Inquiry* (January 1, 2004), 307.
21. On the constitutional duty to consult and accommodate see *Haida Nation v British Columbia (Minister of Forests)*, [2004] SCC 73; and for a recent review see Maureen Killoran, Dan Kolenick, and Thomas Gelbman, "Navigating Murky Waters: Emerging Trends in Aboriginal Consultation and Project Approval" 52, no. 2, *Alberta Law Review* (2014), 207.
22. RSY 2002, Section 68.
23. RSC 1985, Ch. 36 (2nd Supp.).

24. RSC 1985, Ch. O-7.
25. COGOA, Section 5.1(1) and (3).
26. Ibid., Section 5.2(1).
27. Ibid., Section 5.2(3).
28. Indigenous and Northern Affairs Canada, *Benefits Plan Guidelines for the North*, http://www.aadnc-aandc.gc.ca/DAM/DAM-INTER-HQ-NOG/STAGING/texte-text/ben_bpgn_1378407673593_eng.pdf.
29. Ibid., 9.
30. Ibid., 10.
31. Ibid., 11–12.
32. Ibid., 1.
33. See for example: Indian and Northern Affairs Canada, Northern Oil and Gas Branch, "Decision Report, Taglu Field Development Plan," http://www.aadnc-aandc.gc.ca/DAM/DAM-INTER-HQ/STAGING/texte-text/og_ben_tfddr_1310052685085_eng.pdf.
34. Accord Act (Canada-Newfoundland and Labrador Atlantic Accord Implementation Act, SC 1987).
35. "The Atlantic Accord," February 11, 1985, http://www.cnlopb.ca/legislation/statutes.php. See especially Paras. 50–53.
36. Accord Act, Section 44.
37. *Hebron Project: Canada—Newfoundland and Labrador Benefits Plan* (April 2011), http://www.cnlopb.ca/pdfs/hda_benplan.pdf .
38. *Report of the Public Review Commission for the Hebron Development Application* (St. John's, Newfoundland: Canada-Newfoundland Offshore Petroleum Board, February 2012), http://www.cnlopb.ca/pdfs/hebpubrev.pdf.
39. *Staff Analysis, Hebron Benefits Plan* (Canada-Newfoundland Offshore Petroleum Board, April 2012), http://www.cnlopb.ca/news/pdfs/sahebbenplan.pdf.
40. "Hebron Development Application, Decision Report 2012.01" (Canada-Newfoundland Offshore Petroleum Board), http://www.cnlopb.ca/news/pdfs/hebdecision042012.pdf?lbisphpreq=1.
41. Ibid., Condition 2012.01.03.
42. Accord Act, Sections 31–32.
43. Ibid., Section 138.
44. For example, the chronology for the Hebron Project is as follows: Hebron Benefits Agreement, August 20, 2008; Hebron Fiscal Agreement, August 20, 2008; Hebron Benefits Plan, April 2011; Report of the Hebron Public Review Commissioner, February 2012; CNLOPB Staff Analysis, April 2012; and Board Hebron Decision Report, May 31, 2012.
45. *Hebron Project: Canada—Newfoundland and Labrador Benefits Plan*, April 2011, 2-2. In formal terms the two procedures are distinct. The Hebron Benefits Agreement (Section 5.13) for example provides that "The Parties agree that nothing contained in this Agreement shall be applied or construed so as to limit or affect the authority of

the Board to approve and administer a Benefits Plan for the Hebron Project," http://www.hebronproject.com/docs/benefits/finalexecutedbenefits.pdf.

46. Denstedt and Thrasher, "The Accord Acts Twenty Years Later," 302–303. While this level of political support should provide assurance for the proponent that its benefits plan will be approved by the board, the authors also discuss one example in the Nova Scotia offshore (the Deep Panuke project) where that proved not to be the case. Denstedt and Thrasher perhaps overemphasize (e.g., pp. 303 and 305) the point that as a matter of law the province has no veto over the terms of a benefits plan on the grounds that approval of benefits plan, unlike the approval of a development plan, does not require the approval of either government (Accord Act, Section 45[6]).

47. The working interest owners are Chevron Canada, ExxonMobil, StatoilHydro Canada, Petro-Canada, and the Oil and Gas Corporation of Newfoundland and Labrador.

48. *Saint John's (City) v Canada-Newfoundland Offshore Petroleum Board*, [1998] NJ 233 (TD).

49. The court noted (ibid., at Para. 11) that all parties proceeded on the basis that the board had the ability to attach conditions to its approval even though the legislation did not explicitly so provide. The court also concluded (ibid., and referring to Section 134 of the Accord Act) that the board could revoke its authorization for any work or activity in the event that it determined that there was noncompliance with any condition.

50. *Canada-Newfoundland Offshore Petroleum Board Decision* 97.02, Section 3.2 (St. John's, Newfoundland: CNOPB, December1997), 13, http://www.cnlopb.ca/news/pdfs/d97_02e.pdf.

51. See discussion in Denstedt and Thrasher, "The Accord Acts Twenty Years Later," 299, noting that the board's power of enforcement, while real, is likely limited to the "blunt instrument" of revoking a license or authorization where the conditions of such interests have been breached.

52. But see *Mil-Davie Inc v Hibernia Management and Development Co*, 2003 FCT 297. In this case a shipyard owner brought an action against the Hibernia project on the grounds that the project had breached Section 45 of the Accord Act by awarding a significant contract without a bidding process. The Hibernia group applied for summary judgment on the basis that Section 45 could not support a cause of action, but Justice Noël dismissed the application.

53. *Canada-Newfoundland and Labrador Benefits Plan Guidelines* (February 2006), http://www.cnlopb.ca/pdfs/guidelines/benplan.pdf .

54. Ibid., "Appendix II," 1.

55. Ibid., "Appendix I," 1.

56. Ibid., "Appendix I," 3.

57. Ibid., Section 5.0, "Assessment of the Capacity of the Newfoundland and Labrador and Canadian Economies to meet the Requirements of the Project," 14.

58. Ibid., Section 6.2.1, "Pre-development Monitoring and Reporting," 17.

59. *Hibernia Management and Development Company and Petro-Canada v Canada-Newfoundland and Labrador Offshore Petroleum Board*, 2008 NLCA 46 (hereafter *Hibernia Management*).

60. *Mobil Investments Canada Inc. and Murphy Oil Corporation v Canada*, May 2012.

61. The evolution of the guidelines is discussed in *Hibernia Management*.

62. The preceding two paragraphs are based on my blog post on the arbitral award: Nigel Bankes, "From Regulatory Chill to Regulatory Concussion," May 6, 2013.

63. The trial court's decision is discussed by Denstedt and Thrasher, "The Accord Acts Twenty Years Later," 305–308.

64. *Hibernia Management*, Para 65. References are to the leading judgment given by Justice Welsh. Justice Barry gave separate concurring reasons, and Justice Rowe dissented.

65. Ibid., Paras. 67–69.

66. Ibid., Paras. 83–92.

67. Ibid., Paras. 93–99.

68. Ibid., Paras. 99–109.

69. Ibid., Para. 163, suggesting that the board's approach would only be supportable to the extent that there was a close connection between R&D expenditures and the needs of the particular project.

70. Ibid., Paras. 170–173, referring to the terms of the underlying political accord that should inform the interpretation of the legislation: *Mobil Oil Canada Ltd. v Canada-Newfoundland Offshore Petroleum Board*, [1994] 1 SCR 202.

71. Ibid., Para. 181.

72. In addition, the tribunal unanimously concluded that the application of the R&D guidelines to the existing projects of the investors was not a breach of the fair and equitable treatment standard.

73. See synopsis available here: *Mobil and Murphy v Canada*, http://www.international.gc.ca/trade-agreements-accords-commerciaux/topics-domaines/disp-diff/mobil.aspx?lang=eng; and Les Whittington, "Oil Giants Win $17M from Ottawa under NAFTA," *Toronto Star*, March 13, 2015, http://www.thestar.com/news/canada/2015/03/13/oil-giants-win-17m-from-ottawa-under-nafta.html.

74. See case summaries: http://www.international.gc.ca/trade-agreements-accords-commerciaux/topics-domaines/disp-diff/murphy.aspx?lang=eng; and http://www.international.gc.ca/trade-agreements-accords-commerciaux/topics-domaines/disp-diff/mobil2.aspx?lang=eng. The Mobil arbitration is proceeding as *Mobil Investments Canada Inc. v Canada* (ICSID Case No. ARB/15/6).

6

CHINA

Leslie Zhang Weihua (张伟华)

Introduction

Local content as a policy in the oil and gas industry was said to have originated from the North Sea early in the 1970s and took the form of import restrictions and the creation of national oil companies. Its major objectives were transfer of technology, provision of local job opportunities, creation of backward and forward linkages, and increase in ownership and control."[1] As an academic study revealed, "resource-rich countries are increasingly inserting requirements for local content (local content provisions) into their legal framework, through legislation, regulations, contracts, and bidding practices."[2] This observation is reiterated by Victor Galante, who noted that "the phrase 'local content' has begun to appear more frequently in the laws, internal policies, and tender protocols of governments and companies around the world, particularly with regard to the oil and gas industry. The main argument to justify the local content rules is the enhancement of the local industry."[3] This trend is especially evident in the hydrocarbon-rich developing countries in Africa, South America, and Asia.[4]

Unlike some other countries' centralized approach to promulgating local content laws or regulations, the requirements regarding local content in China are provided by various laws and regulations, governmental policies, and production sharing contracts. This follows a management mechanism in which the Chinese petroleum industry is not regulated by a sole governmental agency but by various agencies following the dissolution of the Ministry of Petroleum Industry in 1988.[5] Chinese local content requirements are scattered in laws, regulations, policies, production sharing contracts, employment and labor licenses, and tax incentives. Scattered laws and regulations, emphasis on employment, and the dominance of NOCs in the petroleum industry characterize Chinese local content policies. This section further explores different aspects of the governing landscapes on local content policies and regulations in China.

Local Content Policy and Baseline

Policy background

Although China does not have systematic, specific local content laws and regulations, it adopts local content requirements in its practices relating to the petroleum industry and petroleum operations. From government policies, NOC guidelines, and other various limitations or incentives that will be further explained in this section, we can conclude that China's long-term purpose is to maximize the potential of its domestic manufacturing capacities, service providers, and human resources.

All the laws, regulations, or guidelines are based on the presumption that China will develop its own complete petroleum industry supporting capacities, especially regarding the frontier exploration areas, advanced technology, and development of highly trained technicians.[6]

Overview of key prevailing local conditions

The Chinese petroleum industry has advanced significantly in the past 50 years. Compared with other developing countries, China, as one of the largest petroleum-producing countries, has already built a well-developed petroleum industry. The CNPC, SINOPEC, and the CNOOC are each among the top 50 oil companies in the world.

The US Energy Information Administration (EIA) reports that China surpassed the United States at the end of 2013 as the world's largest net importer of petroleum and other liquids, in part because of China's rising oil consumption. China's oil consumption growth accounted for about 43% of the world's oil consumption growth in 2014.[7]

Key Players

There are a number of key players in the Chinese petroleum industry.

Regulatory agencies

As discussed above, there is no sole centralized regulatory agency that is designated to oversee the petroleum industry. The petroleum industry is regulated by different councils and ministries, the most important of which are introduced in the following text.

The central government. The highest regulatory agency to oversee the energy sector (including oil and gas) is the State Energy Council. Its members include the prime minister, vice premier, deputy secretary of the State Council, and the heads of many different ministries.[8] The State Energy Council is responsible for national energy strategy planning, national energy security decisions, and national energy development strategy.[9]

The National Development and Reform Commission (NDRC). The NDRC is the leading governmental agency that formulates and implements strategies of national economic and social development, annual plans, and medium- and long-term development plans.[10] The NDRC, through its Energy Management Division, the National Energy Administration (NEA), has numerous responsibilities. A few of these responsibilities include the following:

- Formulating and implementing energy development plans and industrial policies
- Promoting institutional reform in the energy sector
- Administering energy sectors, including coal, oil, natural gas, power (including nuclear power), and new and renewable energy
- Taking charge of energy conservation, comprehensive utilization of resources in the energy sector, and various other functions.[11]

For the petroleum industry, the investment plan, the overall development plans (ODPs) for oil and gas fields, and other petroleum investment activities are regulated by the NEA. Under such a management system, the local content policy is also mainly managed by the NEA.

The Ministry of Land and Resources (MLR). The MLR is responsible for the management of all petroleum licenses and permits.[12] It is also responsible for administrative review in accordance with relevant regulations, as well as the development of policies regarding the management, protection, and rational utilization of land, mineral, and marine resources. It also is responsible for formulating the technical criteria, rules, standards, and measures for land, mineral, and marine resources.

The Ministry of Commerce (MOC). Key missions of the MOC include formulating the strategies, guidelines, and policies of developing domestic and foreign trade and international economic cooperation; drafting the laws and regulations governing domestic and foreign trade, foreign investment in China, foreign assistance, overseas investment, and foreign economic cooperation; and devising relevant departmental rules and regulations.[13] Under such mandates, the MOC was previously authorized to approve the production sharing contracts entered by NOCs and foreign contractors.[14]

The State Administration of Foreign Experts Affairs (SAFEA). SAFEA is responsible for certifying foreign experts to work in the Chinese mainland and organizing overseas training for Chinese technical and managerial professionals.[15] All foreign technical and managerial professionals in various industries (including the petroleum industry) must first obtain permissions from SAFEA. In such regards, SAFEA plays an important role in Chinese local content regulatory landscape.

The State Administration of Taxation (SAT). SAT was established at the central government level as an organization in charge of taxation, directly under the State Council.[16] In terms of the local content tax policies, SAT passed some incentive tax policies relating to petroleum operations' importation of equipment that cannot be produced by Chinese local manufacturers.

The State-Owned Assets Supervision and Administration Commission (SASAC). The SASAC was established by the central government in 2003. Its primary function is to supervise (on behalf of the state) the state-owned enterprises (SOEs). The NOCs and other SOEs engaging in oil and gas operations must include the management of the SASAC as a shareholder. The SASAC usually sets up the annual performance parameters for the SOEs and ranks the SOEs according to their performance appraisals.[17]

National oil companies

The customarily recognized "Big Three" national oil companies are the CNPC,[18] Sinopec,[19] and the CNOOC.[20] There are also several state-owned enterprises engaging in petroleum operations, but only on small scale, and only the Big Three NOCs[21] are empowered to enter into production sharing contracts with foreign investors in conventional petroleum resources.[22] As Chinese NOCs originating from the regulatory agency dissolved by the central government, the three NOCs more or less have regulatory powers in onshore or offshore petroleum operations.[23]

Oil and gas companies operating in the country

Along with the CNPC, Sinopec, the CNOOC, and Yanchang Petroleum Group, there are other kinds of petroleum investors, as categorized below.

Foreign investors. To attract foreign investments and advanced technology, foreign investors are allowed to make investments in petroleum industry in China by entering into production sharing contracts with the NOCs.[24] To the extent limited by the applicable laws and regulations,[25] foreign investors are not only allowed to make investments in the upstream sector, but also are allowed to invest in the middle and downstream sectors.

Other Chinese SOEs. The NOCs are comparatively monopolized in the conventional petroleum industry. Only the CNPC, Sinopec, the CNOOC, and Yanchang Petroleum Group are allowed to have oil and gas licenses. Other Chinese SOEs, such as Sinochem Group and ZhenHua Oil, are investing in the petroleum industry by seeking overseas opportunities.

Privately owned enterprises (POEs). In the past, POEs were not allowed to participate in the upstream sector. With the government's policy to promote shale gas development in the past several years, POEs were encouraged to invest in the shale gas industry. At the same time, many POEs have invested in the petroleum industry overseas.

Service providers operating in the country

The service companies owned by NOCs. Each NOC employs integrated or controlled service companies to provide petroleum services.[26] Before the 1990s, there were few privately owned petroleum services companies in the market; almost all the service companies in the market were owned or controlled by the NOCs. The service companies owned by the NOCs dominated the petroleum services market until the 1990s and the start of the 21st century.

Privately owned service companies. Most of the privately owned service companies in Chinese petroleum industry were established in 1990s.[27] Privately owned service companies now hold 10% to 15% of the market share in the petroleum services sector. Most of the privately owned service companies are focused on providing the low to medium end of services provided in the petroleum industry.

In order to be qualified to work for the NOCs, privately owned service companies must first obtain the necessary qualification certificates from the NOCs.

International service companies. As discussed, 80% of the market share in the petroleum services sector is held by subsidiaries of NOCs that provide different petroleum services. With 10% to 15% market share taken by privately owned service companies, the international service companies only account for 5% to 10% of the market share in the petroleum services sector in China. Although the international service companies do not hold a substantial market share, they provide high-end petroleum services to Chinese companies operating in the upstream sectors.

Industry Knowledge

The affiliated research institutes controlled by the government

There are many research institutes controlled by the government that conduct energy policy research, comparable analysis on energy legislations, and various academic or semiacademic studies on energy policies. Those research institutes more or less act as think tanks for the central government. Numerous laws, regulations, and policies promulgated by the government are based on the research outcome of those research institutes. Not all of these research institutes are market oriented, and many of them are subsidized by the government.[28]

The affiliated research institutes controlled by the government usually do not provide academic education for students, but some of them are qualified as postdoctoral research centers, where the students with doctoral degrees can participate in research projects.

The affiliated research institutes controlled by the NOCs

All NOCs have their own affiliated research institutes that provide assessment of strategic options, project analysis, opportunities capture, and other decision-supporting work. According to the CNPC's Research Institute of Petroleum Exploration and Development website:

> The Research Institute of Petroleum Exploration and Development (RIPED) was established in 1958. It is the R&D center of China National Petroleum Corporation (CNPC) and PetroChina Company Limited (PetroChina). Its research activities cover almost all the areas in upstream petroleum business including strategic development planning for CNPC/PetroChina global upstream business, R&D for major fundamental applied theories and techniques, and E&P technical support for the company's oil/gas projects around the world.[29]

Besides RIPED, the CNPC owns many other research institutes that focus on different sectors of the petroleum industry. For example, the CNPC also owns the Tubular Goods Research Institute (TGRI), which focuses on the research of tubular goods. According to the TGRI's website, the institute was founded in 1981. It is located in the southern suburb of Xi'an and functions directly under the authority of the CNPC. It is a polytechnic center, integrating scientific research, quality and technical supervision, and engineering technical services relating to petroleum tubular goods. Decision-making support is offered by its advisory staff for oil pipe engineering. Further R&D support is offered through a testing center for quality and safety of oil tubing, as well as a center for technical support and service of major pipeline projects and oil and gas field exploration and development projects.[30]

The CNOOC, Sinopec, and other SOEs engaging in the petroleum industry also have their respective research institutes that conduct different research functions.

Similar to those affiliated research institutes controlled by the government, some of the affiliated research institutes controlled by the NOCs are also qualified as postdoctoral research centers.

The affiliated research institutes controlled by the universities

There are several universities in China focusing on petroleum-related education. Among these universities, China University of Petroleum, Northeast Petroleum University, Southwest Petroleum University, Yangtze University (formerly known as Jianghan Petroleum College), Xi'an Shiyou University, Chongqing University of Science & Technology, and Guangdong University of Petrochemical Technology are well-known names. In addition to those public universities, there are also private universities and colleges in China focusing on providing petroleum-related education. Whether public or private, the universities offer academic education ranging from undergraduate degrees to doctoral degrees in various petroleum-related disciplines. At the same time, these universities also provide continuing education for employed professionals.

Many of these universities and colleges have affiliated institutes and training centers under their auspices. The government or NOCs often subsidize the affiliated institutes and training centers operated by the universities and also fund many projects initiated by the universities.

The universities provide seminars and training courses to the public and to industry professionals. Such seminars and training courses range from basic to comprehensive.

The privately subsidized research institutes

In China, most institutes on petroleum-related research are official or semi-official. Only a few privately subsidized research institutes focus their research on petroleum.

The institutes established by the government, NOCs, or universities have more influence than those institutes subsidized by the privately owned enterprises.

The institutes using social media

Recently, institutes have begun using social media to spread opinions, views, and research outcomes. The privately subsidized research institutes are adept at using social media to disseminate information and gain influence; some institutes only publicize their research reports, opinions, and views on social media. Some private research institutes have successfully attracted numerous social media followers.[31]

Legislative Hydrocarbon Regime Concerning Local Content

Relevant policies, legislative framework, and key provisions

The Constitution.[32] The constitution provides that

> [a]ll mineral resources, waters, forests, mountains, grassland, unreclaimed land, beaches and other natural resources are

> owned by the state, that is, by the whole people, with the exception of the forests, mountains, grass lands, unreclaimed land and beaches that are owned by collective in accordance with the law. The state ensures the rational use of natural resources and protects rare animals and plants. Appropriation or damaging of natural resources by any organization or individual by whatever means is prohibited.

The provisions in the Constitution set the foundation of the petroleum management system, and anyone who wants to understand the whole petroleum system needs to study the Constitution first.

Hydrocarbons law. There is no uniform hydrocarbons law in China. Instead, different aspects of the oil and gas industry are governed by different bodies of laws issued by different governmental authorities.

Although there is no hydrocarbons law in China, the Chinese National Congress promulgated the Mineral Resources Law of the People's Republic of China. Its purpose is "developing the mining industry, promoting the exploration, development, utilization and protection of mineral resources and ensuring the present and long-term needs of the socialist modernization programme."[33]

Based on the principles provided in the Constitution, the Mineral Resources Law of the People's Republic of China further provides that the mineral resources belong to the state, and the rights of state ownership in mineral resources are exercised by the State Council. There is also a clause regarding oil and gas in the Mineral Resources Law of the People's Republic of China that provides "the competent departments authorized by the State Council may conduct examination of and grant approval to mining of such specified minerals as oil, natural gas, radioactive minerals and issue mining licenses."[34]

In China, the Constitution and the Mineral Resources Law of the People's Republic of China set the tone that the oil and gas industry is a highly regulated industry.

The hydrocarbon regulations. To better understand the hydrocarbon legislation and the requirements made therein for local content, there are two important regulations in relation to Sino-foreign cooperation in the petroleum industry that shall be emphasized. The two regulations have laid out the whole framework concerning the cooperation between NOCs and foreign investors, which has played an important role in the development of the Chinese petroleum industry. The first regulation is Regulation of the

People's Republic of China on Sino-Foreign Cooperation in the Exploitation of Onshore Petroleum Resources (中华人民共和国对外合作开采陆上石油资源条例), Order No. 606 of the State Council, often shortened to Order 606. It was issued in 1993 and amended in 2001, 2007, 2011, and 2013. Order 606, Article 21 provides that "a petroleum contract may stipulate the personnel required for petroleum operations, and Chinese citizens may be given preference in the employment of such personnel by the operator."

The second regulation is Regulation of the People's Republic of China on the Exploitation of Offshore Petroleum Resources in Cooperation with Foreign Enterprises (中华人民对外合作开采海洋石油资源条例), Order No. 607 of the State Council, often shortened to Order 607. It was issued on January 30, 1982 and amended in 2001, 2011, and 2013. Identical to Order 606, Article 13 of Order 607 provides that "a petroleum contract may stipulate the personnel required for petroleum operations, and Chinese citizens may be given preference in the employment of such personnel by the operator."

Other laws and regulations relating to local content. There are some important laws and regulations in relation to the Chinese local content requirements that merit attention.

The first regulation is the Rules for the Administration of Employment of Foreigners in China, which sets the whole legal framework for foreigners working in China.[35] Basically, the principle is that any employer in China shall apply for employment permission if it intends to employ foreigners and may do so after obtaining official approval. The regulation also requires that the post to be filled by the foreigner (as recruited by the employer) shall be a post of "special need" that cannot be filled by any domestic candidates for the time being but violates no government regulations.[36] These provisions are typical local content requirements, and they also apply in the oil and gas industry. In the regulation, Article 9 gives some special consideration to offshore oil operations, stipulating:

> Foreigners who meet any of the following conditions are not required to hold employment licenses and employment permits: foreign contract workers who engage in offshore oil operations through a Permit for Foreigners to Engage in Offshore Oil Operations in the People's Republic of China, and who have special skills and do not need to go ashore.

In China, an employment contract with a foreigner has a limited term. The employer and its foreign employee should conclude a labor contract, the term of which cannot exceed five years.

Chinese laws and regulations also put limitations on the import of goods, which is typically the requirement of the local content policy. For certain goods, the Department for Foreign Trade[37] under the State Council may, based on the need to monitor imports and exports, implement an automatic import and export licensing system for certain goods subject to free import and export. It also publishes the catalog listing such items according to the Foreign Trade Law of the People's Republic of China,[38] which stipulates the following:

> The State may impose restrictions on the import or export of goods and technologies in any of the following circumstances: (1) where the import or export shall be restricted in order to safeguard the national security or public interest; (2) where the export shall be restricted on account of domestic shortage in supply or effective protection of exhaustible domestic resources; (3) where the export shall be restricted due to the limited market capacity of the importing country or region; (4) where the import shall be restricted in order to establish or accelerate the establishment of a particular domestic industry; (5) where the restriction on the import of agricultural, animal husbandry or fishery products in any form is necessary; (6) where the import shall be restricted in order to maintain the State's international financial status and the balance of international payments; (7) where, as the international treaties or agreements to which the People's Republic of China is a contracting party or a participating party require, the import or export shall be restricted.

Based on the above-mentioned laws and regulations, it could be concluded that the Chinese government, by using various laws and regulations, encourages the use of local labor and goods and sets the necessary threshold concerning the use of foreign labor and goods.

International treaties. There are three international treaties in which China is a member (or has observer status) that are worthy of discussion in this text.

- *New York Convention.* China joined the New York Convention on the Recognition and Enforcement of Foreign Arbitral Awards in 1987.
- *International Center for Settlement of Investment Disputes (ICSID).* China became a member state to the International Center for Settlement of Investment Disputes in 1993.

- *Energy Charter Treaty (ECT)*. China is now an observer to the Energy Charter Conference and is considering joining as a member to the Energy Charter Treaty.

Foreign investors' rights are protected under the three international treaties mentioned above.

Relevant contracts

There is a common misconception that China only has a production sharing contract (PSC) regime in its petroleum industry. In reality, the oil and gas regime in China is a combination of licensing and PCSs (manifested by the petroleum contract). The governmental granting authority in China grants the petroleum license to a domestic (PRC) company (more often than not, an NOC), and usually registers the domestic company as the license holder. In the past, some foreign contractors under their petroleum contracts with an NOC were registered as the license holder in the exploration stage, although the NOC still held the license for the foreign contractor(s). The rationale for that approach was that the foreign contractors under the petroleum contract functioned as the operator during the exploration stage. Currently, however, the granting authority rarely registers the foreign contractors as the license holder even in the exploration stage. In any event, the NOC still holds all the licenses during the exploration, development, and production stages.[39]

China's model PSCs provide for the preferential employment of Chinese personnel, goods, and services. For example, the model offshore PSC provides that the contractor shall, in accordance with the petroleum regulations, give preference to the employment of qualified Chinese personnel in the performance of petroleum operations. For this purpose, the contractor must submit in advance to the NOCs a plan for the employment of Chinese personnel, listing the posts and number of persons involved. The NOC shall, in accordance with the plan, provide or assist in recruiting Chinese employees as candidates for such employment. For the performance of petroleum operations, the contractor is obligated to employ competent Chinese personnel and to employ those who have become qualified after being trained in accordance with the training program. The contractor must give preference to employing the Chinese personnel who have participated in the training program provided by the contractor.

A model PSC also addresses the training of Chinese personnel and transfer of technology. It stipulates that the foreign investor must transfer operatorship to the NOC as early as possible. It requires that the operator, through the training of Chinese personnel, gradually increase the percentage of Chinese personnel, including the key personnel, while reducing the percentage of expatriate employees accordingly. By the end of the fifth year of the production, the operator must endeavor to make the total number of the Chinese personnel no less than 90% of the total personnel. The total number of the Chinese personnel shall not include those personnel rendering living services of any third party.

The model PSC contains numerous other clauses giving preference to Chinese labor and goods. There are several examples:

> The Contractor, in carrying out the Petroleum Operations, if the conditions of such factors as quality, price, delivery time and services are competitive, shall give preference to the goods, equipment and services provided within the territory of People's Republic of China....
>
> In accordance with the Petroleum Regulations, the NOCs shall have the right to assign its staff to participate in the master designs and engineering designs made by the Contractor for the purpose of the implementation of the Contract. Engineering design companies within the territory of the People's Republic of China shall be given preference in entering into the subcontracts for the aforesaid master designs and engineering designs provided that their technical level, price, and delivery time are competitive....
>
> In the course of the implementation of the Contract, those scientific research projects which are required by the Petroleum Operations but not carried out by the Parties, with the approval of the JMC, may be commissioned to, and carried out by, any Third Party, and the Parties shall enter into subcontracts or service contracts with relevant scientific research departments within the territory of the People's Republic of China, provided that they are competitive. The aforesaid required expenses shall be included in the budget specified herein and charged to the Joint Account. All inventions and experimental or research results developed from the aforesaid research projects carried out by a Third Party delegated by the Operator shall also be shared by and belong to the Parties who, subject to the provisions hereof, shall not disclose any of them to any other Third Parties. The Operator shall endeavor to incorporate the provisions herein in the subcontracts or service contracts signed with a Third Party.

The constitution, petroleum-related laws and regulations, various policies, and model PSCs constitute the full chain of the local content requirements.

Positive and Negative Experiences and Expectations Regarding the Regulatory Framework

Positive experiences regarding the regulatory framework

The existing regulatory framework regarding the Chinese petroleum industry has proven to be an effective tool in establishing China's fully integrated petroleum industry.

In 1949, the annual output of crude oil in China was only 120,000 tons, but imported crude oil was 2 million tons. In 2015, the annual output of crude oil in China reached 215 million tons, and the annual output of natural gas reached 127.1 billion cubic meters.

Crude oil output of China is now ranked fourth in the world. Natural gas output ranks seventh in the world, and China's three petroleum giants—the CNPC, Sinopec, and the CNOOC—rank among the top 50 oil companies across the world.

The second positive experience regarding the previous regulatory framework is that it provided opportunities for Chinese oil companies to develop and expand their competitiveness in the global market. Nowadays, the CNPC, Sinopec, and the CNOOC operate globally, with corporate footprints all over the world. These companies are not only serving the Chinese domestic market, but also the global market.

The third positive experience regarding previous regulatory framework is that it successfully attracted foreign capital and technology, which have been important for the development of the Chinese petroleum industry.

Negative experiences regarding the regulatory framework

Although the current regulatory framework brought enormous benefits to the Chinese petroleum industry and the Chinese people, it is now slightly outdated and in need of some changes and innovations.

The first negative experience of the current regulatory framework is that it overprotects the NOCs and other SOEs in the petroleum industry. As a result, there is no effective competition in the petroleum industry, especially in the upstream sector. The lack of competition in the petroleum industry is now being regarded as the most serious obstacle to deepening the reform in this industry.

The second negative experience of the current regulatory framework is that it should put more emphasis on the development of advanced technology and innovations. Although China has established a well-developed petroleum industry, the industry still needs to be upgraded and renovated. The whole regulatory framework also needs to be reexamined, and necessary changes should be made.

The third negative experience of the current regulatory framework is that privately owned enterprises should be more actively encouraged to participate in the petroleum industry. Privately owned enterprises already contribute to the oil services sector, and more involvement in the upstream sector should be expected.

In a nutshell, the current regulatory framework should be modified to reflect the new trends and requirements.

Current Issues

Deregulation in petroleum industry

The Chinese government now is implementing an "opening up" policy in the petroleum industry in an effort to attract more investors in the upstream sector and in the oil services sector.[40] The overall changes in the Chinese petroleum industry will include amending the regulatory framework and outdated policies to allow more competition in this industry. The government not only encourages foreign investment in the upstream sector but also welcomes various kinds of investments in the petroleum industry.

Encouraging the shale gas industry

The Chinese government has expressed an uncharacteristic enthusiasm within the shale gas industry and independently introduced many subsidies to promote the exploration, development, and production of shale gas. To encourage the development of shale gas resources, the government allows POEs and IOCs to participate in shale gas developments. It also urges (by publicizing specific guidelines) financial institutions to provide financial support to the entities that conduct petroleum operations in unconventional oil and gas development. With an aggressive annual production target of 1.05 Tcf in 2025, the shale gas industry brings more opportunities for POEs and international investors in Chinese upstream investment.

The independence of petroleum transportation network

Currently, most of the petroleum pipelines are owned by the NOCs. The government already requested a feasibility study regarding the separation of petroleum pipelines from the NOCs. The main petroleum pipelines will be operated independently from the NOCs. The independence of the petroleum transportation network is deemed one of the most important reform actions adopted by the government.

Market-determined price for oil and gas products

In the past, the Chinese government (by authorizing NDRC) has controlled the price of oil and gas products. Even today, the refined oil price is still regulated by the government through the oil product pricing system.

The consensus is that the government should let the market decide the price of oil and gas products, rather than do so by administrative order.

Establishment of oil and gas distribution market

In the past, the government had a license system to regulate the oil and gas distribution market. Without the licenses granted by the government, a company is not permitted to import crude oil or refined oil from the overseas market. Currently, the government is granting more companies the right to import crude oil or refined oil from the overseas market. More players will be entering into the oil and gas distribution market. And the government is considering allowing the establishment of new oil and gas stations by private companies or other kinds of investors.

Diversified ownership of the SOEs/subsidiaries of SOEs

The Chinese government is encouraging various investors to participate in the diversified ownership campaign that aims to diversify the ownership of SOEs/subsidiaries of SOEs.

Other than those nationally monopolized industries, other industries are open to various kinds of investors. The diversified ownership approach gives private companies many more opportunities to participate in the petroleum industry.

The many changes occurring and under consideration are opening the door to investors other than NOCs having an incentive to participate in the Chinese petroleum industry.

Conclusion

The local content requirements in the Chinese petroleum industry are scattered in different laws and regulations and are regulated by various governmental agencies. Due to the de facto monopoly positions of the NOCs, the local services industry in the petroleum sector is less directed by the government. Therefore, there are no uniform laws or regulations governing the local content requirements in China.

The Chinese government focuses more on the employment issues in the local content policy and adopts a more relaxed policy in selection of petroleum contractors. Although the government is using its incentive policy to encourage domestic companies to develop their own technology in deepwater operations and other advanced territories, there are no express limitations on the selection of petroleum services providers in laws and regulations. The reason behind this approach is partly due to the fact that China already has developed a thriving petroleum industry over the last 50 years, and the use of the local goods and labor is not excessively emphasized. Also, with the natural monopoly positions of NOCs in the petroleum industry, the Chinese government does not need strict laws and regulations specializing in local content. Actually, with the current regulatory framework, China has successfully fostered its own petroleum industry.

With further lessening of the governmental regulations in the Chinese petroleum industry, more liberal policy in local content could be expected. Such changes would strengthen the position of private companies in the oil and gas sector in China.

Notes

1. Rabiu Ado, "Local Content Policy and the WTO Rules of Trade-Related Investment Measures (TRIMs): The Pros and Cons," *International Journal of Business and Management Studies*, 2, no. 1 (2013).
2. "Local Content Laws & Contractual Provisions," Columbia Center on Sustainable Development, Columbia University, http://ccsi.columbia.edu/work/projects/local-content-laws-contractual-provisions/.

3. Victor Galante, "Brazil: Local Content in Brazil (Global Energy)," *Mondaq.com*, http://www.mondaq.com/x/240372/Oil+Gas+Electricity/Local+Content+In+Brazil.
4. The host countries in Africa, South America, and Asia have endeavored to develop local content laws in the past 10 years. For example, the Nigerian Oil and Gas Industry Content Development Act was enacted on April 22, 2010, and the Ghana Local Content and Local Participation Bill 2013 was passed in November 2013.
5. During the years 1988 to 1993, the Ministry of Energy regulated the petroleum industry for the Chinese government. Since its dissolution in 1993, there has been no centralized regulatory agency governing the Chinese petroleum industry.
6. According to the 12th energy development plan promulgated by the NEA, development of deepwater oil drilling, offshore platform construction, and shale gas exploration technology in the petroleum industry will be encouraged.
7. "China," Beta, Energy Information Administration (May 14, 2015), http://www.eia.gov/beta/international/analysis.cfm?iso=CHN.
8. The other members from different governmental agencies are: director of the Central Finance, foreign minister, director of the National Development and Reform Commission, minister of Science and Technology, minister of Security, minister of Finance, minister of Land and Resources, minister of Environmental Protection, minister of Transportation, minister of Water Resources, minister of Commerce, deputy governor of the People's Bank, chairman of the SASAC (State-owned Assets Supervision and Administration Commission), director of State Administration of Taxation, director of State Administration of Work Safety, chairman of the China Banking Regulatory Commission, deputy chief of general staff of the PLA, and head of the National Energy Administration.
9. As of its establishment in 2008, the State Energy Council has convened two plenary meetings to discuss critical issues in Chinese energy developments.
10. More detailed functions of the NDRC are given at: "Main Functions of the NDRC," http://en.ndrc.gov.cn/mfndrc/.
11. Other functions include: (1) guiding scientific and technological advancement; (2) organizing and carrying out the R&D of important equipment and guiding the assimilation and innovation of imported complete sets of major equipment; (3) organizing and coordinating key energy-related demonstration projects and promoting the deployment of new products, new technologies, and new equipments; (4) approving, reviewing, or examining fixed-asset investment projects of the energy sector within national plans and the scale of annual plans in accordance with the authority stipulated by the State Council; (5) conducting energy forecasting and precaution and participating in energy operation coordination and emergency preparedness; (6) formulating and implementing national oil reserve plans and polices; (7) taking the lead in launching international energy cooperation; (8) participating in the formulation of policies related to energy such as resources, finance, and taxation, environment protection, and addressing climate change; (9) making recommendations on energy price adjustment and imports and exports aggregate; and (10) undertaking the daily work of the National Energy Commission. More information is available at "National Energy Administration," http://en.ndrc.gov.cn/mfod/200812/t20081218_252224.html.
12. More detailed functions of the MLR can be found at: "Responsibilities of the Ministry of Land and Resources," http://www.mlr.gov.cn/mlrenglish/about/mission/200710/t20071015_656461.htm.

13. More detailed functions of the MOC can be found at: "Ministry of Commerce, People's Republic of China," http://english.mofcom.gov.cn/column/mission2010.shtml.
14. As the new Chinese administration is conducting deregulations on governmental approvals, the right to sanction the PSCs now is assigned to the NOCs, and only a filing with the Ministry of Commerce is required.
15. More detailed functions of the SAFEA are available at: "State Administration of Foreign Experts Affairs: Responsibilities," http://en.safea.gov.cn/2013-08/15/content_16897568.htm.
16. More detailed functions of the SAT are available at: "State Administration of Taxation, May 8, 2012," http://www.chinatax.gov.cn/2013/n2925/n2959/c307242/content.html.
17. More detailed functions of the SASAC are available at: "State-Owned Assets Supervision and Administration Commission of the State Council," http://en.sasac.gov.cn/.
18. The China National Petroleum Corporation was established on September 17, 1988 from the Ministry of Petroleum Industry; it is mainly in charge of oil and gas upstream operations. It is a state oil company endowed with certain governmental administrative functions.
19. The China Petrochemical Corporation (Sinopec Group) is a very large petroleum and petrochemical enterprise group, established in July 1998 from the former China Petrochemical Corporation. Sinopec Group is a state-owned company solely invested by the state, functioning as a state-authorized investment organization in which the state holds the controlling share.
20. The China National Offshore Oil Corporation (CNOOC), the largest offshore O&G producer in China, is a state-owned company operating directly under the State-Owned Assets Supervision and Administration Commission of the State Council of the People's Republic of China.
21. Shaanxi Yanchang Petroleum (Group) Corp., Ltd. (abbreviated as Yanchang Petroleum Group), directly attached to Shaanxi People's Provincial Government, is one of the four qualified enterprises for oil and gas exploration in China. (The others are the CNPC, Sinopec, and the CNOOC.) Yanchang Petroleum Group is not an NOC but is owned by the local government.
22. Previously, a company called CUCBM (China United Coalbed Methane Corporation, Ltd.) could also enter into production sharing contracts with foreign investors in CBM only. In 2015, the CUCBM became a wholly owned subsidiary of the CNOOC.
23. For example, the three NOCs can grant oil and gas services qualification permits to services providers pursuant to their selection standards.
24. The rights to cooperate with foreign contractors were granted by "Regulations of the P.R.C. on Sino-Foreign Cooperation in the Development of Continental Petroleum Resources" and "Regulations of the People's Republic of China on the Exploitation of Offshore Petroleum Resources in Cooperation with Foreign Enterprises," which were promulgated by the State Council.
25. The NDRC issues an annual foreign investment categorization, dividing all industries into three types. The first is comprised of preferred investment projects; the second consists of the restricted investment projects; and the third consists of the forbidden investment projects.
26. For example, the CNPC has owned many service companies that provide a wide range of services, including seismic, drilling, logging, construction, and other services. The

CNOOC also owned a full chain of service companies, including an offshore drilling company, offshore construction company, offshore supply service company, offshore catering service company, and other service companies.

27. Ma Jian, "The Study on the Privately-Owned Petroleum Service Companies in the Internationally Competitive Environment," *China Petroleum Enterprise* (August 2011).

28. See the official website of Chinese Energy Research Institute National Development and Reform Commission, one of the affiliated research institutes controlled by the NDRC, at http://www.eri.org.cn/.

29. See the website for the Research Institute of Petroleum Exploration and Development (RIPED), http://riped.cnpc.com.cn/ripeden/index.shtml.

30. "Brief Introduction," CNPC Tubular Goods Research Institute, http://tgrc.cnpc.com.cn/tgrcen/briefintroduction/column_common1.shtml.

31. For example, there is a website called *Oil Observer* that is operated by a private company and has successfully attracted more than 70,000 followers. The website offers industry insights, analytical research reports, seminars and training, and transaction opportunity matching. See http://www.oilobserver.com/.

32. The Constitution is the ultimate source of law in China. It lays down the foundation rules or principles for the Chinese legal regime, including the nature of the regime, the framework of government, and the basic rights and obligations of Chinese citizens.

33. See Article 2, Mineral Resources Law of the People's Republic of China.

34. See Article 16, Mineral Resources Law of the People's Republic of China.

35. The regulation was promulgated jointly by the Ministry of Labor, Ministry of Public Security, Ministry of Foreign Affairs, and the Ministry of Foreign Trade and Economic Cooperation of the People's Republic of China on January 22, 1996.

36. See Article 6, Rules for the Administration of Employment of Foreigners in China.

37. The Department for Foreign Trade is now the Ministry of Commerce.

38. This law was adopted at the 7th Meeting of the Standing Committee of the Eighth National People's Congress on May 12, 1994. It was revised at the 8th Meeting of the Standing Committee of the Tenth National People's Congress and promulgated by Order No. 15 of the president of the People's Republic of China on April 6, 2004.

39. Previously all oil and gas licenses in the Chinese upstream sector had been granted to Chinese national oil companies, and privately owned enterprises did not have the option to participate in the exploration, development, and production of oil and gas fields within Chinese territory. IOCs could only participate in the upstream sector by executing PSCs with NOCs.

40. The NDRC is currently taking the lead to initiate the "the overall reform plan in petroleum industry," which aims to encourage various kinds of investors in the petroleum industry. The monopoly in the upstream sector is expected to be broken by further deregulation.

7

CYPRUS

Deborah Primett, George Z. Georgiou, and Romanos V. Loizides

Introduction

The Republic of Cyprus currently depends heavily on imported petroleum products from non-EU countries to satisfy its energy needs. Petroleum products are mainly imported to the island by the Electricity Authority of Cyprus, oil marketing companies, and the cement industry, with the country currently spending approximately 7% of its GDP on such imports.[1] The net imports of primary energy in Cyprus are in the range of 3,000 tonnes of oil equivalent, which amounts to 0.004 tonnes per inhabitant.[2]

In fact more than 90% of Cyprus's gross energy consumption is satisfied by imported petroleum products, with the balance of the island's energy demands being met by solar and wind generation.[3] However, natural gas has been discovered in Cyprus's exclusive economic zone (EEZ), which has been divided into 13 exploration blocks comprising 51,000 square kilometers offshore to the south and southwest of the island. The discovery may mean the days of Cyprus relying on imported primary energy products are numbered.

Oil and gas exploration in Cyprus dates back to 1938, when Iraq Petroleum Company Limited started the first surveys in Cyprus. The most recent developments in this area started in 2003, when Cyprus delimited its EEZ and its borders with Egypt and Lebanon. In 2006, the discovery of hydrocarbons by neighboring countries such as Egypt and Israel prompted

Cyprus to conduct two-dimensional surveys in all 13 blocks of its EEZ. Based on available seismic data, in February 2007 the Republic of Cyprus launched its first international tender for three-year oil and gas exploration licenses. After extensive negotiations, an exploration license for Block No. 12 was granted to Noble Energy International on October 24, 2008, for an initial period of three years.[4]

In December 2011, the Republic of Cyprus in conjunction with Noble Energy announced the discovery of substantial natural gas deposits of around 7 trillion cubic feet (Tcf) in Block 12 in the EEZ. Since then, further explorations have been carried out in other exploration blocks by the Italian-Korean consortium of Eni-KOGAS and the French company Total E&P Cyprus B.V. The Republic of Cyprus has begun to formulate a strategy to invest in the extraction of oil and gas with the aim of becoming a significant potential natural gas producer and exporter.

This recent discovery, along with Cyprus's geographical location at the crossroads between Europe, Asia, and Africa, may potentially enable Cyprus to transform itself into an energy hub in the eastern Mediterranean. However, Cyprus needs to build a national industry from scratch. Furthermore, "relationships of conflict" currently exist in the immediate geographical area (Cyprus/Turkey and Israel/Lebanon) that are likely to impact Cyprus's ability to maximize the natural gas deposits found in its EEZ. Thus Cyprus will have to overcome many challenges before it is able to achieve prominence as an energy hub.

Local Content Policy and Baseline

Policy background, including short- and long-term goals

As a member of the European Union since 2004, Cyprus has developed a strategic energy plan that is aligned with the European energy strategy and European Community *acquis communautaire* (the cumulative body of EU law). Cyprus adopted national legislation based on Directive 94/22/EC on the conditions for granting and using authorizations for the prospection, exploration, and production of hydrocarbons and Directive 2003/55/EC concerning common rules for the internal market in natural gas. Cyprus implemented the Hydrocarbon (Prospection, Exploration and Exploitation) Law of 2007 (No. 4(I)/2007), called the Hydrocarbons Law, and the Hydrocarbon (Prospection, Exploration and Exploitation) Regulations of

2007 and 2009 (No. 51/2007 and No. 113/2009), called the Hydrocarbons Regulations.

According to a presentation by the Natural Gas Public Company, the republic's strategic energy plan is based on the following principles:[5]

- Security of energy supply
- Increasing the competitiveness of the Cypriot economy
- Adherence to environmental commitments

The republic's short-term goal is to explore viable, feasible, and efficient ways to reap the possible benefits of the exploitation of its hydrocarbon resources to the maximum extent possible, with a view to bolstering its economic recovery.

At present Cyprus's long-term goals are to fulfill its own energy needs while at the same time becoming a significant producer and exporter of natural gas. Moreover, the republic aims to establish a competitive and secure market for supplying other countries with natural gas at reduced prices. In conjunction, the republic aims to develop measures to safeguard the environmental integrity of the island by protecting the environment from industrial pollution.

Overview of key prevailing conditions in Cyprus

The Republic of Cyprus has many advantages that may assist along the road to achieving its goals. There is a highly educated, qualified, and multilingual workforce on the island, as well as a stable business environment, comparatively low taxation, and a stable, reputable political system.

Cyprus also has well-developed socioeconomic infrastructure. Its economy has been heavily dependent on the financial services and tourism sectors for many years. But Cyprus has sought to develop into other sectors as the financial and tourism sectors have been greatly affected by the global economic crisis. The discovery of natural gas reserves in Cyprus's EEZ offers an opportunity for Cyprus to develop into the energy markets in an effort to find its way back to prosperity.[6]

However, there are impediments to developing an international energy industry in Cyprus. One problem is that it has no interconnections with the trans-European electricity or gas transmission networks at present. It also lacks the necessary infrastructure to enable it to participate in the single European energy market. Large capital investments would be required before it could market its hydrocarbon reserves.

Cyprus has therefore sought cooperation from countries in the region to support it in exploiting its natural gas reserves. Cyprus has signed memoranda of understanding with Greece, Israel, Egypt, and Jordan in relation to the exchange of information, expertise, and experience in hydrocarbons, energy infrastructure, energy network management, liquefied fuel reserves, and on the protection of the environment.

The most recent agreement was entered into between the Republic of Cyprus and Egypt in February 2015. The two countries agreed to explore methods for both the upstream and downstream exploitation of output from the Aphrodite offshore exploration block and possible liquefaction of natural gas for export using Egypt's existing infrastructure.

One prospective export option being considered by the Cypriot government is the pumping of natural gas from Block 12 to Egypt through a subsea pipeline network, facilitated by a floating production, storage, and offloading platform (FPSO) for the mutual benefit of the two countries.

Other potential export options being considered by the Cypriot government include the construction of an onshore liquefied natural gas (LNG) terminal, an eastern Mediterranean subsea pipeline pumping gas to Europe via Greece, a floating liquefied natural gas (FLNG) facility, and the deployment of a compressed natural gas (CNG) marine transport system.

Key Players

The Cypriot government and its agencies

The main regulator of the upstream sector is the government of the Republic of Cyprus and other bodies that implement and manage state policy in the oil and gas industry.

The Council of Ministers (which is comprised of all of the ministers of the government) has a number of powers in the upstream stage under the domestic oil and gas legislative framework. Some of these powers include the following:

- It may determine the areas, within the territory of the republic, to be made available for carrying out the activities of prospecting, exploring for, and exploiting hydrocarbons.
- It has the authority, following a submission of an application, to grant an authorization for hydrocarbons operations.

- It may renew or extend the authorization for such a period as provided in the contract between the government of the republic and the holder of an authorization.
- Under special circumstances, it may revoke, suspend, or cancel the authorization granted.
- It may object to a transfer of an authorization to another entity.
- The operations and activities related to the production of hydrocarbons can be carried out only after the Council of Ministers approves the development and production plan.
- It may issue regulations with the purpose of harmonizing the Cypriot legislative framework with EU directives.

The Ministry of Energy, Commerce, Industry, and Tourism. The Cypriot Ministry of Energy, Commerce, Industry, and Tourism (MCIT) is the main executive ministry responsible for the implementation of government energy policy. It is assisted by the Energy Service of the MCIT, which has overall responsibility for the energy policy (including the oil and gas sector policy) and targets of the Republic of Cyprus. Its responsibilities include the following:[7]

- "Monitoring and coordinating the supply and availability of sufficient energy capacity for domestic needs.
- "Monitoring and participating in the formation of the European Policy for energy issues.
- "Suggesting ways to implement the European Acquis, assisting in the preparation of Laws, Regulations, Rules etc. and implements programs for their promotion.
- "Assisting the Government in the formation of the national energy policy for Cyprus in coordination with all other bodies involved."

The MCIT also has the responsibility for initiating the process of licensing by inviting applications for licenses for defined blocks within the EEZ of Cyprus. Moreover, the MCIT participated in negotiations that took place for the grant of exploration licenses to the oil and gas companies currently operating in Cyprus.

Other entities involved in national energy policy in Cyprus include the Natural Gas Public Company (DEFA_ and the Cyprus Organization for Storage and Management of Oil Stocks. Additional information about these entities may be found in the "Legislative Hydrocarbon Regime in Cyprus" section.

Ministry of Agriculture, Natural Resources, and the Environment (MANRE). This ministry supervises the implementation of government policy and the adoption of European policy and legislation on the environment. It chairs the Committee for Environmental Impact Assessment and promotes the enforcement of regulations relating to the Water Pollution Control Law and the Management of Solid and Hazardous Waste Law. It also encourages environmental awareness and serves as an information resource.

One of the main functions of the MANRE during the upstream stage of operations is to review the environmental impact study submitted by the entity in its application for authorization to explore for hydrocarbons. The department reviews the likely impact of exploration activity on the environment and the plans to minimize its effect.

National Hydrocarbons Company. Following the discovery of hydrocarbons in 2011, the Cypriot government set up a state-owned company to deal with the exploration and exploitation of Cypriot hydrocarbon resources. The Cyprus National Hydrocarbons Company (CNHC) officially began operating on January 1, 2013. Following a decision by the Council of Ministers on October 16, 2013, the CNHC was renamed the Cyprus Hydrocarbons Company (CHC), and a new board consisting of seven nonexecutive members was appointed.

The CHC has been established to undertake the following activities:[8]

- Marketing of hydrocarbons
- Involvement in construction of necessary infrastructure, including pipelines, LNG plants and terminals, loading facilities, etc.
- Involvement in all aspects of hydrocarbons E&P
- Involvement in operation and management of facilities necessary to support oil and gas sales
- Implementation and management of hydrocarbon-related contracts , new or future
- Ensuring that Cyprus's energy needs are met from hydrocarbons found in the EEZ

In February 2015, the CHC signed a memorandum of understanding (MOU) with the Egyptian Natural Gas Holding Company (EGAS) to facilitate cooperation between the two countries for the development of one of the blocks of the Cyprus EEZ by exploiting the gas infrastructures and facilities in Egypt.

Oil and gas companies operating in Cyprus

Although it is still battling its way out of a recession, Cyprus has attracted significant interest from investors worldwide. Cyprus's hydrocarbons discovery has caught the attention of global energy companies, and some are shifting their eastern Mediterranean base of operations to Cyprus. Since Cyprus's oil and gas industry is still in its developmental stages, energy companies currently operating in Cyprus are concentrating on searching for potential underwater natural gas fields and drilling exploratory wells. In 2011 exploration Block No. 12, known as Aphrodite, was awarded to Noble Energy and Delek for hydrocarbon exploration during the first round of hydrocarbon licensing undertaken by the Cypriot government. In 2012 a second round of licensing was undertaken involving a number of large oil and gas companies. As a result, the Cypriot government granted exploration licenses to Eni-KOGAS, the consortia led by the Italian company Eni and the Korean company KOGAS, for exploration in Blocks 2, 3, and 9, and to the French company Total for exploration in Blocks 10 and 11.

Four years after their initial exploration in Block 12 of Cyprus's EEZ, Noble Energy declared a gross reserve of natural gas amounting to about 4.5 Tcf as being commercially viable. This means Cyprus has moved from the exploration to the exploitation phase. In March 2015, Noble Energy presented its proposed development plan to the Cypriot government for the exploitation of natural gas from Block 12.

In January 2015, after evaluating exploration survey data, Total declared that it could not locate any targets to drill in Blocks 10 and 11. Then in March 2015, the Cypriot government signed an agreement with Total on further exploration works in the island's EEZ, which Total will undertake to further assess the prospects of Block 11.

In March 2015, Eni-KOGAS concluded its second exploration well drilling in a new geological formation in Block 9 in search of commercially exploitable hydrocarbons. After both drillings proved unsuccessful, Eni-KOGAS submitted a request to the Republic of Cyprus to proceed with the reevaluation of the geological model over a two-year period while postponing drilling in the interim. This would give them time to study the geological data collected from the two wells and reevaluate their exploration program in Blocks 2, 3, and 9.[9]

Service providers operating in Cyprus

The island's political stability, strategic location, and its potential as a future energy hub has attracted the interest of some of the world's largest oilfield services companies. Halliburton, Schlumberger, and Weatherford International are using Cyprus as their base of operations for the eastern Mediterranean. These service companies are currently assisting and servicing the technical needs of the multinational oil and gas companies that are operating in Cyprus.

Halliburton provides services and products to upstream oil and gas customers across the world and won the contract for drilling, cementing, and supporting the Eni-KOGAS wells in Cyprus. It is also currently providing support to Eni-KOGAS in its exploratory activities offshore in Block 9.

Schlumberger also provides services from exploration through production, and at present it is carrying out logging and testing for Eni-KOGAS in Cyprus.

Industry Knowledge

Over the last two years, oil and gas courses have been introduced in almost all of the private universities in Cyprus. Further, private organizations are offering professional training in this field, as there are few qualified professionals currently supporting and promoting the development of the hydrocarbon industry in Cyprus.

The University of Nicosia offers programs in English for obtaining a bachelor's degree in the fields of energy, oil and gas management, and oil and gas energy engineering. It also offers a master's degree program in the field of oil and gas energy engineering.

Other private universities offer undergraduate programs in the field of mechanical engineering with a specialization in oil and gas engineering, and civil engineering programs with specialization in petroleum geotechnical studies. Further master's degrees are available in the field of oil and gas engineering, with specializations in (a) oil and gas engineering, with emphasis on downstream, (b) offshore engineering, with emphasis on the midstream, and (c) petroleum engineering, with emphasis on the upstream. All programs are offered in English.

Besides private universities in Cyprus offering knowledge in the field, there is also the Solon Kassinis Energy Training School and Research Center at the Institute of Professional Studies at UCLan-Cyprus. The center offers training on the oil and gas industry and offers a diploma in oil and gas with specialization for (a) drilling technicians, (b) electrical engineering technicians, (c) mechanical engineering technicians, (d) welding and pipeline technicians, and (e) health, safety, and environment.

Finally, the Levantine Training Center (LTC) at ShipCon Limassol offers professional training in relation to the oil and gas industry.

Legislative Hydrocarbon Regime in Cyprus

Regulatory bodies (including functions and roles)

Natural Gas Public Company (DEFA). In order to develop the internal gas market, the Cypriot Natural Gas Public Company (whose Greek acronym is DEFA) was set up by the state as the entity exclusively responsible for the import, storage, distribution, transmission, supply, and trading of natural gas, and to manage the distribution and supply system for natural gas in Cyprus.

According to DEFA, its responsibilities include the following:[10]

- "Buying, Importing, Holding, Using, Distributing, Selling, Supplying Natural Gas, in any Form.
- "Operation of the Natural Gas Transmission and Distribution Network.
- "Signing of Treaties/Conventions with the Republic of Cyprus or any other Country or Company or Organisation etc.
- "Negotiating, Buying, Selling, Managing, Storing, Importing, Exporting, Re-exporting etc. any goods, tangible or intangible, including Natural Gas."

DEFA was involved in the discussions with the Israeli contract team for concluding a bilateral agreement between Cyprus and Israel on natural gas matters. These discussions led to the signing of a trilateral MOU between Cyprus, Greece, and Israel for cooperation in the fields of energy and sea resources in August 2013. In April 2015, DEFA received an authorization from the minister of the MCIT to start negotiations with Noble Energy and Delek, the partners in Block 12, for the supply of Cypriot gas to the domestic market. Moreover, DEFA has been charged with running the tender for the temporary import of natural gas (the so-called interim solution) from other parties.[11]

Cyprus Energy Regulatory Authority (CERA or RAEK). The CERA is an independent institution led by three persons appointed by the Cypriot government. Its main functions related to the gas market are defined in the Market Regulation on Natural Gas Law of 2004, Law 183 (I)/2004, as follows:[12]

- "To encourage, promote, and safeguard the healthy and substantial competition in the Electricity and Natural Gas Markets.
- "To protect Consumers' interests.
- "To promote the development of economically viable and efficient Electricity and Natural Gas Markets.
- "To ensure the Security, Continuation, Quality and Reliability of Electricity Supply.
- "To take into consideration the Protection of the environment."

A significant function of the CERA relates to advising the MCIT on matters concerning electricity and oil and gas. Moreover in 2012, it conducted a series of consultations with DEFA and various other stakeholders in order to prepare proposals for strategic guidelines in developing the natural gas sector in Cyprus.

Cyprus Organization for Storage and Management of Oil Stocks (COSMOS or KODAP). COSMOS is a nonprofit semigovernmental organization founded for the purpose of harmonizing local legislation with European Directive 68/414/EEC, which imposes an obligation on member states to maintain minimum stocks of crude oil and/or petroleum products.

Its mission "is to maintain and manage the minimum stocks of crude oil and oil products of Cyprus, which have to be available at all times." However, it "cannot interfere with oil prices in the local market, as its main purpose is to use the oil stocks only in cases of shortages or supply difficulties, and only after an Order by the Minister of Commerce, Industry and Tourism."[13]

Relevant policies, legislative framework, and key provisions

Hydrocarbon exploration and exploitation activity in the territorial waters, the continental shelf, and the EEZ of the Republic of Cyprus are primarily governed by the following laws and regulations:

- The Hydrocarbon (Prospection, Exploration and Exploitation) Law of 2007 (No. 4(I)/2007) (the Hydrocarbons Law) and the Hydrocarbon (Prospection, Exploration and Exploitation) Regulations of 2007 and 2009 (No. 51/2007 and No. 113/2009) (the Hydrocarbons Regulations), implementing EU Directive 94/22/EC on the conditions for granting and using authorizations for the prospection, exploration, and production of hydrocarbons
- Laws on environmental protection, health, and safety, including Solid and Hazardous Waste Law, Law 215(I)/2005, the Assessment of Impact on the Environment from Certain Plans and/or Programs Law (No. 102(I)/2005), which is harmonized with EU Directive 2001/42/EC
- The Law to Provide for the Proclamation of the Continuous Zone by the Republic of Cyprus (2004)
- The Law to Provide for the Proclamation of the Exclusive Economic Zone by the Republic of Cyprus (2004)
- The United Nations Convention on the Law of the Sea (UNCLOS 1982)

Cyprus has signed a number of agreements with Egypt, Israel, and Lebanon in line with UNCLOS 1982, including agreements that relate to the delimitation of Cyprus's EEZ.

The Hydrocarbons Law is supplemented by the Hydrocarbons Regulations, which set out the criteria applicable to licensing applications for prospection, exploration, and exploitation of hydrocarbons in Cyprus.

Ownership. According to the Hydrocarbons Law, hydrocarbons in Cyprus are deemed to be and always to have been vested in the Republic of Cyprus. This is true regardless of where hydrocarbons are located within the republic's domain. The Republic of Cyprus has sovereign rights in this area, including the right to grant licenses to investors for hydrocarbon exploration and production in its EEZ.

Licenses. Licenses in Cyprus are awarded through licensing rounds. The authority to issue licenses for prospection, exploration, and exploitation of natural gas rests with the Council of Ministers of the Republic of Cyprus. According to the Hydrocarbons Law and the Hydrocarbons Regulations, the Council of Ministers may issue the following types of licenses:

- *A prospection license for up to one year.* This license does not permit drilling activity, but it allows evaluation of potential by identifying geological structures by means of gravity, magnetic, and seismic surveys.
- *An exploration license.* This type of license is initially valid for three years and allows the holder to undertake gravity, magnetic, and seismic surveys and conduct exploratory drilling. It is renewable for another two periods of two years, and on each renewal, 25% of the initial license is relinquished. In the event of a discovery, the licensee has the right to be granted an exploitation license for the discovery.
- *An exploitation license, granted for an initial period of up to 25 years, with the option to renew for another period of up to 10 years.* This type of license, in case of a discovery of commercially exploitable reserves during exploration, is granted only after the Council of Ministers approves a development and production plan.

Licensing is initiated by the Ministry of Energy, Commerce, Industry, and Tourism (the MCIT), which invites applications for licenses for defined blocks within the EEZ of Cyprus. These applications are evaluated by an advisory committee, which then makes recommendations to the MCIT. The MCIT then seeks the approval of the Council of Ministers for the relevant license application(s).

In granting any of the above types of licenses, the Council of Ministers may take the following into consideration:

- National security and the public interest
- The technical and financial capacity of the applicant
- The means by which the applicant intends to carry out the activities specified in the license
- The financial consideration that the applicant is offering in order to obtain the license
- Any lack of efficiency or responsibility that the applicant may have demonstrated in a previous license or authorization of any form in any country in the world

Successful applicants must enter into an exploration and production sharing contract (EPSC) in the form of the "Model Exploration and Production Sharing Contract (MEPSC)" published by the MCIT. On signing the EPSC, the successful applicant will be granted the exclusive right to conduct exploration, development, and production operations, in accordance with the terms of the license granted. Such authorizations can be transferred only with the consent of the Council of Ministers.

The annual surface fees payable by the licensees under the MEPSC are currently as follows:

- €25 per square kilometer during the exploration period
- €30 per square kilometer during the first renewal period
- €35 per square kilometer during the second renewal period

An additional annual fee of €500 per square kilometer is payable by the licensees to the Republic of Cyprus during the exploitation period.

Health, safety, and the environment. Operations involving hydrocarbons must be undertaken in a proper, safe, and workmanlike manner, and in accordance with good oilfield practices. The operator must ensure that any environmental pollution or damage to the water, ground, or atmosphere resulting from its E&P operations is kept at an absolute minimum. Specifically, operators must not discharge waste into the Mediterranean Sea. Waste management systems and contingency plans must be implemented in order to deal with accidental spillages. Operators must also comply with the following:

- The Hazardous Waste Law (Law 215 (I)/2002) regarding hydrocarbon waste
- The International Convention of Pollution from Ships (MARPOL 73/78) regarding the prevention of pollution of the marine environment by ships from operational or accidental causes
- The Convention for the Protection of the Marine Environment and the Coastal Region of the Mediterranean (Barcelona Convention 1995)

Moreover, the Ministry of Agriculture, Natural Resources, and the Environment (MANRE) introduced the Strategic Environmental Assessment (SEA), which governs the licensing of hydrocarbons within the EEZ of the Republic of Cyprus in order to ensure environmental protection and sustainable development. The SEA is governed by the Assessments of the Effects on the Environment of Certain Plans and/or Programs Law, Law

140 (I)/2005 (as amended), which is harmonized with Directive 2001/42/EC. The overall objective of the SEA is to identify, describe, and evaluate the likely significant environmental effects of implementing hydrocarbon exploration and exploitation activities.[14] Operators, whether by exploration or production licenses, must conduct a preliminary environmental study prior to the initiation of any exploration or production work, in line with the results and recommendations prescribed by the SEA.

Relevant contracts

As mentioned previously, the conditions and requirements of any license granted in Cyprus are contained in an exploration and production sharing contract (EPSC). A standard form of the model EPSC (MEPSC), published by the minister of Energy, Commerce, Industry, and Tourism, covers exploration, appraisal and development, and production activities. The MEPSC grants a right for the licensee to progress from exploration to appraisal and development upon making a commercial discovery. The MEPSC provides for governmental oversight of the licensee's activities, as the licensee is subject to general inspection, supervision, and review by the MCIT.

The main characteristics of the MEPSC are as follows:

1. The contractor is granted the exclusive right to conduct hydrocarbon operations[15] in the contract area at its sole risk, cost, and expense and in accordance with applicable law[16] and the terms and conditions set out in the EPSC.
2. From the first day of production (and during and when hydrocarbons are being produced), the contractor is entitled to use money from produced oil or gas to recover capital and operational expenditures.[17] Moreover, the contractor is entitled to take a percentage share[18] of the remaining money[19] from produced oil or gas, in consideration for its investment in the hydrocarbons operations.
3. The corporate tax is included in the share of the republic's profit oil, and thus profits made by the contractor are effectively exempted from tax.
4. The contractor must make a declaration of commerciality within the period of the duration of the exploration license.
5. If the contractor declares that the discovery is commercially exploitable, the contractor must submit a proposed development and production plan within four months of making the declaration to the minister of the MCIT for approval by the Council of Ministers.

6. A contractor is obligated to relinquish specific blocks or parts of an exploration block when renewing the licensing period.
7. The MEPSC may be terminated by the minister and the license revoked by the Council of Ministers if a material breach or recurrent breach of the MEPSC takes place on the part of the contractor.
8. In the event of any difference or disagreement or dispute between the republic and the contractor regarding the interpretation or implementation of any provisions of the license agreement, or if one of the parties wishes to challenge the decision reached by the expert pursuant to Article 36(2), the dispute will be referred to arbitration.
9. Under the MEPSC, the licensee's rights and obligations are to supply all necessary funds and purchase or rent all equipment and materials necessary for the performance of the licensee's operations.
10. Prepare and implement works programs in accordance with best international industry practices and to prepare and submit budgets.
11. The MEPSC stipulates that preference will be given to local and European Economic Area (EEA) enterprises and goods.
12. The contractor shall assure priority of employment for Cypriot and EEA personnel and contribute to the training of the personnel in order to allow them to have access to any position of a skilled worker, foreman, executive, and manager.
13. The contractor, if requested in writing, is obligated to sell to the republic a portion of the contractor's share of profit oil and profit gas for the domestic consumption of the republic at a price to be agreed between the contractor and the minister or any person authorized by the minister.
14. Have the right to free access and operation in the contract area.
15. Prepare and submit development and production plans to the minister for approval.
16. Provide sales and marketing assistance to the minister.
17. Provide necessary guarantees, including performance guarantees.
18. Lift, dispose of, and export the licensee's share of hydrocarbons.
19. Take out and maintain in force all insurance of the type and in the amounts customarily used in the international hydrocarbon industry.

Positive and Negative Experiences and Expectations Regarding the Regulatory Framework

Regulatory bodies in Cyprus do not themselves possess significant experience in the energy sector, and therefore, they are likely to need to rely on external expertise in order to supervise or monitor activities effectively. The Republic of Cyprus is currently being advised by independent professional advisors, mainly from abroad, in an attempt to set up and administer a model for managing the economic aspects of gas reserves.

A legal and regulatory framework exists to some extent in Cyprus, but it will be constantly improving and developing as Cyprus develops its experience in this sector.

At present there is not much to be said about positive or negative experiences regarding the regulatory framework in Cyprus, as it has not really been tested. The operations carried out in Cyprus at the moment are still at the exploration phase. At this preliminary stage, one of the positive aspects is that there is definitely an intention to promote local recruitment, acquisition of local goods and services, transfer of know-how, implementation of other methods of training local skilled workers, and development of the domestic industrial infrastructure.

This is evidenced by the requirements that have to be fulfilled by the contractors set out in the MEPSC during the licensing phase, as mentioned previously.

Current Issues

Infrastructure—pipeline network

Infrastructure needs to be developed for the effective exploitation of hydrocarbon reserves. Currently, there are ongoing discussions regarding how Cyprus can collaborate with neighboring countries for the export of natural gas. Noble Energy has recently presented its proposed development plan for the exploitation of natural gas from Block 12 in the Cyprus EEZ.

The development plan currently includes the following:

- Installing an FPSO platform over the Block 12 field to collect and process the gas.
- Exporting gas to local and nearby markets. The most likely way is through a pipeline network, pumping gas to Egypt through existing LNG production facilities. Another option under consideration is construction of a pipeline network with Jordan, for the use of the local market.

Another proposed infrastructure project is the construction of a subsea "East Med" pipeline. This project would incorporate a subsea pipeline from offshore Cyprus to Cyprus, then to mainland Greece via Crete, and finally to Italy via mainland Greece and the IGI (Italy–Greece Interconnector) pipeline to serve the EU market.[20]

Infrastructure—LNG terminal Cyprus

As part of its efforts to secure an uninterrupted supply of natural gas, initially for power generation and then for other uses, DEFA has planned and launched the work for the construction and development of a transmission and distribution natural gas pipeline network. Initially, this network will consist of three pipelines that will supply the three power stations of the Electricity Authority of Cyprus in Vasilikos, Dhekelia, and Moni. This initial network will serve as the backbone for the development of the future network in the cities and industries, and is estimated to have a total length of about 80 km. The initial cost estimate for this phase of the project is approximately €60 million (this project has already secured EU funding of €10 million under the European Economic Program for Recovery, following a relevant proposal submitted by the Ministry of Energy, Commerce, Industry, and Tourism).[21]

The idea of constructing an LNG terminal at Vasilikos was an option that has been evaluated but currently abandoned, mostly due to the major increase in the cost of the project. A floating LNG terminal seems to be the leading option for a liquefaction and purification facility, as it appears to be the most economical and feasible option at the time of writing.

Fall in the price of oil

The recent fall in the price of oil has affected the activities of oil companies and the oil industry as a whole. The fall in the price of oil internationally has affected many companies, making them more risk-adverse. Major oil companies have already taken measures to reduce capital expenditures by trimming their upstream business efforts, as the drop in oil prices has made exploration, production, and LNG projects less economical.

These developments have significantly affected the oil and gas companies operating in Cyprus. Total SA announced plans to slash capital spending for 2015 by 10% and exploration spending by 30%. Moreover, due to the decline in oil prices, Noble decided to freeze its drilling program for Block 12 and announced that it was interested in giving away 30% of its share in Block 12 in order to reduce its exposure.[22] Noble finally ended its Block 12 obligations through sale of half of its interest to BP cheaply, and relinquishment of the remaining half interest in May 2016.[23]

Geopolitical problems

Turkey threatens stability in the region. Since the allocation of Blocks 5 and 6 to Eni-KOGAS in December 2013, Turkey has been carrying out seismic surveys within an area that trespassed into parts of Blocks 4, 5, and 6, violating the UN Convention on the Law of the Sea (UNCLOS 1982) and the right of Cyprus to exploit its EEZ. Moreover, Turkey has taken steps to dissuade participants in the Cyprus bidding process. It has banned Eni from operating in Turkey, and it has also sought to impede survey vessels operating in the Cyprus EEZ. As a result, hydrocarbon projects in Cyprus appear less attractive to foreign oil companies.

Skilled labor

Lack of skilled labor and equipment on the island may result in difficulties for oil companies seeking to expand their activities rapidly in Cyprus.

Concerns over health, safety, and the environment

Oil and gas companies and service providers are involved in heavy industrial activities and occupy large spaces just one kilometer away from residential areas. This causes difficulties with local residents who are concerned with environmental health and safety risks of the nearby operations. In 2014, these concerns led to complaints to the ombudsman and to an online petition demanding the facilities be removed from the designated area.

Conclusion

The Cypriot government is developing a comprehensive strategy for exploiting its potentially large offshore gas reservoir and transforming its domestic energy sector. A successful strategy will contribute to establishing a stable public revenue flow and to lower electricity prices.[24] While the action plan for the exploitation of its hydrocarbons reserves exists, implementation and streamlining are yet to take place.

Cyprus will not only benefit from the revenue flow from the future exploitation of its gas reserves, but also from the direct foreign investment in significant infrastructure projects such as the pipeline network. The emergence of this new industry will definitely create the need for a number of support services that will lead to the opening of hundreds of jobs.

Despite the fact that Cyprus is still in the early stages of development in this industry, good management and decision making will assist the industry to overcome the technical, geographical, and economic challenges arising from the exploitation of Cypriot hydrocarbon reserves.

Notes

1. CYSTAT (Statistical Service of the Republic of Cyprus), Ministry of Finance, offers several energy publications at http://www.mof.gov.cy/.
2. Ibid.
3. "Commission Staff Working Document: Country Report Cyprus," SWD(2015) 32 Final (Brussels: European Commission, February 26, 2015), http://eur-lex.europa.eu/legal-content/EN/TXT/PDF/?uri=CELEX:52015SC0032&qid=1429620042872&from=EN.
4. "Hydrocarbon Exploration," Ministry of Energy, Commerce, Industry and Tourism, http://www.mcit.gov.cy/mcit/mcit.nsf/All/FE3EB5707ADA0E6EC225771B0035B0D2?OpenDocument.
5. "The Energy Market in Cyprus: Natural Gas, Changing the Picture," Natural Gas Public Company (DEFA), 4, http://www.defa.com.cy/presentations/4Dec12UBIFRANCEGREECE_atKEVE_.pdf.
6. Solon Kassinis, "Natural Gas—Fundamentals, Geopolitics and the Way Forward for Resource Monetization," (2014), 22.
7. "Energy Service," Ministry of Energy, Commerce, Industry, and Tourism, http://www.mcit.gov.cy/mcit/mcit.nsf/dmlenergyservice_en/dmlenergyservice_en?OpenDocument.
8. See "Objectives," Cyprus Hydrocarbons Company (CHC), http://chc.com.cy/about-us/objectives/; and "Responsibilities," CHC, http://chc.com.cy/about-us/responsibilities/.

9. Charles Ellinas, "Cyprus Gas Export Options," Cyprus Gas News, April 13, 2015, http://www.cyprusgasnews.com/archives/8002/cyprus-gas-export-options-2/.
10. "Scope of Work," National Gas Public Company (DEFA), http://www.defa.com.cy/en/defa-company/purpose.
11. Lefteris Adilinis and Fiona Mullen, "Expanding Cyprus Gas Options," In-Cypress.com, May 6, 2016, http://in-cyprus.com/expanding-cyprus-gas-options/.
12. "Cyprus Energy Regulatory Authority (CERA): Company Overview," *Cyprus Profile*, http://www.cyprusprofile.com/en/companies/overview/cyprus-energy-regulatory-authority-cera.
13. "Mission," COSMOS (KODAP), http://www.kodap.org.cy/index.php/en/the-organization/mission.
14. "Energy Service."
15. According to the "Model Exploration Production Sharing Contract," Republic of Cyprus, MCIT, 10, http://www.resourcecontracts.org/contract/ocds-591adf-1348839751/download/pdf, "hydrocarbon operations" are defined as follows: "Exploration Operations, Development and Production Operations and all other related activities carried out under this Contract, including the lifting of Hydrocarbons from the Contract Area but excluding any storage, transportation or processing beyond the Delivery Point."
16. As it relates to the legislative framework, discussed previously in the text.
17. This is known as *cost oil* or *cost gas*.
18. The production sharing contract signed with Noble Energy in the first licensing round provides for a 65% share to the Republic of Cyprus and a 35% to Noble Energy.
19. This is known as *profit oil* and/or *profit gas*.
20. Kassinis, "Natural Gas."
21. "Energy Service."
22. Charles Ellinas, "The Price of Oil and Its Impact," *In-Cypress.com*, 2015, http://in-cyprus.com/price-oil-impact/.
23. Charles Ellinas, John Roberts, and Harry Tzimitras, *Hydrocarbon Developments in the Eastern Mediterranean: The Case for Pragmatism* (Washington, DC: Atlantic Council, July 31, 2016), 11, https://issuu.com/atlanticcouncil/docs/hydrocarbon_developments_in_the_eas.
24. "Commission Staff Working Document: Country Report Cyprus," 2015.

8

EGYPT

Loic Conan and Sayed El Araby

Introduction

Egypt lies to the north of Sudan, with Libya to the west and the Gaza Strip to the east, comprising a surface area of about 1,250,000 square kilometers. Egypt's population was around 90 million at the end of 2015. Arabic is the mother language of Egypt, and English and French are the most commonly spoken foreign languages. A main revenue resource of the country comes from transit fees from the Suez Canal, a major shipping lane. Other revenue sources include agriculture, trade, tourism, industry, and minor exports of some crude oil and petroleum products.

The first discovery of oil in Egypt was made in the 18th century in the western onshore part of the Gulf of Suez. Since the advent of the petroleum industry, Egypt has followed a variety of oil regimes. Until 1958, the tax/royalty system was employed, followed by participation agreements during the 1960s. Since 1972, Egypt has employed production sharing contracts, following Indonesia's pioneering example.

In Egypt, the People's Assembly (or the president in the absence of the People's Assembly) is the body that authorizes the minister of Petroleum and Mineral Resources to sign a production sharing agreement (PSA) with the investor once the relevant law is issued. Briefly, the production sharing regime in Egypt operates as follows:

- The international oil company (IOC) incurs all exploration risks.

- If the IOC is successful in finding hydrocarbons in sufficient quantities, it can recover its costs from a share of production (customarily up to 40%).
- The remaining share of production, called *profit oil and gas*, is split between the international oil company and the national oil company, with an average ratio of 75% for the latter and 25% for the former, subject to the potential of the area and risks associated with the exploration/production of its deposits.
- Royalty and income tax are normally paid to the government from the national oil company's share.
- There are also bonuses related to signature, transfer of interest, and meeting production thresholds.

The Egyptian government has signed more than 500 agreements with multinational companies since the production sharing regime began.

Local Content Policy and Baseline

The national oil company of Egypt is called the Egyptian General Petroleum Corporation (EGPC).[1] Its main purpose is to increase or at least sustain production to satisfy the local needs for petroleum products while minimizing the importation of products for domestic use. The EGPC is one of the largest economic organizations in Egypt and has an influential and strategic role in securing sufficient petroleum products to meet the country's daily needs at prices far below actual market costs. The most important challenges spring from the continuous increase in domestic consumption, which increases the need to secure sufficient foreign currency to cover the cost of importing petroleum to meet the balance of those consumption demands. The EGPC is tasked with developing strategies in order to cope with industry changes and price volatility. During the fiscal period 2013 to 2014, EGPC succeeded in securing about 35.4 million tons of petroleum products for domestic consumption via production from refining and processing companies. It met the remaining demand for gas, oil, LPG, and fuel oil with imports. In addition, the government revisited the existing price of natural gas in order to make it fair and more profitable for the upstream companies, especially those operating in deep waters, since local production remains cheaper than importing this same energy. The development of approximately 15 trillion cubic feet (Tcf) of stranded gas through ad hoc service agreements has been another way Egypt has boosted production of brownfields. In the same vein, the government has increased its investments in infrastructure to provide facilities to the operators. Egypt has also exerted some effort to develop the potential of the younger workforce in different disciplines by supporting knowledge-sharing experiences with senior staff.

Key Players

Host government/national oil companies

EGPC. As mentioned previously, Egypt's hydrocarbon activities are managed by the Ministry of Petroleum and Mineral Resources through national oil companies, the foremost being the Egyptian General Petroleum Company (EGPC). Established in 1958, the EGPC is tasked with managing upstream oil activities and implementing the three prongs of the government's oil policy:

1. Maximizing the declining domestic oil production
2. Obtaining additional foreign investment
3. Operating the country's large refining system

In addition to sharing in production through partnerships with international oil companies, the EGPC directly produces 20% of Egypt's output.

EGAS and Ganope. Two other important national oil companies, EGAS and Ganope, are tasked with implementing the government's oil policy. The Egyptian Natural Gas Holding Company (EGAS) was established in 2001 and is in charge of natural gas activities. Ganoub El Wadi Holding Company (Ganope) was established in 2002 and is responsible for both natural gas and crude oil activities in Upper Egypt. EGAS was established to cope with the boom of natural gas discoveries in the Mediterranean and Nile delta region. As for Ganope, it was established due to the increase of development projects in Upper Egypt. A national oil company distinct from the EGPC was needed to improve the management of these specific activities. While the EGPC remains the most important in size, these three national oil companies assume similar functions, and all report to the Ministry of Petroleum and Mineral Resources.

Oil and gas companies operating in the country

The involvement of the international oil companies (IOCs) is important, and it primarily occurs through production sharing arrangements with the relevant national oil company. Key oil partners are BP and Eni, along with Apache, an IOC. BP, Eni, and Apache are also the main gas players, as well as BG, Dana Gas, Petronas, and Shell. At least 70% of Egypt's natural gas is extracted from the Nile delta basin and offshore Mediterranean Sea blocks, where Eni recently made the Zohr discovery. The Zohr gas field is the largest gas discovery made in Egypt and the Mediterranean Sea to date. Given the intensive capital needs of offshore deepwater gas exploration and production, most investments in the region have been made by foreign operators. Local players are present in small, mature fields, such as in the Gulf of Suez, the Western Desert, and Nile delta. Private local players account for about 10% of total Egyptian production.

Service providers operating in the country

The Egyptian petroleum market is considered one of the most competitive markets in the Middle East. Around 120 multinational services companies are listed as registered or approved on the EGPC/EGAS/Ganope vendors list. Some are state-owned companies such as Petrojet, Enppi, Sino Tharwa Drilling Company, Egyptian Drilling Company, Petroleum Air Services, and Petroleum Marine Services Company. These companies face a very competitive environment that leads to performing services and providing materials at the lowest price possible. As explained further below, local sourcing of goods and services is encouraged by the host country agreement, provided certain standards are met.

Industry Knowledge

The EGPC has been carefully following and studying the influence of the changes in the petroleum industry, with the support of established petroleum training and research centers in cooperation with local universities and international oil companies with local assets. EGPC is aware of the important role of human resources management in the petroleum industry. It recognizes the need for personnel in the national oil companies who are qualified, talented, and capable of meeting private sector standards. The EGPC has established several training centers to allow opportunities for personnel to acquire relevant experience and to become acquainted with the latest technologies. The EGPC has also concluded cooperating agreements with multinational training and education centers to provide courses and training programs for different levels of staff and management. In addition, the EGPC participates in conferences and seminars inside and outside of Egypt.

Legislative Hydrocarbon Regime Concerning Local Content

Regulatory bodies

The EGPC, EGAS, and Ganope are the three main regulatory bodies for the Egyptian oil and gas industry (see the "Host government/national oil companies" section).

Relevant policies, legislative framework, and key provisions

Legal sources on local content exist at different levels. The top tier is the Constitution, adopted in January 2014. It addresses local content as follows:

- *Article 20.* The state shall encourage and develop technical and technological education, as well as vocational training, and expand all their types in accordance with international quality standards and labor market needs.
- *Article 32.* The state's natural resources belong to the people. The state shall preserve and effectively exploit them.

The level below the Constitution is comprised of the legislative sources, with Law No. 66 of 1953 (as amended), Concerning Mines and Quarries, and Law No. 8 of 1997, the Investment Law. Both list local content objectives. Decree 758 of 1972, Executive Regulations of Law No. 66/1953, Concerning Mines and Quarries, contains some details on local content that are also found in production sharing agreements. The details are described later in this chapter.

Equally important legally are the production sharing agreements entered into by the IOCs and the government. The PSAs further describe the applicable local content tools and policies.

Relevant contracts

In the case of Egypt's local content, concrete policies are to be found in the granting instrument, which in most cases is a production sharing contract.

Petroleum rights allocation system/NOC equity participation. If a commercial discovery is made during the exploration period, a *joint operating company* is established— a company equally owned between the national oil company and the international oil company or consortium. If one of the international oil companies subsequently seeks to assign its participating interest to a third party, the national oil company party to the relevant production sharing contract has a preemptive right to acquire the interest and to act as the international oil company. In this case, assuming there is a joint operating agreement among the international oil companies in place, the national oil company will join the existing joint operating agreement. Sometimes there is no joint operating agreement because one international oil company acts as sole contractor member. In this case, the production sharing contract (PSC) provides that a joint operating agreement based on the most recent Association of International Petroleum Negotiators (AIPN) model is to be agreed between the national oil company and the international oil company. The Egyptian PSC model is designed for the national oil company to ultimately own 100% interest in the area (through exercising its right of preemption over time on each occasion an assignment is proposed).

Employment rights and training. The Egyptian PSA states the preference for employment of qualified Egyptian nationals. International oil companies and national oil companies are expected to carry out training programs for all Egyptian employees, with the expectation that qualified Egyptian national employees will gradually replace nonexecutive expatriate staff as openings are available. There is also an annual training budget allocated to this purpose. This budget will depend on the value of the concession awarded but usually ranges between US$100,000 and US$250,000. Another incentive to hire local staff is the rule on the minimum fraction of expatriate salaries (25%) that is to be paid in local currency.

Domestic sourcing of goods and services. The PSC gives priority to local contractors if services performance meets international standards and prices are not more than 10% higher than the costs of a foreign supplier. Likewise, preference is given to locally manufactured material, equipment, machinery, and consumables if guidelines are met concerning quality standards, timely delivery, and pricing of not more than 10% above the price of same imported items.

Tariffs. Rules on tariffs mirror tender procedures, i.e., there is a right for the international oil company to freely import and be exempted from custom duties, but these exemptions do not apply to any imported items if the same items are manufactured locally and meet quality, safety, price, and timely delivery expectations.

Positive and Negative Experiences and Expectations Regarding the Regulatory Framework

Petroleum rights allocation system/NOC equity participation

To date, Egyptian national oil companies (the EGPC, EGAS, or Ganope) have mostly been acting as regulators, only acting as operators in a marginal role. This is predominantly due to their debt level, which constitutes a significant barrier to development and hinders employment of personnel with the required technical skills. Indeed, the national oil companies have had to cope with a distressed budget because of the government's heavy energy subsidies, which amounted to approximately US$20 billion in 2013. This has resulted in the inability of the national oil companies to fulfill their financial obligations to the international oil companies, and likewise to comply with the terms of the production sharing contracts. The amount of debt peaked in June 2014, with arrears of US$7.5 billion. It was still US$2.9 billion at the end of August 2015.

The delay in paying international oil companies has led to delays in investments, which in turn has led to further declining production. The government plan to progressively phase out subsidies within the next five years should offer some relief to the national oil companies. However, subsidies policies are highly sensitive political topics and are subject to change. Also, the drop in barrel price should help decrease the debt. This mechanically reduces the national oil companies' dues, at least for crude oil not subject to long-term arrangements, as opposed to gas sales agreements subject to the price-reopener provision. But for the near future, it seems the national oil companies will fall short of becoming fully business-oriented organizations, impeded by bureaucracy and lack of efficiency.

Employment rights and training

The proper implementation of the employment rights and training policies is monitored through EGPC channels, via the cost-recovery mechanism. The effectiveness of these policies is difficult to assess, given the lack of data and audits on targets and outcomes. That said, it seems that with respect to training programs, the national oil company systematically requests the cash amount from the international oil company rather than allowing the latter to show that the equivalent amount has been spent on training programs. The cash is understood to cover the national oil companies' debt instead. Either way makes it difficult to control whether the cash paid to the national oil company under the training program clause is actually spent for this purpose. Recently, senior officials for the national oil companies and the Ministry of Petroleum and Mineral Resources have acquired their managerial skills within the private sector.

Domestic sourcing of services and goods

Unlike operatorship that is very capital intensive, Egypt has been quite successful in raising local participation in sourcing of services and goods. Some state-owned oilfield services companies have shown competitiveness in construction and engineering projects. To name the most well-known, Petrojet, Enppi, and Petroleum Marine Services have even managed to extend their reach outside Egypt and are now present in the Middle East/North Africa region. International contractors retain a residual role in the most sophisticated projects requiring advanced technology, such as LNG. In such projects, domestic contractors may still be involved through subcontractor roles.

Current Issues and Challenges

Stabilization

In the mid-2000s, EGPC introduced one of the most attractive features in the PSC: the stabilization clause. This clause is designed to protect the investor against any laws or regulations that could adversely impact the economics of the project. In this context, the parties may agree on some changes to restore the balance of the contract toward terms more nearly approaching those of the effective date of the PSC. In light of the combined global oil price drop, local foreign currency shortage, and political unrest, this has been reassuring to investors.

Decommissioning

Some PSCs in force since the 1970s do not contain any provision on decommissioning. These early PSCs will expire soon, leaving a legal vacuum for the concerned areas. The latest adopted PSC model will remedy this shortcoming for the coming blocks to be tendered. The newer PSCs include a decommissioning clause, which obligates both the national oil company and the international oil company to restore the area of operations as it was before operations commenced, after removing all facilities at contract termination. The budget for decommissioning is agreed through a specific percentage of production allocated for such purpose.

EGPC debt

As discussed previously, the main challenge faced by international oil and gas companies in Egypt is the large debt they are owed by the EGPC and the other national oil companies. Problems arise when the national oil company, in an effort to satisfy local consumption, decides to purchase the profit oil share from the international oil company. (Under the PSA, the national oil company is granted a preemption right, since normally the share is free for export.) The EGPC has been reducing this debt by securing financing from banks. The government strongly believes the subsidy reform is key to resolving the longstanding balance of payments problem. Fuel subsidy reforms have been implemented since July 2014.

Conclusion

Local content experience in Egypt has so far led to mixed results. The legal framework is in place, but realities such as the level of debt and the bureaucratic mindset have made its implementation challenging. The NOCs have not been able to use their preferential rights to develop into a commercial entity. The outlook is brighter with state-owned oilfield services companies than for the national oil companies operating in the upstream sector.

Note

1. See the "Key Players" section for a more detailed introduction.

9

REPUBLIC OF EQUATORIAL GUINEA

Cristina Sánchez Cosme

Introduction

Since the beginning of the development of the oil and gas industry in Equatorial Guinea, the state has been trying to encourage the inclusion of national companies and human capital in the operation. This country has the advantage of having closely observed the experiences and challenges of its neighbors in the region, which have been producing precious resources for a significant amount of time. Equatorial Guinea's first exploration license was awarded in 1983, but it was not until 1990 that the first discovery of a commercial deposit was made. This successful deposit was discovered in the Alba field, offshore Bioko Island. The country is truly a young player in the industry in comparison to its neighbors Cameroon, Nigeria, and Gabon. According to the 2015 census, the population of Equatorial Guinea is more than 1.2 million,[1] and 77% of its GDP comes from the petroleum industry. Given the overwhelming contribution to the country's economy, it is important for the government of Equatorial Guinea to ensure that its citizens play an important role in the development of the industry.

In an effort to complement the Hydrocarbons Law of 2006, the Ministry of Mines, Industry, and Energy (MMIE) published Ministerial Order Number 1/2014, through which the National Content Regulation was adopted. The evolution of strategies with regard to national content in Equatorial Guinea has been slow, but the efforts to develop and implement the measures to include local content can be seen year by year. The ministry's greatest challenge with implementing the regulation is resistance to change in the status quo. Some resistance comes from the international oil companies, which resist changes due to the potential for increased costs and lower production efficiency, among other reasons.

Local Content Policy and Baseline

The development of public policy behind national content is featured in several texts. Public policy is a key premise in the Hydrocarbons Law of 2006 and in the National Content Regulation of 2014. It is also an important part of both the Petroleum Regulation of 2014 and the Contract Model of 2006. In an effort to develop human capital and train people from Equatorial Guinea for jobs in the sector, the state has signed nonbinding cooperation agreements with several countries in the African continent, including the Republic of Côte d'Ivoire (Ivory Coast), the Republic of Ghana, and the Democratic Republic of São Tomé and Príncipe. The idea behind this is to create a strong regional and continental bond, facilitated by the cultural and historical similarities between the nations. At a national level, the country is promoting the creation of legislation that facilitates the contribution of other Africa countries before seeking solutions outside the continent.

Chapter XX of the Hydrocarbons Law talks exclusively about the measures aimed at strengthening national content. It is focused on providing general provisions in regard to the contribution of contractors to the National Technological Institute of Hydrocarbons of Equatorial Guinea (ITNHGE), also called the Technological Institute of Hydrocarbons. It also focuses on measures to be taken to ensure national investment, contribution to social works within the country, and training of national human capital. In the Hydrocarbons Law, *contractor* refers to any person with whom the state has entered into a contract that authorizes the performance and regulates the conduct of petroleum operations. Other parties that have to comply with local content are not really mentioned under Chapter XX of the Hydrocarbons Law. It was not until the National Content Regulation was created, more than a decade after the Hydrocarbons Law was enacted, that the ministry had the necessary tools to proactively promote public policy.

In regards to short-terms goals, the state wants to assure that part of the revenue coming from petroleum operations remains in the country, and that more citizens become part of the sector's labor force. In the long term, the state's goal is the total inclusion of its citizens in all levels of operation. The state also is eager to see domestic companies participating in the progress achieved in the industry. The involvement of national companies in providing services to companies in the sector is expected. However, this can only happen with an effective transfer of technology. On the other hand, certain social work programs are seen as part of the long-term development of the country's civil infrastructure. Generally, the Ministry of Mines, Industry, and Energy approves construction work as part of contractor's social contribution, as per local content regulations. The construction of schools in rural areas and of sanitary infrastructure is particularly important for the state.

One of the main obstacles that both the private sector and the Ministry of Mines, Industry, and Energy have in regards to the implementation of the local content policy for labor is the lack of individuals from Equatorial Guinea specialized in any job necessary for operations in the hydrocarbon sector. In order to encourage national content development and to bring the level of skills in the national workforce up to international standards, the government created the Technological Institute of Hydrocarbons, as mentioned previously. The institute started its operations in late 2007. Little by little the state, with the support of IOCs, has been able to develop a qualified workforce for certain areas within the industry.

Considering that Equatorial Guinea has 1.5 billion barrels of proven reserves, the growth of the sector, even when prices are low, will be continuous. The drop in price is clearly an obstacle to the development of the sector. Since the oil and gas field in the country is not saturated in terms of activity, and there are currently only a few main players, the country has the capacity of welcoming additional investors willing to take risks in exploration. As proof of this, the state awarded five new licenses for different blocks at the end of 2015. Moreover, as a result of government policy, Equatorial Guinea has positioned itself at the forefront in the development of downstream and midstream projects in the region. For example, projects such as Ophir's floating LNG platform, the Bioko oil terminal, and the Riaba integrated petrochemical complex are all pioneering initiatives for the continent. Given the country's small size and its large ambitions, the need for an adequate workforce to fulfill its goals is readily apparent.

Key Players

The Ministry of Mines, Industry, and Energy is the entity that supervises and regulates all matters related to petroleum operations in Equatorial Guinea. Originally, an operating company called GEPSA (Guineo-Española de Petróleos, SA) represented the commercial interests of the state. GEPSA was owned and operated as a joint venture by the state and Hispanoil. Afterwards, GEPSA was dissolved. Due to the significant progress of the industry during the 1990s, the government created its national company in 2001 under the name of Guinea Ecuatorial de Petróleos (GEPetrol), through Presidential Decree 9/2001. GEPetrol[2] is focused on the management of the state's participation in production sharing contracts. It is the state's hydrocarbons sales agent, and it has the authority to participate in E&P activities inside and outside of Equatorial Guinea. However, the Ministry of Mines, Industry, and Energy is the entity that grants licenses, regulates, and administers the industry. At present, GEPetrol participates in 30 blocks, and it is the operator of Block P and Block N.[3]

In 2005, President Teodoro Obiang Nguema Mbasogo signed Decree 45/2005 for the creation of a natural gas state company called the National Society of Natural Gas of Equatorial Guinea (Sonagas). This company manages gas assets for both the industrial and residential markets. It is also in charge of the distribution, commercialization, and exportation of natural gas. Per the most recent version of the Hydrocarbon Law, both GEPetrol and Sonagas have the right to at least 20% participation in any concession.

International oil companies have been key in the evolution of the country's economy. Companies such as Nobel Energy, Marathon Oil Corporation, ExxonMobil, Hess Equatorial Guinea, Ophir Energy, Atlas Petroleum International, and Tullow Oil PLC stand out due to their contributions and work within the sector in Equatorial Guinea. Moreover, it is worth mentioning several service companies working together with the IOCs, such as Schlumberger, Halliburton, Wood Group PSN, and Baker Hughes. These multinational companies have the same legal obligations as any oil and gas company in regard to local content.

Among the key agents, Luba Freeport Ltd. should be mentioned. It is a logistics company that provides services to operating companies in the hydrocarbons sector that are established in Equatorial Guinea. It is a joint venture company between GEPetrol and a private investor. This company is attractive as it functions as a "one-stop shop" and is free of custom taxes, saving time and money. Among the services Luba Freeport Ltd. offers to companies in the hydrocarbons sector are commercial property rental space, accommodations, medical facilities, and catering.

Industry Knowledge

With respect to knowledge and training of national employees, companies have focused efforts on the development of their human capital. Since the Technological Institute of Hydrocarbons opened its doors in 2007, it has expanded to several campuses around the country. This institution was created in order to address the necessity of training young nationals so they could participate in the sector. The government is currently developing the main campus of the Technology Institute of Petroleum of Mongomo. This institute will contribute to the activities of the Technological Institute of Hydrocarbons.

The National University of Equatorial Guinea (NUEG) is the only academic institution for higher education. At present, it offers training for careers in petroleum engineering. However, this institution was founded in 1995 and does not have the same resources in its curriculum as equivalent institutes found overseas. This shows the importance of the Technological Institute of Hydrocarbons, which offers more complete training for young workers entering the job market. Moreover, it is very common to see locals who have trained and studied outside of Equatorial Guinea. Many of them have completed their degrees at universities overseas. Some of the most common destinations include countries such as Russia, China, Cuba, the United States, and Spain.

Legislative Hydrocarbon Regime Concerning Local Content

Before further discussing the Ministry of Mines, Industry, and Energy as the regulatory body overseeing all petroleum operations, we have to mention the responsibilities of the Ministry of Work and Social Security. This ministry oversees all labor matters in the country, including employment with the IOCs and services companies.

One aspect of the social and public policy considerations behind the Labor Law, as amended in 2012, emerges in its encouragement to contract nationals and reduce the numbers of imported workers. Moreover, it demands that local companies implement professional training and education programs for their employees.

An important aspect of local policy is established by the Constitution,[4] as amended in 2012. It addresses, as one of the basic aspects of Equatoguinean society, the importance of work protection, "through which a man develops the personality that creates National wealth towards social being." This means that there is a constitutional right to work in Equatorial Guinea. Following on from this, the national work legislation has been written and interpreted in a very protective manner.

The Regulating Law for the National Work Policy of 1992 establishes certain parameters through which the Ministry of Work and Social Security can ensure that the professional development of Equatoguineans remains a priority for any employer. This law establishes the procedural obligations necessary for an employer to obtain work permits for foreign employees from the Ministry of Work and Social Security. In order to obtain such work permits, the employer has to prove that there has been a check through the National Work Office and that no adequate candidates have been found to fill that position. This means that the employer has to prove that it has carried out all possible efforts to hire local trained employees, and that it has not been able to fill the jobs. At this point, the employer is allowed to employ foreign workers, as long as the required degrees or qualifications are present. In any case, under no circumstances is the employer allowed to hire more than 10% of foreign employees. In terms of companies in the hydrocarbons sector, the legislation allows an exception of 30% of foreign employees.

In 2010, the Department of National Content was created in affiliation with the MMIE. In 2014, the National Content Ministerial Order was announced. This law promotes and expands the application of Chapter XX of the Hydrocarbons Law and covers all aspects of financial and human capital associated with local content in Equatorial Guinea. There are certain key ideas in this text, such as a local presence, training, networking, equal conditions, investments, and compliance.

In order to carry out business within the Republic of Equatorial Guinea, a company must establish a legal entity in the jurisdiction. Decree 127/2004 establishes that the participation of Equatoguinean partners in companies created by foreigners or foreign companies established in the country cannot be lower than 35% of the share capital. Moreover, it establishes that national participation shall never be lower than three national partners.[5] The state wants to obligate international companies to establish themselves in the country and become clear participants in the industry.

While it is unlikely that the hiring practices of IOCs will go unnoticed, service companies can sometimes fly under the radar. This dynamic can even be seen when the service companies are involved in million-dollar contracts. Therefore, certain companies might enter and exit Equatorial Guinea without necessarily complying with the national content regulation or other norms. In order to combat this problem, the MMIE has developed the Company Registry. Companies are not only obligated to register, but they also have to apply for a license. Without this license, which has a fixed cost, companies in the sector cannot operate in Equatorial Guinea.

The MMIE would also like to eliminate situations in which financial documents are kept outside the country, and fiscal audits are carried out overseas. In many cases, it is difficult for local partners to have full access to the books of a company, even when this partner is the state. It is even more complicated to force a company to share information and documents when their offices are not in the country. In general terms, often the state and other parties of interest are not able to carry out audits correctly due to a lack of documentation.

Under the National Content Regulation, the MMIE will work as a platform to connect different players in the sector. Before E&P companies contract with any service company, they have to refer their requirements to the ministry. This gives the ministry evidence of all the operator's partners (including service companies). The participating companies have an obligation to update this requirement list with the ministry every six months, and in turn, the ministry will update the database of national companies accordingly. The same conditions apply to both local and international companies. The fiscal identification number is particularly important. If a local company does not have all the necessary permits in order to operate commercially in Equatorial Guinea, they will not be given access to the registry, and they will not be able to provide services to either multinational companies or the government.

It is expected that local service companies will not have the same expertise level as international companies that have been in the industry longer. This does not mean that they should be left behind. This is where the idea of strategic alliances between local, regional, and international companies has been developed in order to improve balanced participation in oil and gas projects. This means that a group of companies with mixed origin can combine their resources to contribute a certain percentage of human capital toward a specific service. Moreover, Article 17 of the National Content Regulation establishes that "companies in the hydrocarbon sector engaged in exploration and production, as well as contractors and subcontractors, shall offer split contracts, with the intention of creating smaller contracts, and to the extent possible, invite domestic firms to participate, in accordance with their capabilities and resources." This aims to combine the capacity of many local companies. The core idea is to avoid overlooking local companies because they might not be able to compete with international service providers.

The Technological Institute of Hydrocarbons and the internal training programs of each company try to develop human capital through other means. For example, the National Content Regulation included a mandatory annual internship program provided by IOCs and others for graduates of the UNGE. This institution offers a degree in technical petroleum engineering, among other careers, which could be useful to oil and gas companies at administrative levels. Internship plans and the education and training plans for employees are not new to the local legislation. In fact, such plans are in line with the regulatory expectations of the Ministry of Work and Social Security.

Equatorial Guinea has been blessed with precious natural resources, including hydrocarbons. Unquestionably, such resources have had a positive impact on the development of national infrastructure, services, and education. However, there remains a long way to go. In such a profitable sector, the government wants to see a direct impact on those who are most in need in the country. Therefore, social responsibly commitments are at the top of the hierarchy in regards to national content. The Department of National Content has the task of identifying the social necessities of the country in order to assign the financing for various projects to different companies. These companies are obligated to contribute a certain amount of capital per year, according to the total value of their contracts in Equatorial Guinea. The most visible and relevant contributions are in education and sanitary services. Year by year, more projects are being developed for building primary schools and clinics in remote areas of the country. It is made clear to contractors and subcontractors that they must hire national construction companies to carry out the construction of these projects. They are only allowed to hire foreign construction companies if no local company capable of doing the job is available.

With respect to compliance, the most effective tool the Department of National Content has is its ability to carry out annual audits. The department can carry out inspections under the ministerial order and the corresponding law. These audits are done at the expense of the company. Moreover, a mechanism has been created for national companies to present complaints to the Ministry of Mines, Industry, and Energy in cases in which international companies do not comply with the ministerial order. If international companies fail to follow all or part of the National Content Regulation, the ministry has the power to impose sanctions, fines, and suspensions. In the case of national companies, they could be suspended from participating in the contractors' listing of the Ministry of Mines, Industry, and Energy for a period of six months. The possible sanctions are not specifically established within the set of laws. In certain cases a limit is set, but the sanctions will depend on the recommendation of the Department for National Content to the Ministry of Mines, Industry, and Energy.

Chapter XXI on national content of the Regulation of Petroleum Operations in Equatorial Guinea delineates certain obligations for national companies, contractors, associates, and any other entity involved in exploration or production activities. This chapter makes reference to the materials, equipment, machinery, and goods that must be used in operations. Products should be locally sourced, as long as they meet the same quality standards as the foreign products, and as long as the prices do not exceed imported product costs by 10%. Moreover, it establishes an obligation to contract local services under the same conditions. Lastly, it establishes that Equatoguinean nationals, whether as an entity or as individuals, will be hired under no lesser conditions than those of foreign contractors and suppliers. It is important to remember that services that are covered by local content have been extended to all the necessities required by operating companies. This includes legal and tributary consultancy, supply services, language courses, document handling and translations, travel agencies, helicopter services, waste management, and transport.

Relevant legislation concerning local content policy includes the following:

- Law 8/2006, dated November 2, 2006, established the Hydrocarbons Law of Equatorial Guinea.
- Ministerial Order 1/2014, dated September 26, 2014, approved and adopted the National Content Regulation.
- Decree 116/2004, dated September 15, 2004, created the National Technological Institute of Hydrocarbons of Equatorial Guinea (also called the Technological Institute of Hydrocarbons).
- Ministerial Order 4/2013, dated June 20, 2013, approved and adopted the Regulation for Petroleum Operations.

Positive and Negative Experiences and Expectations Regarding the Regulatory Framework

With respect to previous experiences and expectations both in the private and public sector, the only rule that could be compared to the National Content Regulation is Chapter XX of the Hydrocarbons Law. While the chapter contains all the essential elements of the National Content Regulation, they are only outlined in general terms. The regulation intends to make the law feasible. It is expected that the National Content Regulation and the Ministry of Mines, Industry, and Energy fill in the gaps in the law. The ministerial department expects to apply final control over the rules and obligations that are contained both in laws and in contracts with E&P companies in the sector.

Current Issues and Challenges

Equatorial Guinea has a leading position in the petroleum sector at a global and African level. However, it encounters many problems, including the diversification of its economy, unemployment, and low levels of technology, among many others. In regard to unemployment, in 2009 about 22.3% of the population was unemployed.[6] This number is even higher today, due to the number of dismissals resulting from a decrease in the activities of several companies in the country. It is not a new development that oil prices are very low, and it will likely continue to decrease before stabilizing. Dismissals have been a regular occurrence since the beginning of the oil price slump, and given that operator and service companies are some of the major employers in the country, the social and economic pressures have had a noticeable impact. Moreover, intense training programs are also affected due to the lack of financial resources. It puts IOCs coming to the country in an extremely difficult situation, as they are forced to invest in training and education as part of their operation costs in the country.

Apart from all this, one of the main problems is the inclusion of local service companies. At present, they are nonexistent. Financial resources are not the only hindrance; there also is a lack of liquidity, knowledge, and technology. Consequently, local companies cannot compete with foreign investors. As a core market, the local services sector is quite limited, and actually needs to develop dramatically.

In the past months there has been progress in regard to the implementation of the public policy content. Previously, there was a disconnect between legislation and implementation. However, little by little, the Ministry of Mines, Industry, and Energy is utilizing tools in order to comply with the National Content Regulation. For example, the ministry is carrying out general audits on all companies in the sector.

Conclusion

Equatorial Guinea is known for its contribution to the hydrocarbons sector and its derivatives. Even when the country is going through an economic recession and the price of a barrel of oil has dropped considerably, the Ministry of Mines, Industry, and Energy still attracts the interest of investors, both new and familiar. The country has economic stability, and it is a relatively easy market to enter. By 2014, 10 blocks were offered through tenders, and another 4 through direct negotiations. This small country located in central Africa still has a lot to offer. There is a dichotomy between the drop in price and the new unexploited resources. However, new discoveries and opportunities within the sector continue to materialize in Equatorial Guinea. There are new opportunities for the inclusion of local companies and Equatoguineans who would like to be a part of the workforce that moves the economy of the country. The National Content Regulation of 2014 not only gives clear directions to companies in regards to national content, but it also gives the state control in managing how nationals are included in the hydrocarbons sector.

Notes

1. "Censo de Población 2015: República de Guinea Ecuatorial," September 24, 2015, http://www.inege.gq/publicaciones/Censo2015_Folleto_R_Preliminares.pdf.
2. GEPetrol manages the state's participation as a shareholder. However, the ministry approves and supervises operations even when GEPetrol is a party of the PSC.
3. GEPetrol has participation in all of the blocks awarded in Equatorial Guinea.
4. The Constitution is also known as the Fundamental Law.
5. Decree 127/2004, Article 2, states that "the national participation will never be of less than 3 shareholders."

6. "Africa: Equatorial Guinea," *The World Factbook*, CIA, https://www.cia.gov/library/publications/resources/the-world-factbook/rankorder/2129rank.html#ek.

10

GHANA

Thomas Kojo Stephens and Theophilus Acheampong

Introduction[1]

Petroleum agreements usually contain clauses geared at ensuring that the IOCs make a substantial direct input to the local economy. Thus, there will be provisions geared at such areas as employing and training of nationals by the IOC; promoting research and development, local industries and services; and ensuring transfer of technology. This is what is collectively known as *local content*, and in some jurisdictions as *domestic* or *national content*. It refers to the level of involvement of the local sector in the oil industry.

Local Content Policy and Baseline

Ghana discovered oil in commercial quantities in the Jubilee field in 2007 and began production in commercial quantities in December 2010. The importance of local content was not lost on the country. Ghana's Ministry of Energy issued a policy document on February 26, 2010, entitled "Local Content and Local Participation in Petroleum Activities—Policy Framework." The local content policy stated, "It is the desire of the government and people that the control as well as the benefits from the oil and gas discovery and production will remain with Ghanaians."[2] Further, in the preamble of Ghana's Model Petroleum Agreement, it states:

> The Parties recognize that Ghanaian nationals should as soon as reasonably possible be engaged in employment at all levels in the petroleum industry, including technical, administrative

> and managerial positions, and that to achieve this objective an adequate program of training must be established as an integral part of this Agreement.[3]

The policy document had an initial figure of 90% local content to be achieved over the next 10 years, which was wildly optimistic considering the country's industrial capacity and resource base. It stated:

> [T]he Government of Ghana is committed to deploying an effective local content and local participation policy as the platform for achieving the goals for the oil and gas sector with full local participation in all aspects of the oil and gas value chain of at least 90% by 2020.

In furtherance of giving impetus to local content, the Petroleum (Local Content and Local Participation) Regulations, 2013 (LI 2204) was passed on November 19, 2013 after consultations with both national and international stakeholders. The Local Content Regulations took effect February 1, 2014, and apply to local content with respect to petroleum activities.[4] Thus, the purpose of the Local Content Regulations, as stated under Regulation 1, is to:

> (a) promote the maximization of value-addition and job creation through the use of local expertise, goods and services, business, and financing in the petroleum industry value chain and their retention in the country;
>
> (b) develop local capacities in the petroleum industry value chain through education, skills transfer and expertise development, transfer of technology and know-how and active research and development programs;
>
> (c) achieve the maximum local employment level and in-country spend for the provision of goods and services in the petroleum industry as specified in the First Schedule;
>
> (d) increase the capability and international competitiveness of domestic businesses;
>
> (e) create petroleum and related supportive industries that will sustain economic development;
>
> (f) achieve and maintain a degree of control for Ghanaians over development initiatives for local stakeholders;
>
> (g) provide for a robust and transparent monitoring and reporting system to ensure delivery of local content policy objectives;

(h) provide for the submission of the local content plan and related sub-plans by contractors, sub-contractors, licensees and any other allied entity involved in the petroleum industry including

i. the provision of goods and services;

ii. the transfer to the Corporation [national oil company] or the Commission and Ghanaians of advanced technology and skills related to petroleum activities;

iii. a recruitment and training program; and

iv. supervision, coordination, implementation and monitoring of local content.

The overall approach to local content in Ghana is governed by the interaction of the Petroleum (Local Content and Local Participation) Regulations, 2013 (LI 2204), the Petroleum (Exploration and Production) Act, 2016 (Act 919) and the specifics of the petroleum agreement. The Petroleum Commission—the regulatory body of the industry—broadly defines local content as the "quantum or percentage of locally produced materials, personnel, financing, goods and services rendered in the petroleum industry value chain and which can be measured in monetary terms."[5] *Local participation*, on the other hand, is contextualized as the "level of Ghanaian Equity Ownership in the oil and gas industry. To qualify as a Ghanaian/ indigenous company, the company must have at least 51% of its equity owned by a Ghanaian with 80% management and senior positions occupied by Ghanaians."[6]

Overview of Key Prevailing Local Conditions

There are four main areas in which Ghana aims to achieve added value with regard to local content:

1. Employment and training of nationals
2. Local industries/goods and services
3. Research and development
4. Technology transfer/capacity development programs

At the time of commercial discovery, there were high public expectations that the oil industry would provide a flood of jobs within the economy. Some individuals, however, juxtaposed these expectations against the grim realities. Attafuah for instance noted:

> The political elite, as well as the leadership of other sectors of civil society, must temper public enthusiasm with realism. The oil industry in developing countries is typically dominated by expatriates with specialized and technical skills, and there is no reason to assume that the situation in Ghana will be any different. It is doubtful how many Ghanaian companies are sufficiently positioned financially and technically, to compete effectively against their multinational counterparts in the provision of the highly specialized oilfield, seismic, geophysical, drilling, packing (cementing wells) and logging services required in the oil sector. It is imperative that we conduct a meticulous reality check on our current state of competence and assess our levels of preparedness vis-à-vis the external competition.[7]

Some industry officials also tapered down expectations as they warned that the Ghanaian industry was still at a relatively early stage and could only absorb a limited number of Ghanaians. Gert-Jan Smulders, at that time Tullow's country manager for Ghana, noted, "We need to manage public expectations. Tullow is committed to grow and maximize the number of national staff, but the size of the operation isn't such that you'll get hundreds of people employed."[8] The expected flood of jobs has not materialized, primarily because the oil and gas industry is an enclave activity, with spillovers primarily the result of direct state interventions. This is coupled with the fact that Ghana's resources are modest and located in deepwater areas; the technology involved requires relatively few, but highly skilled and experienced, people.

The conventional wisdom is that if local goods and services are patronized, the economy will be fueled and avenues for opportunities and growth will trickle down to a great many in the populace—the multiplier effect. The reality of the matter, however, is that a great swath of local industries and personnel are not yet versed in petroleum operations so as to be active participants. The primary benefit that the country can hope to harness in the foreseeable future will mainly be restricted to the revenue obtained from oil production itself, as well as from the gas by-products. The World Bank notes:

> Unlike gas, oil discovery will primarily impact the economy through the budget. As located offshore and demanding equipment and expertise not yet available in Ghana, the extraction of oil will not generate significant backward linkages (in terms of demand for domestic inputs and upstream activities) in the short to medium term. And unlike gas, forward linkages and related downstream activities are also expected to be minimal, but through their impact on the budget.[9]

In Ghana's circumstances, the key first and foremost is to seek to harness the numerous spillover benefits that come from the oil industry. Kastning is most apt when he notes:

> As there is only a very limited number of jobs in the highly technical upstream production of oil…, it is of enormous importance to create a strong mid- and downstream oil industry. Mid- and downstream means every sector of the oil and gas industry in addition to the actual exploration and production. By supporting these sectors, not only are jobs created, but moreover value is added to the resources and dependence on the traded barrel price is being decreased.[10]

Ghana has sought to harness the by-product from the industry—gas—to set up a viable gas industry. It embarked on the Western Corridor Gas Infrastructure Development Project to utilize the gas from the oil producing fields. Thus the project inter alia refers to the offshore pipeline, the onshore pipeline, the gas processing plant, and the natural gas liquids (NGLs) export system. The Ghana National Gas Company[11] was formed and tasked with the responsibility of harnessing the gas from the fields for utility. Ghana has sought to use gas from the oil-producing fields primarily to provide lean gas for power generation as well as liquefied petroleum gas (LPG) for domestic purposes. There is a very huge demand for LPG, and it is used primarily in Ghana as a form of energy for cooking. It has also become popular with taxi drivers, who use it to power their vehicles, as it is much cheaper than petrol. Natural gas offers opportunities that justify the investment, as it has great potential to create a whole range of benefits within the economy.

The petroleum sector in and of itself has limited capacity to be the engine of growth for the economy in terms of providing job opportunities. The opportunities will flow from linkages of the local industry to the oil industry. The opportunity to form effective linkages is hampered by the economies of scale required in the industry. There is not sufficient work to justify setting up Ghanaian companies to service an industry with sporadic demand. Further, the oil industry is highly specialized, requiring people with specific expertise, which Ghana lacks in its workforce. Panford notes that with respect to employment and training, the country faces enormous challenges.[12] He states, "[T]here has been considerable lag between Ghana's needs and what educational institutions produced and almost without exception, local institutions are not equipping Ghanaians with the requisite skills for the petroleum and high technology industries."[13]

The country must evaluate the strength of its local industry in order to manage expectations. The local industry is in poor shape to interlink with the oil industry. As Panford notes:

> An even more daunting challenge is how in the last few decades, due to outright or benign neglect, Ghana's meager industrial base has virtually collapsed. In addition, Ghana's business climate and capacity to support a new oil sector is currently inadequate. An undisputed fact about the economy is that most locally owned businesses are under-capitalized and not competitive enough to meet the standards of MNOCs.[14]

There is the further constraint that the practice of the IOCs has traditionally been to rely on the big oil services companies, especially when operating in the developing world. Thomas Manu, then director of exploration and production at GNPC and currently its deputy chief executive officer, noted in 2008 at the time of oil discovery, "Plans to create jobs by encouraging local companies to service the oil industry are likely to prove tough to implement. I worry about our capacity to benefit from this resource. How much Ghanaian goods and services will be used in this whole thing?"[15]

The harsh reality is that Ghana's local industry cannot unilaterally lift itself by its bootstraps. Belderbos et al. note that local content requirements are not very successful in developing the local sector.[16] Drawing from this conclusion, Heum et al. note, "This means that local content policies should appreciate and encourage foreign firms to collaborate with local companies. In turn, this should be expected to give impulses and generate dynamics that will have positive impacts on the development of indigenous firms."[17] These policies must first and foremost identify the local firms that have the potential to engage in effective partnership with the IOC. Panford is most apt when he notes, "The absence of competitive businesses that can supply oil contractors may be resolved by creating a carefully planned, public-private process targeted specifically at identifying and assisting locally-owned businesses that have potentials to succeed as local suppliers to MNOCs."[18] Heum et al. are of a similar view but go on to opine very rightly, "Attention should be on how to facilitate their participation in the domestic petroleum activities without compromising quality, health, safety and environmental standards."[19]

Ghana has tried to further its local content agenda through the use of joint venture (JV) contracts. Thus it is mandated that a nonindigenous Ghanaian company cannot conduct business in the industry without incorporating a joint venture with an indigenous Ghanaian company.[20] A separate legal entity must be incorporated, and business must be transacted through that entity. The transfer of shares to a Ghanaian is not regarded as a joint venture regardless of the shareholding structure. The commission requires that the Ghanaian partner be part of the management and operation of the business. The intent of the JVs is to ensure that there is minimum Ghanaian interest through equity participation and to help prepare indigenous Ghanaian companies to be competitive in the industry through technology transfer and capacity building.

At the point of registration to undertake a petroleum activity, JV companies are required to submit to the Commission a detailed plan for transfer of technology and know-how to the Ghanaian partner. They are further required to furnish the commission at that juncture with all forms of agreements between the parties, including the JV agreement, board resolutions, and the technology and skills transfer plan. There has naturally been some Ghanaian participation, and the intent is that the local entities will imbibe the culture, practices, and business acumen of their foreign counterparts. Currently, JVCs include McDermott Marine Construction Ghana Limited (MMCGL); a JVC between Hydra Offshore (Ghanaian) and McDermott Marine Construction; and another between Ghana National Petroleum Corporation, 30%, and Technip (TP Offshore International), 70%.

However, apart from collaborations with international firms, it can be argued that there is a need for an altogether different approach whereby reputable foreign goods and services companies supplying the oil sector could be induced to set up operations/subsidiaries locally and manufacture or assemble in Ghana, at the very least. Heum et al., when commenting on the Norwegian experience, note:

> Further, politicians were in principle open to the contribution from foreign firms, while being quite determined to use the opportunity to develop domestic industrial competence. This was partly done by requiring the oil companies to set up fully operating subsidiaries in Norway, where the Norwegian authorities encouraged the recruitment of Norwegians.[21]

They further note, "In a globalized industry a local subsidiary of a multinational can be just as effective in using local inputs and developing capacity and competence in the host country as a company in which domestic citizens hold a majority of the shares."[22] It can be argued that this would be a better and faster way to upgrade local expertise and facilitate real participation as Ghanaian nationals would work in these companies and gain the needed expertise. It can be further argued that it would be beneficial to the country if build, operate, and transfer (BOT) agreements, for instance, were entered into primarily with companies with responsibility for maintenance—as opposed to construction—in the industry. In essence—and in the particular context of Ghana—it has been argued that there should be a shift in emphasis in local content from the litmus test of involvement of locally owned companies to one in which participation by the local industry is viewed holistically. The counterargument would be that Ghana's circumstances are vastly different from those of Norway in terms of the level of expertise of its local industries as well as the sheer size of the industries of the two countries and that, in some aspects, it may be better for Ghana to forgo trying to achieve local content that might well turn out to be an exercise in futility or a largely cosmetic venture, in exchange for securing better fiscal terms in the agreement.

The key, however, to pursuing a local content agenda is effective partnership with international firms in order to realize the local content agenda. It would require effective collaboration with foreign firms, with the subsequent synergies that will enhance their capacity and ability to be players in the industry. The onus should be put on the IOCs to help further this goal. Thus, in the specific case of Ghana, it is the case that each development plan has detailed schedules and plans for transitioning jobs to Ghana. The development plan specifies which services will be carried out locally. However, it should be the case that when the development plan is being discussed, the onus should be placed on the IOC to ensure that local content is significantly achieved. The plan of development stage should be viewed as the catalyst for local content in regards to each petroleum agreement, and the ball should be placed in the court of the IOC to come out with a workable plan. With this approach as well, the chances of success are much higher. Heum et al. encapsulate this when they note, "The point is not to instruct oil companies as to what has to be done, but to challenge them to come up with schemes, which they expect to work, which they can commit themselves to, and for which they can be held responsible."[23] Commenting on the Norwegian experience, they note, "As part of the concessionary procedure, the oil company had to come up with plans as to how the local content could be enhanced on a competitive basis."[24]

Key Players

Host governments and their agencies

The main governmental bodies engaged in the petroleum industry are the Ministry of Petroleum, the Petroleum Commission, the Local Content Committee formed under the Petroleum Commission Act, 2011 (Act 821),[25] and the Ghana National Petroleum Corporation (GNPC).

The Ministry of Energy. The Ministry of Energy formulates, implements, monitors, and evaluates energy sector policies.[26] Acting for and on behalf of the government, the country's objectives and goals are articulated through policy developed and implemented by the ministry, headed by the minister.

The Petroleum Commission. The Petroleum Commission has regulatory, technocratic, and advisory functions. The activities of the commission are governed by the Petroleum Commission Act, 2011 (Act 821). The ministry has direct oversight of the Petroleum Commission. Among the functions enumerated in Section 3 of its governing Act are the following:

- The Commission recommends to the minister national policies related to petroleum activities.[27]
- The Commission monitors and ensures compliance with policies, laws, regulations, and agreements.[28]
- The Commission monitors petroleum activities and carries out inspections and audits.[29]
- The Commission receives applications and issues permits for specific petroleum activities,[30] more particularly, all permits and approvals for the conduct of IOC service companies.
- The Commission promotes local content and local participation in petroleum activities.[31]
- The Commission also receives all work permit applications for companies that wish to engage expatriates to work in the petroleum sector. The Commission evaluates and furnishes to the director of the Ghana Immigration Services, its recommendations for approval.[32] The Commission has set up a Work Permits Committee to assist in

this process and to serve as an initial screening point for applications for work permits for expatriate staff.[33]

The Local Content Committee

The Petroleum Commission Act provides for the establishment of the Local Content Committee to deal with the local content and local participation program.[34] It is more particularly tasked with overseeing the implementation of the Local Content Regulations.[35] The Committee is tasked with overseeing, coordinating, and managing the development of local content;[36] preparing guidelines to include targets and formats for local content plans as well as reporting;[37] and for making appropriate recommendations to the Commission for smooth implementation of the regulations.[38]

Further, the Committee is tasked with setting minimum standard requirements for local content in local content plans where applicable,[39] undertaking public education,[40] local content monitoring and audit,[41] and performing any other functions conferred on it by the Commission in accordance with the provisions of applicable laws.[42] The Committee, in implementing the provisions of the regulations, is tasked with ensuring that there is "measurable and continuous growth in local content in all petroleum activities."[43] The Committee submits quarterly reports of its activities to the Commission.

Ghana National Petroleum Corporation (GNPC) The national oil company (NOC) of Ghana, GNPC, became fully operational in 1985. The object of GNPC, as stated in its governing act, is to "undertake the exploration, development, production and disposal of petroleum."[44] Under the Petroleum Act, GNPC is the entity vested with the exclusive right to engage in petroleum operations, and no oil company can enter into a petroleum agreement for exploitation of petroleum in Ghana unless in partnership with GNPC.[45] The Interpretation section of the Petroleum Act defines *petroleum agreement* as "an agreement entered into between the Republic, the Corporation [GNPC] and a contractor." The GNPC has a subsidiary (exploratory arm) known as GNPC Exploration and Production Company Limited (Explorco).

GNPC officials are often seconded to some of the IOCs in order to understudy. Article 21.3 of Ghana's Model Petroleum Agreement[46] provides as follows:

> Contractor shall, if requested by GNPC, provide opportunities for a mutually agreed number of personnel nominated by GNPC to be seconded to Contractor for on-the-job training or attachment in all phases of its Petroleum Operations under

> a mutually agreed secondment contract. GNPC shall present recommendations and consult with Contractor as to the number of proposed secondees, the nature of their secondment and the estimate of expenses thereof and shall obtain the approval of the JMC [Joint Management Committee] thereof.

Oil and gas companies operating in the country

Oil and gas companies operating in Ghana include the following:

Kosmos Energy Ghana HC—West Cape Three Points

Tullow Ghana Limited—Deepwater Tano

Eni Ghana—Offshore Cape Three Points

Hess Ghana—Deepwater Tano Cape Three Points

Oranto Petroleum Int. Ltd.—Offshore Saltpond

Anadarko Petroleum Corporation

AMNI International Petroleum—Central Tano

Sahara Energy Fields—Shallow Water Cape Three Points

Brittania-U—South West Saltpond

A-Z Petroleum—Deepwater Cape Three Points West

Heritage Oil—Offshore South West Tano, East Keta Ultra Deepwater

UB Resources—South West Cape Three Points

Medea Development—East Cape Three Points

AGM Petroleum—South Deepwater Tano

CAMAC Energy—Expanded Shallow Water Tano

Springfield Energy—West Cape

Three Points Block 2

Swiss African Petroleum—Onshore/Offshore Keta Block

Eco Atlantic Oil and Gas—Deepwater Cape Three Points West

Service providers operating in the country

The service providers operating in the country include the following:

- Schlumberger
- Halliburton
- Baker Hughes
- GE Oil and Gas
- FMC Technologies
- Maersk Drilling
- Opus Offshore
- Tenaris
- Intertek
- Seaweld Engineering
- Technip
- MODEC
- Seadrill

Industry Knowledge

The World Bank has provided funds to local institutions for capacity building. The Oil and Gas Capacity Building Project (OGCBP) officially began in March 2011. It was targeted to train more than 200 public officials over the next four years in key subject areas, with a view to improving policy formulation and regulatory capacity in the oil and gas sector. The beneficiary institutions include the Ministry of Energy, Ministry of Finance, Petroleum Commission, GNPC, EPA, Economic and Organized Crime Office (EOCO), and the Ministry of Justice, through the office of the Attorney General. It also includes selected technical and vocational institutions, such as the Kwame Nkrumah University of Science and Technology (KNUST), which has been supported to improve petroleum and petrochemical engineering, teaching, and research capabilities. Support has also been provided to vocational training institutions through curriculum development of programs focused on technical skills and development for the oil and gas industry. This includes support to the Takoradi Technical Institute, Kikam Technical Institute, and the Regional Maritime University.

The government of Ghana, led by the Ministry of Energy and the Ministry of Trade and Industry (MoTI), works in collaboration with the IOCs for purposes of furthering development of the local economy. For example, the government of Ghana partnered with Jubilee field companies Tullow Oil Ghana, Kosmos Ghana, Anadarko, Petro SA, and the GNPC to set up the Enterprise Development Center (EDC) in May 2013. The EDC is a five-year project to provide support to Ghanaian micro, small, and medium enterprises (MSMEs) and is under the general supervision of the Ministry of Petroleum and the Ministry of Trade and Industry. The EDC provides a range of services such as business training, capacity building programs, advisory services, and access to markets and information.

In collaboration with Sribima Maritime Training Center (SMTC Global), the Oil and Gas Safety Training Center has been established at the Regional Maritime University (RMU). Its purpose is to conduct mandatory safety courses for the offshore oil and gas industry. The course involves helicopter underwater escape training (HUET), along with emergency breathing system (EBS), first aid, sea survival, fire fighting, and self-rescue training. The training is geared toward all upstream workers, policy makers, and lawmakers visiting the oil rigs, among others.[47]

There is also the USAID-funded Ghana Supply Chain Development Program, which uses PYXERA Global's approach to local content and supplier development to seek to increase sustainable participation of Ghanaian small and medium enterprises (SMEs) in procurements in the industry.[48] PYXERA notes that it leverages "the unique strengths of corporations, governments, social sector organizations, educational institutions, and individuals to enhance the abilities of people and communities to solve complex problems and attain mutually beneficial goals."[49] PYXERA Global seeks to enhance the capabilities of local companies to successfully become suppliers to multinational corporations.[50] The Ghana Supply Chain Development Program seeks to ensure the engagement of qualified SMEs in procurement opportunities, create strategic partnerships with local business associations, and analyze gaps between the capabilities of local SMEs and the needs of the oil and gas sector. The program provides advanced-level training and certification, along with access to guidance for business service providers and the SMEs.[51]

The University of Ghana, Legon, also offers a course of study leading to the award of a master of laws degree (LLM) in oil and gas law.

Legislative Hydrocarbon Regime Concerning Local Content

The legislative hydrocarbon regime dealing with local content consists mainly of the Petroleum (Exploration and Production) Act, 2016 (Act 919) and the Petroleum (Local Content and Local Participation) Regulations 2013 (LI 2204). The Petroleum Act repealed the Petroleum (Exploration and Production) Act, 1984 (PNDCL 84), which was considered outmoded. The Petroleum Act has provisions covering local content mainly in sections 60–69 of the Act.

The repealed Petroleum Act had provisions regarding local content captured under Section 23,[52] specifically Sections 23(10), (12), (13), and (14). The Petroleum (Local Content and Local Participation) Regulations, 2013 (LI 2204) was passed on November 19, 2013 specifically to govern matters related to local content within the petroleum industry and to provide details, guidance, and rules as to local content. According to Regulation 2, "These Regulations apply to local content with respect to petroleum activities."

Regulatory bodies

Petroleum Commission. The Petroleum Commission was established in July 2011 by the *Petroleum Commission Act*, 2011 (Act 821) to regulate and manage the exploitation of petroleum resources and to coordinate the policies in relation to them. The commission is the regulator of Ghana's upstream petroleum sector and is mandated to regulate, manage, and coordinate all activities in this sector for the overall benefit and welfare of Ghanaians. It states its mission as follows: "To promote, regulate and manage the efficient conduct of upstream petroleum operations and all allied activities and the utilization of petroleum resources on a sustainable basis for the overall benefit of the citizens of Ghana."[53]

Under Section 3(f) of the *Petroleum Commission Act*, the Petroleum Commission is tasked with the responsibility to "promote local content and local participation in petroleum activities… to strengthen national development."[54] Section 8(2)[55] of the *Petroleum Commission Act* mandates that there be established "a Local Content Committee to deal with the local content and local participation programme." The committee is chaired by a member of the board of the Petroleum Commission.[56]

Relevant policies, legislative framework, and key provisions

In 1970, oil was discovered in commercial quantities, albeit small, in the Saltpond field, and production began in 1978. The since-repealed *Petroleum (Exploration and Production) Act* (PNDCL 84) was passed in 1984, and it contained local content provisions under Part III, "Rights and Obligations of Contractors and Sub-Contractors." Section 23 stated:

> (10) A contractor or sub-contractor shall, in accordance with the Regulations and with the terms of a petroleum agreement or petroleum sub-contract, as the case may be, ensure that opportunities are given as far as is possible for the employment of Ghanaians having the requisite expertise or qualifications in the various levels of the operations....
>
> (12) A contractor or sub-contractor shall, as far as practicable, in accordance with the Regulations and the petroleum agreement or sub-contract use goods and services produced or provided in Ghana for his operations in preference to foreign goods and services.
>
> (13) A contractor or sub-contractor shall, in consultation with the Corporation, prepare and implement, in accordance with the Regulations and in accordance with the terms of any such petroleum agreement or petroleum sub-contract, plans and programmes for training Ghanaians in all job classifications and in all aspects of petroleum operations.
>
> (14) A contractor or sub-contractor shall while carrying out petroleum operations prepare and implement plans for the transfer to the Corporation of advanced technological know-how and skills relating to petroleum operations but this provision shall not be interpreted to disable the contractor or sub-contractor from protecting their competitive position in the petroleum industry or requiring the Corporation also to take steps to protect such competitive position.

In 2010, production of oil began from the Jubilee Field. The Petroleum (Local Content and Local Participation) Regulations 2013 (LI 2204) set out a detailed framework for local participation in the industry. The provisions under the Local Content Regulations largely expounded upon the Petroleum (Exploration and Production) Act 1984, which was repealed in 2016 by the Petroleum (Exploration and Production) Act, 2016 (Act 919). The act contains provisions covering local content as well, but naturally they are not as detailed as what is contained in the regulations. Thus, the broad framework is outlined in the Petroleum Act and the details in the regulations, though the latter came into existence first. The four thematic local content areas are captured under the Model Petroleum Agreement under Article 20 and 21, titled Purchasing and Procurement, and Employment and Training respectively, and are incorporated in each specific agreement.

The Petroleum (Exploration and Production) Act, 2016 (Act 919). The Petroleum Act outlines the provisions dealing with local content as follows. Section 60 deals with Employment and Training of Ghanaian Citizens; section 61, Use of Ghanaian Goods and Services; and Section 62, Technology Transfer; which are the broad parameters for the local content agenda. Section 63 covers the Local Content Plan and mandates that a licensee, contractor, or subcontractor prepare and implement a local content plan, which covers the broad parameters outlined. The plan must be submitted to the Petroleum Commission for approval. The Petroleum Act, Section 64, titled Establishment of the Local Content Fund, establishes a local content fund. The object of the fund is to provide financial resources for indigenous Ghanaian companies and citizens engaged in petroleum activities.[57] The money from the fund is to be applied to education, training, research, and development in petroleum activities by the aforementioned, as well as Ghanaian institutions of learning.[58] The monies from the fund also provide loans on a competitive basis to both small- and medium-scale enterprises to support their participation in petroleum activities.[59] The monies for the fund are to accrue from contributions from a contractor as agreed to in the petroleum agreement,[60] contributions from a subcontractor of 1% of the total sum payable by the contractor for every contract,[61] monies approved by Parliament,[62] as well as grants.[63]

The Petroleum Act mandates that monies to the fund be paid into a bank account opened by the board of the Petroleum Commission, with the approval of the controller and accountant general.[64] Any withdrawals from the fund can only be done under the signature of either the chairperson of the board and the chief executive of the Petroleum Commission,[65] or the chairperson of the board and one other member of the Local Content Committee.[66] The fund is administered by the minister for energy and the Local Content Committee.[67] They are mandated for the purpose of administering the fund to formulate policies to generate money for the fund,[68] determine the allocations to be made towards the fund's objectives,[69] and determine the fund's annual targets.[70] The minister approves the annual budget of the fund,[71] and the Local Content Committee may invest a part of the fund that it considers appropriate in the manner approved by the minister for energy in consultation with the minister for finance.[72] It is also instructive to note that the fund is exempt from the payment of tax.[73] The Local Content Committee is mandated to keeps books of account and proper records.[74] The Commission must within one month of receipt of an audit report from the auditor general submit an annual report covering operations of the fund for the year to which the report relates.[75]

Petroleum (Local Content and Local Participation) Regulations, 2013 (LI 2204). Details to the local content provisions, are manifested in the *Local Content Regulations.* It must be reiterated that contrary to what one would assume, the regulations came into being in 2013, before the Petroleum Act in 2016. As seen in the regulations, Ghana has sought to have direct participation by locals in the industry. In this vein, the regulation titled Interest of a Citizen in Petroleum Operations mandates that an indigenous Ghanaian company be given first preference in the grant of a petroleum agreement or license with respect to petroleum activities, subject to its fulfilling conditions specified in the regulations.[76] Regulation 4(2) mandates that there be at least a 5% equity participation of an indigenous company other than the corporation to be qualified to enter into a petroleum agreement or license. The minister is given the discretion to vary this requirement in circumstances where the 5% requirement cannot be satisfied. This regulation is perceived by many as simply an avenue for the well-connected to use the back door to rope in cronies to have interests—usually carried—in the petroleum agreements. It is argued that the evidence of this is that the GNPC already takes a mandatory stake, therefore the logical step would have been to increase the GNPC's stake or find another way to allow the public to benefit. It has been suggested that instead of having this 5% requirement provision, it might be a better option to mandate that the IOCs offer a certain percentage of shares on the local stock exchange,[77] so that whoever is interested in being part of the process would have the opportunity to do so, as opposed to a privileged few.

There is also the requirement under Regulation 4(6) that a nonindigenous Ghanaian company that intends to provide goods and services to the industry must incorporate a joint venture company with an indigenous Ghanaian company and afford that indigenous Ghanaian company an equity participation of at least 10%. Critics again argue that the intent is to ensure that a small group of well-connected Ghanaians benefit.

As mentioned previously, the focus of the Local Content Regulations is to attain local content in four main areas, as follows:

1. Employment and training of nationals
2. Local industries/goods and services
3. Research and development
4. Technology transfer/capacity development programs

Employment and training of nationals. In respect to employing and training nationals, under Regulation 17[78] of the Local Content Regulations, a contractor, subcontractor, licensee, or other allied entity is required to submit an employment and training subplan. The subplan must include a forecast of its hiring and training needs,[79] specification of the skills needed,[80] anticipated skill shortages in the Ghanaian workforce,[81] specific training requirements,[82] and the anticipated expenditure to be incurred in implementing the employment and training subplan as forecasted.[83] They are further required to provide a time frame within which they will offer employment opportunities for the Ghanaian workforce for each phase of the petroleum activity to enable Ghanaians prepare for such opportunities.[84] They must also provide information on efforts made and procedures adopted for the accelerated training of Ghanaians.[85] Where Ghanaians are not employed because of lack of expertise, it must be ensured to the satisfaction of the Commission that every reasonable effort is made to provide training to Ghanaians in that field, locally or elsewhere.[86] As part of the employment and training subplan, it is also required that a succession plan for any employment position that is occupied by a non-Ghanaian be submitted to the commission. This stipulation is to ensure that the minimum local content levels are achieved.[87] Contractors, subcontractors, licensees, and other allied entities are also mandated to employ only Ghanaians in junior or middle-level positions.[88] The Local Content Regulations define junior or middle-level positions to "include the position of foreman, supervisor or any corresponding position designated as such."[89]

Local goods and services. With respect to local goods and services, a contractor, subcontractor, licensee, or other allied entity is required to establish and implement a bidding process for the acquisition of goods and services, giving preference to indigenous companies.[90] Where the total value of the bid of a qualified Ghanaian company does not exceed the lowest bid by more than 10%, the contract must be awarded to that indigenous company.[91] When bids are evaluated and judged to be equal, the bid that contains the highest level of local content must be selected.[92] Where a nonindigenous company is required to provide goods and services to a contractor, subcontractor, licensee, or other allied entity, it must incorporate a company in Ghana and operate it from there.[93] Further, where practicable, it is required to provide the goods in association with an indigenous Ghanaian company.[94] Contractors, subcontractors, licensees, and other allied entities are also mandated to provide a quarterly report on contracts or purchase orders estimated to exceed the cedi equivalent of US$100,000 and intended to be tendered for or executed in the next quarter.[95]

The *Local Content Regulations* mandated the establishment of a Common Qualification System[96] to serve as the sole system for the registration and prequalification of local content in the industry.[97] The system's purpose is for the verification of the contractor's capacities and capabilities,[98] the evaluation of the application of local content that is submitted,[99] tracking and monitoring of performance and provision of feedback,[100] and the ranking and categorization of service companies based on capabilities and local content.[101]

Regulation 27(2)[102] of the *Local Content Regulations* stipulates that the insurable risks relating to petroleum activities shall be insured through an indigenous brokerage firm, or where applicable, a reinsurance broker. Where a person seeks to obtain an insurance service relating to petroleum activity out of the jurisdiction, the person or entity must obtain the written approval of the National Insurance Commission.[103] In granting such approval, the National Insurance Commission must ensure that Ghanaian local capacity has been fully exhausted.[104]

Further, it is required that contractors, subcontractors, licensees, or allied entities that require legal services in the country shall retain only the services of a Ghanaian legal practitioner or firm of Ghanaian legal practitioners whose principal office is located within the jurisdiction.[105] Critics point out that this provision is largely a white elephant, as the Ministry of Petroleum itself and the other government entities engaged in the industry rely at least to some extent on foreign legal representation. Contractors, subcontractors, licensees, or allied entities are further required to provide a legal services subplan to the Commission. The plan must include a comprehensive report on legal services utilized by the entity in the preceding six months by expenditure,[106] where applicable; a forecast of legal services required during the ensuing six months[107] and the projected expenditure for the services;[108] and the annual services budget for the ensuing year, quoted in the local currency (Ghana cedis) and US dollars.[109]

Concerning financial services, a contractor, subcontractor, or other allied entity that requires financial services relating to petroleum activity is mandated to retain only the services of a Ghanaian financial institution or organization.[110] There is, however, the caveat that they may, with the approval of the Commission, engage the services of a foreign financial institution or organization.[111] They are also required to submit a financial services subplan to the Commission, which must specify the financial services utilized in the preceding six months by expenditure;[112] a forecast of financial services required in the ensuing six months and the projected expenditure;[113] and a list of financial services utilized in the preceding six months, along with the nature and expenditure incurred.[114] Contractors, subcontractors, licensees, or other allied entities must also maintain a bank account with an indigenous Ghanaian bank and transact business through banks in the country.[115] An *indigenous Ghanaian bank* is defined under the legislation as "a bank that has one hundred percent Ghanaian or a majority Ghanaian shareholding."[116]

Research and development. In respect to research and development, Regulation 20[117] of the Local Content Regulations mandates the submission of a program for research, development, and budget to the commission. This program must be targeted toward the promotion of education, practical attachments, training, and research and development in the country in relation to the entity's overall work program and its activities. The program must outline a revolving three- to five-year program for petroleum-related research and development initiatives to be undertaken in the country,[118] as well as details of the expected expenditure to be incurred in implementing the research and development subplan.[119] They are further required to provide for public calls for proposals for research and development initiatives associated with their activities, as well as criteria for selecting proposals that qualify for support.[120] The research and development subplan must be updated annually[121] and must be submitted to the commission for review and approval.[122]

Transfer of technology. With respect to transfer of technology, the emphasis is more on transfer of new technological programs, advanced know-how, and skills related to petroleum operations and the like, as opposed to the transfer of physical equipment, which is captured under a separate provision altogether. Contractors, subcontractors, licensees, or other allied entities are also required to support and facilitate technology transfer regarding formation of joint ventures, partnering of license agreements between indigenous Ghanaian companies or citizens, and foreign contractors and service or supply companies.[123] The minister is required to consult with relevant government agencies to propose fiscal incentives. Such incentives are designated to assist foreign companies that aim to develop technological capacity and skills of Ghanaian citizens,[124] as well as indigenous Ghanaian companies that establish factories and production units in the country.[125] The government agencies are required to collaborate with the Petroleum Commission,[126] and it is the Commission that proposes the criteria for obtaining the fiscal incentives.[127]

Regulation 46 of the Local Content Regulations, Offences and Penalties, encapsulates the sanctions for breach of the Petroleum Regulations, which are primarily in the form of administrative penalty units.

Positive and Negative Experiences and Expectations Regarding the Regulatory Framework

The provisions in the Local Content Regulations were passed after consultation with both national and international stakeholders, and optimistic targets were discussed as to where certain levels of local content would be achieved.

The problem, however, is that unlike Norway[128] and other developed countries where the local economy was relatively developed, with varying degrees of expertise in various economic sectors, Ghana faces significant challenges. As Al-Kasim notes,

> Before the advent of petroleum activities, Norway already possessed considerable technological expertise in several areas relevant to petroleum operations. Examples include concrete technology, dam building, marine operations, and advanced methods of numerical analysis.[129]

Ghana's economy cannot boast of that level of expertise. Ghana's 2008 policy document on local content identified three main constraints on this desire to be actively engaged in the industry: finance, human resource capacity, and technology. These constraints naturally serve to dampen the potential level of participation by local industries. Thus, with regards to local content in these two facets—local industries and services—at least in the foreseeable future, one can expect to see local participation on an altogether small scale and on the lower tiers of development.

A challenge with the current regulatory framework is that although on paper it appears that aggressive strides have been made to ensure high levels of local content, there arises the issue of implementation. In Ghana, there is often a gap between the law on the books and the law in action, a situation that is exacerbated where private actors or parochial interests stand to benefit handsomely from such a state of affairs. An inherent problem with Ghana's system is that Ghana is largely a grabber friendly country, as opposed to producer friendly. A World Bank report on Ghana for instance, notes:

> However, as numerous political economists point out, there is a slightly more complex governance picture beneath the surface that suggests that Ghana is more "grabber friendly" than "producer friendly." World Bank (2007), Booth et al.

> (2005), Nugent (1999) and others all show that Ghana does indeed exhibit the political characteristics of a "factional democracy."[130]

Grabber friendly would imply that as opposed to generating revenue and income from production ventures, the focus is on taking from what already exists. This could be in the form of patronage from the government or seeking avenues to cash in on existing ventures that are lucrative.

Furthermore, although there are laws on the books in relation to local content, and much publicity is given to the topic, in practice, there is a lack of a concerted, dedicated, and genuine commitment to local content within the government. The focus has been, and remains, the revenue that is generated directly from the industry. Local content in practice is often conveniently ignored, relegated to the background, or cosmetically applied. There is a need to lay out a blueprint for a local content agenda that can be aggressively tackled, primarily at the plan of development stage.

It is also instructive to note that it took quite some time before the Petroleum Act of 1984 was repealed and replaced with the Petroleum Act of 2016 (Act 919). Thus, due to the lack of proper sequencing in terms of the enactment of legislation, there is the incongruous situation where the regulations for local content came into existence before the Petroleum Act.

Concerns have also been raised with regard to certain provisions of the Local Content Regulations and their interpretation and implementation by the regulator, to the extent that some provisions might be commercially and practically infeasible or may even provide room for private actors to exploit these provisions. For example, those dealing with the requirement whereby indigenous Ghanaian companies be given 5% equity participation in the grant of a petroleum agreement or a licence largely ignores Ghana's current state of industrial development. It raises the likelihood of local companies fronting for international owners and/or local politicians to the extent that there are significant loopholes in ensuring openness and transparency in the tendering process for the award of oil and gas blocks.

Further, though the *Local Content Regulations* provide a mixed baseline assessment, the detailed component breakdowns are not given on what constitutes the baseline metrics from which an objective assessment of local content penetration can be made. For example, the first schedule that accompanies the Petroleum Regulations states that there should currently be in existence a 20% local capacity in terms of tonnage for drilling modules or packages. There is no way to know, however, if this has been achieved.

Current Issues and Challenges

The aim of the Local Content Regulation sand the local content provisions in the Petroleum Act is to aggressively seek Ghanaian involvement in the industry. Whether or not the legislation is able to translate to activities happening on the ground is the challenge. Developing local capacity requires the concerted effort of all stakeholders. It is primarily dependent on the government's ability to create the enabling drivers to accentuate local content in terms of policy direction and local capability (including technical capabilities), which collectively is a function of the hydrocarbon potential of the country.[131] The prescriptive legislative rules that Ghana has adopted in its approach to local content will in and of themselves not resolve or deepen local content in the oil and gas industry. This is evidenced by other jurisdictions such as Nigeria, Brazil, and Malaysia. For example, the local content levels in Norway, Brazil, Malaysia, and Venezuela are reported to range between 45% and 75%, despite a long history of oil and gas production.[132] In Nigeria, local content targets of 70% have been difficult to attain due to myriad factors, including poor monitoring and supervision of the policy and insufficient regulator capacity.[133]

There is evidence for the existence of some backward linkages in some subsectors of the industry, namely in the real estate, banking, insurance, and hospitality industries, as well as fabrication and construction, well construction and completions, and ICT. Employment of Ghanaians in the service sector has grown annually but is still biased toward nontechnical and nonmanagerial positions. The majority of local business opportunities remain limited to the provision of low-tech ancillary services due to a number of constraints. The local supply chain companies have very little competitive advantage, and most lack leading industry certifications (e.g., ISO/OHSAS). They are further hindered by the relatively high cost of doing business in Ghana, as evidenced by high interest rates and volatile exchange rates, which hurt competitiveness in contract bidding. Even with the existence of supply and fabrication bases, most of the capital equipment and plants needed for the development of the offshore fields are assembled and imported into the country.[134]

Developing and deepening local content takes time and has proven to be a challenge even in established oil and gas provinces with high local capabilities and technical competencies, and supported by an active industrial base, such as Brazil and Norway. This is likely to be the case for Ghana, as the country's industrial base lacks significant capacity to conduct highly technical work in the offshore industry. Developing the skill base will take a considerable amount of time given that the deficit is much more precarious in Ghana than in Brazil or the United Kingdom, for example. There are few coordinated activities, and little interfacing, among the oil and gas companies and the local supply chain, despite attempts by bodies such as the Ghana Oil Club and the Ghana Oil and Gas Service Providers Association (GOGSPA)[135] to bridge this gap. It is estimated that there are more than 150 local Ghanaian companies operating in the upstream sector, of which there are only two major fabrication companies and two piping and casing companies. Total contracts awarded to local companies since 2009 are in the region of $600 million.[136]

The regulations do stipulate that the "contractor shall give preference to materials, services and products produced in Ghana…if such materials, services and products can be supplied at prices grades, quantities, delivery dates and on other commercial terms equivalent to or more favorable" than those from foreign sources. Nonetheless, there is currently no clear mechanism in place to implement this due to the fact that the Common Qualification System (CQS) is not yet fully operational.

Additional information, "Schedule of Local Content Levels to Be Attained from Date of Effectiveness of License of Petroleum Agreement," is incorporated in the Local Content Regulations. It has been argued (some critics would say it can be seen by inspection) that the incremental targets set in the schedule are unrealistic. They suggest that lower and more realistic targets that could genuinely be achieved would give more impetus to the Petroleum Regulations than the current targets, deemed largely unachievable given the country's circumstances and constraints.

It is also worthy of note that the IOCs that entered into the petroleum agreements before the enactment of the Local Content Regulations perceive themselves as not bound by the Petroleum Regulations due to the stabilization clause in their agreements. Although they do engage in local content, they perceive themselves as doing so under the umbrella of the Petroleum Act of 1984, not under the more detailed Local Content Regulations and the attendant penalties prescribed under Regulation 46, *Offences and Penalties.*

Notes

1. The authors would like to thank Seyram Dzikunu (Bentsi-Enchill, Letsa & Ankomah) for her valuable assistance in the preparation of this chapter.

2. "Local Content and Local Participation in Petroleum Activities–Policy Framework," Ministry of Energy, 2015, http://ghanaoilwatch.org/images/laws/local_content_policy.pdf.
3. "Preamble," Clause 6, *Model Petroleum Agreement*, Ghana.
4. Ministry of Energy and Petroleum, "Regulation 2: Application of Regulations," in *Petroleum (Local Content and Local Participation) Regulations, 2013 (LI 2204)*, http://www.reportingoilandgas.org/wp-content/uploads/PETROLEUMLOCAL-CONTENT-AND-LOCAL-PARTICIPATION-REGULATIONS2013.pdf.
5. "Local Content," Ghana Petroleum Commission, http://www.petrocom.gov.gh/local-content.html.
6. Ibid.
7. Kenneth Agyeman Attafuah, "Managing the Political and Social Expectations from Ghana's Oil and Gas Resources," *Ghana Policy Journal*, Vol. 4 (2010), 110, 114.
8. M. Green, "Crude Realities: Can Ghana Escape Africa's Oil Curse?" *Financial Times* (August 28, 2008), 7.
9. World Bank, *Economy-Wide Impact of Oil Discovery in Ghana*. Report No. 47321-GH (World Bank, November 30, 2009), 3, http://siteresources.worldbank.org/INTGHANA/Resources/Economy-Wide_Impact_of_Oil_Discovery_in_Ghana.pdf.
10. Thomas Kastning, "Basic Overview of Ghana's Emerging Oil Industry," Friedrich Ebert Stiftung Foundation, 7, http://library.fes.de/pdf-files/bueros/ghana/10490.pdf.
11. It is a private limited liability company registered under the Companies Code, 1963 (Act 179). It was incorporated on July 27, 2011 and issued with a certificate to commence business on July 28, 2011. It is wholly owned by the government of Ghana. It commenced commercial operations in April 2015.
12. Kwamina Panford, "The Crucial Roles of Ghana's Model Petroleum Agreement: The Public Policy Implications and Requirements," *Ghana Policy Journal*, Vol. 4 (December 2010), 81, 88.
13. Ibid., 88.
14. Ibid.
15. Green, "Crude Realities," 7.
16. René Belderbos, Giovanni Capannelli, and Kyoji Fukao, "Backward Vertical Linkages of Foreign Manufacturing Affiliates: Evidence from Japanese Multinationals," *World Development*, Vol. 29, no. 1, 189–208.
17. Per Heum, Robert Kasande, Ole Frederik Ekern, and Alex Nyombi, "Policy and Regulatory Framework to Enhance Local Content: Yardsticks and Best Practice" (Working Paper 02/11, Institute for Research in Economics and Business Administration, Oslo, January 2011), 7, http://brage.bibsys.no/xmlui/bitstream/handle/11250/166346/A02_11_%281%29.pdf?sequence=1.
18. Panford, "The Crucial Roles of Ghana's Model Petroleum Agreement."
19. Heum et al., "Policy and Regulatory Framework to Enhance Local Content: Yardsticks and Best Practice," 13.
20. Regulation 4(6), Petroleum (Local Content and Local Participation) Regulations (LI 2204) 2013).
21. Heum et al., "Policy and Regulatory Framework to Enhance Local Content," 9.

22. Ibid., 7.
23. Ibid., 16.
24. Ibid., 9.
25. Section 8(2), Petroleum Commission Act (Act 821) (2011), https://s3.amazonaws.com/ndpc-static/pubication/Act+821+-+Petroleum+Commission+Act,+2011.pdf.
26. "Brief Background," Ghana Ministry of Energy, 2013, http://www.energymin.gov.gh/?page_id=54.
27. Section 3(b), Petroleum Commission Act.
28. Ibid., Section 3(c).
29. Ibid., Section 3(e).
30. Ibid., Section 3(h).
31. Ibid., Section 3(f).
32. See Section 16 of the *Immigration Regulations*, 2001 (LI 1691).
33. Petroleum Commission Ghana, "Local Content," http://www.petrocom.gov.gh/local-content.html, 7.
34. Section 8(2), Petroleum Commission Act, "Establishment of Committees."
35. "Functions of Local Content Committee," Regulation 5, Petroleum (Local Content and Local Participation) Regulations (LI 2204) (2013).
36. Ibid., Regulation 5(3)(a).
37. Ibid., Regulation 5(3)(b).
38. Ibid., Regulation 5(3)(c).
39. Ibid., Regulation 5(3)(d).
40. Ibid., Regulation 5(3)(e).
41. Ibid., Regulation 5(3)(f).
42. Ibid., Regulation 5(3)(g).
43. Ibid., Regulation 5(1), "Functions of Local Content Committee."
44. "Objects and Functions of the Corporation," Section 2(1), Ghana National Petroleum Corporation Act (PNDCL 64) (1983), *Ghana Legal*, http://laws.ghanalegal.com/acts/id/516/ghana-national-petroleum-corporation-act.
45. "Exploration, Development or Production of Petroleum," Section 2(1), Petroleum Exploration and Production Law (1984), *Ghana Legal*, http://laws.ghanalegal.com/acts/id/543/section/2/Exploration,_Development_Or_Production_Of_Petroleum.
46. Article 21, "Employment and Training," *Model Petroleum Agreement*.
47. "Oil & Gas: What Is It About?" Regional Maritime University, 2015, http://www.rmu.edu.gh/pages/sections.php?siteid=maritime&mid=4&sid=34.
48. "Ghana Supply Chain Development Program," Pyxera Global, http://pyxeraglobal.org/partners/private/supply-chain-development-program/.
49. "Who We Are," Pyxera Global, http://pyxeraglobal.org/who-we-are/.
50. "Local Content Development," Pyxera Global, http://pyxeraglobal.org/services/local-content-development/.

51. Ibid.
52. "Rights and Obligations of Contractors and Sub-Contractors," Section 23, Petroleum Exploration and Production Law (1984), http://www.petrocom.gov.gh/assets/petroleum_exploration_and_production-law_1984_pndc_law_84.pdf.
53. "Organizational Brief," Ghana Petroleum Commission, http://www.petrocom.gov.gh/organizational-brief.html.
54. Section 3(f), Petroleum Commission Act (Act 821) (2011), and "Local Content Requirement," Regulation 3, Petroleum (Local Content and Local Participation) Regulations (LI 2204) (2013).
55. Ibid., Section 8(2), "Establishment of Committees."
56. Ibid., Section 8(3).
57. "Object of the Fund," Section 65, and "Establishment of the Local Content Fund," Petroleum (Exploration and Production) Act (Act 919) (2016).
58. Ibid., Section 65(2)(a).
59. Ibid., Section 65(2)(b).
60. Ibid., Section 66(1)(a).
61. Ibid., Section 66(1)(b).
62. Ibid., Section 66(1)(c).
63. Ibid., Section 66(1)(d).
64. Ibid., Section 66(2).
65. Ibid., Section 66(3)(a).
66. Ibid., Section 66(3)(b).
67. Ibid., Section 67, "Management of the Fund."
68. Ibid., Section 67(2)(a).
69. Ibid., Section 67(2)(b).
70. Ibid., Section 67(2)(c).
71. Ibid., Section 67(3).
72. Ibid., Section 67(3).
73. Ibid., Section 67(5).
74. Ibid., Section 68(1), "Accounts and Audit."
75. Ibid., Section 69, "Annual Report and Other Reports."
76. Regulation 4(1), Petroleum (Local Content and Local Participation) Regulations (LI 2204) (2013).
77. Seth Asante, Annual Ghana Bar Association Conference, Kumasi, Ghana, September 2015.
78. "Employment and Training Sub-Plan," Regulation 17, Petroleum (Local Content and Local Participation) Regulations (LI 2204) (2013).
79. Ibid., Regulation 17.
80. Ibid., Regulation 17(i).

81. Ibid., Regulation 17(ii).
82. Ibid., Regulation 17(iii).
83. Ibid., Regulation 17(iv).
84. Ibid., Regulation 17(b).
85. Ibid., Regulation 17(c).
86. Ibid., Regulation 17(5).
87. Ibid., Regulation 18(1), "Succession Plan."
88. Ibid., Regulation 19, "Middle and Junior Level Positions."
89. Ibid., Regulation 19(2).
90. Ibid., Regulation 11, "Preference to Indigenous Ghanaian Companies."
91. Ibid., Regulation 12(3).
92. Ibid., Regulation 12(4).
93. Ibid., Regulation 12(5)(a).
94. Ibid., Regulation 12(5)(b).
95. Ibid., Regulation 17(2)(b).
96. Ibid., Regulation 37(1), "Establishment of a Common Qualification System."
97. Ibid., Regulation 38(1), "Object of the Common Qualification System."
98. Ibid., Regulation 38(2)(a).
99. Ibid., Regulation 38(2)(b).
100. Ibid., Regulation 38(2)(c).
101. Ibid., Regulation 38(2)(d).
102. Ibid., Regulation 27, "Insurance and Reinsurance."
103. Ibid., Regulation 28(1), "Approval of Offshore Insurance."
104. Ibid., Regulation 28(2).
105. Ibid., Regulation 29, "Legal Services Sub-Plan."
106. Ibid., Regulation 30(a).
107. Ibid., Regulation 30(b).
108. Ibid.
109. Ibid., Regulation 30(c).
110. Ibid., Regulation 31(1), "Financial Services."
111. Ibid., Regulation 31(2).
112. Ibid., Regulation 32(a), "Financial Services Sub-Plan."
113. Ibid., Regulation 32(b).
114. Ibid., Regulation 32(c).
115. Ibid., Regulation 33(1), "Operation of Bank Account in Ghana."
116. Ibid., Regulation 33(2).

117. Ibid., Regulation 20, "Programme for Research, Development and Budget."
118. Ibid., Regulation 21(a), "Research and Development Sub-Plan."
119. Ibid., Regulation 21(b).
120. Ibid., Regulation 21(c).
121. Ibid., Regulation 21(2)(a).
122. Ibid., Regulation 21(2)(b).
123. Ibid., Regulation 25(1), "Support for Technology Transfer to Indigenous Ghanaian Companies."
124. Ibid., Regulation 25(2)(a).
125. Ibid., Regulation 25(2)(b).
126. Ibid., Regulation 25(3).
127. Ibid., Regulation 25(4).
128. It must also be taken into consideration that Norway has far more petroleum resources than Ghana.
129. Farouk Al-Kasim, *Managing Petroleum Resources: The 'Norwegian Model' In a Broad Perspective* (Alden Press, 2006), 58.
130. World Bank, "Economy-Wide Impact of Oil Discovery in Ghana" 2009, 27.
131. Irena Agalliu, "Review of Ghana's Draft Local Content Regulations," (IHS Global Inc, 2012).
132. Mike Morris, Raphael Kaplinsky, and David Kaplan, *One Thing Leads to Another: Promoting Industrialization by Making the Most of the Commodity Boom in Sub-Saharan Africa* (Lulu Publishers, 2012), 88.
133. UNECA, *Making the Most of Africa's Commodities: Industrializing for Growth, Jobs and Economic Transformation* (Addis Ababa: UN Economic Commission on Africa, 2013), 207, http://www.uneca.org/sites/default/files/PublicationFiles/unera_report_eng_final_web.pdf.
134. Africa Center for Economic Transformation, *Local Content and Value Addition in Ghana's Mineral, Oil, and Gas Sectors: Is Ghana Getting It Right?* (ACET Study, February 2015), 15, http://acetforafrica.org/acet/wp-content/uploads/publications/2016/07/GTF-Local-Content-MOG-in-Ghana-SR-formatted.compressed.pdf.
135. The Ghana Oil and Gas Service Providers Association (GOGSPA) is an advocacy, monitoring, and lobbying association with the mission to strengthen the competitiveness of indigenous companies in the industry, to train and build the capacity of its members, and to work closely with all stakeholders in the country. Its members include both local indigenous companies and foreign companies.
136. ACET, *Local Content and Value Addition in Ghana's Mineral, Oil, and Gas Sectors*, 16.

11

IRAN

Kamran Fatahi

Introduction

After more than a century of oil production, nearly half of Iran's oil reserves have been depleted, whether by the National Iranian Oil Company (NIOC) or by international oil companies (IOCs). However, Iran's associated upstream industry has not grown at the same pace. Current local capabilities are not at internationally accepted levels, especially with respect to R&D efforts, except in a few areas.

Normally new technologies, both hard and soft, were imported to the country by IOCs or service companies. Know-how of working with or utilizing such technologies was transferred to local industries, and little attention was given to the science and innovation behind these technologies. Of course, nationalization of the oil industry in Iran in 1951 ushered in a new era for the country. Some of the E&P activities were performed by the NIOC, and its role continued to expand after the 1973 IOOC[1] consortium contract termination, followed shortly thereafter by Iran's Islamic Revolution in 1979. Recently, attention has been given to the science, technology, and innovation (STI) aspects of the petroleum industry, in line with the Resistance Economy policy of the country. Different programs and initiatives were defined by the Ministry of Petroleum. Several years of international sanctions on the oil industry provided the opportunity for local companies to show their potential capacities and establish a baseline for local service and EPC companies. Iran's agreement with 5+1 countries on the Iran nuclear program at the end of 2015 (Joint Comprehensive Plan of Action, JCPOA), opened a window of opportunity for IOCs and international oilfield service companies. They could access Iran's high-potential market, especially in the prevalent low oil prices. Iran also is looking for new investments in its upstream sector, but it plans to strengthen its local companies in collaboration with international companies, both in know-how transfer and "know-why" transfer (STI mode).

In this chapter, Iran's current local content policy in the framework of the Resistance Economy and corresponding policies in the oil and gas sector are reviewed. Key players in both the public and private sectors are mentioned as well. Then the country's current industrial and academic capacities are addressed. Finally, local content requirements in buyback contracts and the new Iran petroleum contracts are reviewed and discussed.

Local Content Policy, Baseline, and the Resistance Economy Policy

In addition to the direct benefits earned by exploration and production activities, oil countries are seeking the added benefits of domestic economic development. Policies are directed toward benefiting from the oil and gas industry's potential (with oilfield services and EPC projects) as an economic driver of Iran's industrial sector for local content development and enhancement of its infrastructure.

Iran's Resistance Economy policy

After several years of sanctions on Iran's economy, especially during the past decade, Iranian decision makers have followed a more robust economic policy in the post-sanction era. Economic policy, while robust, has been geared toward cushioning the economy against possible long-term external economic shocks. Four main features of this economic policy are the focus on indigenous capabilities, increasing global interaction, developing a knowledge-based economy, and promoting social justice.

On February 19, 2014, Iran's supreme leader announced the general policies of Iran's Resistance Economy. The following are some of the key objectives presented in this document:[2]

- Domestic capacity building with maximal utilization of the country's resources, with a special focus on the participation of lower and middle-income classes in wealth generation

- Promotion of a knowledge-based economy through drafting and implementing a comprehensive, scientific plan for the country and promoting innovation with the ultimate goal of becoming the No. 1 knowledge-based economy of the region
- Increased efficiency in economic activity, improvement of economic competitiveness by empowering the key production elements (labor, infrastructure, etc.)
- Utilization of subsidy reforms to optimize energy consumption in the country, increase employment and domestic production and promote social justice
- Improve Total Factor Productivity especially by empowerment of domestic human resources through skills education and entrepreneurship
- Promotion of domestic production, especially in strategic products and services and the consequent reduction of dependence on imports by diversification of source countries
- Provision of food and medicine security
- Consumption management based on the promotion of consumption of local products parallel to the quality improvement of domestic products and competitiveness
- Comprehensive reform of the financial system to respond to the country's economic needs and economic robustness
- Targeted promotion of exportable goods and services through legal and administrative reform as well as the promotion of foreign investment for export purposes and planning for sustainable share of national exports in target export countries
- Development of economic free zones for facilitating technology transfer and international finance as well as increasing exports
- Increase the economy's resistance through regional and international economic collaboration, especially with neighbors but also through diplomacy
- Reduce vulnerability of oil and gas exports through the selection of strategic buyers and involving the private sector in diversifying sales channels
- Increase oil and gas value-added exports by development [of the] oil and gas value chain and attention to maximum efficient production rate of oil and gas fields

- Increase oil and gas strategic reserves and production to have an impact on international markets with emphasis on cross-border oil and gas fields
- Implement reforms to rationalize government costs, increase tax revenues and reduce dependency on oil and gas export revenue
- Increase the share of the National Development Fund from oil and gas export revenues
- Increase transparency in financial matters and avoid activities that pave the way for corruption

Technology catch-up strategy in Iran's oil and gas industry

Based on the Resistance Economy policy, all governmental executive organizations shall consider transfer of technology in foreign trade and must uphold the policy in signing contracts with foreign partners. Even after a century of oil production, Iran faces some technological problems due to lack of an integrated and comprehensive technology transfer and development program in both the upstream and the downstream oil industry.

With world-class oil and gas reserves and production, the Iranian government is now trying to implement a technology catch-up strategy to make a knowledge-based oil industry. It is hoped this will propel the national economy by empowering local companies and universities.

The development of the local oil industry started very slowly in the years before the Islamic Revolution of 1979. Early development of the local industry occurred primarily through the subcontracting of nonindustrial and civil work to local companies and employment of Iranian nationals in the NIOC and the IOCs active in the country. In that time, the main contractual commitment of the IOCs in regards to local content development was employment of Iranian nationals; almost all the E&P activities were run by the IOCs (IOEPC consortium). Subcontracts to foreign companies were limited except for minor oil production by the NIOC or its joint ventures. Iranian graduates, mainly from Abadan Institute of Technology (AIT) and other local and international universities, were employed by the IOEPC, the IORC, the NIOC, and its JVs with independent oil companies. The IOCs were not interested in transferring the know-how of oil exploration and production to the NIOC or even to their Iranian employees. This was much worse before oil nationalization, where the Anglo-Persian Oil Company (APOC)[3] had the oil concession of the country. J. W. Williamson, in a book written about his observations of Iran's oil activities in 1927, emphasized the role of scientific and industrial methods in oil exploration and exploitation by the APOC. He noted how they had applied science and scientific knowledge in the operations. But local nationals were only being utilized for hard work and construction, and training for locals was only offered for technician-level work. Even though all physical operations were carried out in Iran, from the oil fields to the Abadan refinery, a decision was made to locate an associated oil industry research center in London:[4]

> The decision to establish a research station was taken in 1915. The selection of the site chosen eighteen miles from London was determined by the consideration that the research staff must be in the closest personal contact, not only with the nerve centre of the Company's organisation in London, but also with the most recent developments of scientific thought and activity, the most eminent of scientific societies and the most complete and up-to-date of scientific and technical reference libraries. *These latter things, it is needless to say, are centred in London.*

In view of the necessarily wide and varied range of the scientific investigations to be undertaken, the research station was laid out as a group of separate buildings, each of which was designed and equipped for a specific type of work. Areas of research ranged widely, including pure chemical study of the individual chemical substances present in petroleum and pure physical research on the theory of lubrication. It also included applied research relative to long-term tests on diesel engines and complete refining of crude oil on a scale up to one ton.

After the Islamic Revolution in 1979, all operations from the IOCs were transferred to the NIOC, and a complete shift from the foreign-run oil industry occurred. Gradually local construction companies and consulting firms were established, mainly by former employees of the IOCs and as consortia operating as subcontractors of the NIOC. The speed of development of local companies increased after the Iran-Iraq War. The first engineering company in the upstream sector and integrated EPC contractors were established in response to the subsequent national economic development and reconstruction of the country. Joint ventures between Iranian and international companies were created to undertake big EPC projects in midstream and downstream operations. Buyback contracts in the upstream sector of the oil industry increased project management knowledge and experience of local EPC contractors, general contractors (GC), and field developers.

All efforts to that point, however, had been focused on know-how, not "know-why." Recently, Iran's Ministry of Petroleum defined an initiative with several programs to shift from know-how mode, where knowledge is gained by doing, using, or interacting (DUI), to know-why mode, or science- and technology-based innovation (STI) mode in the oil and gas industry. These two approaches fail when they are not integrated. To this end, the technological catch-up strategy was chosen as the main technology policy (see fig. 11–1).

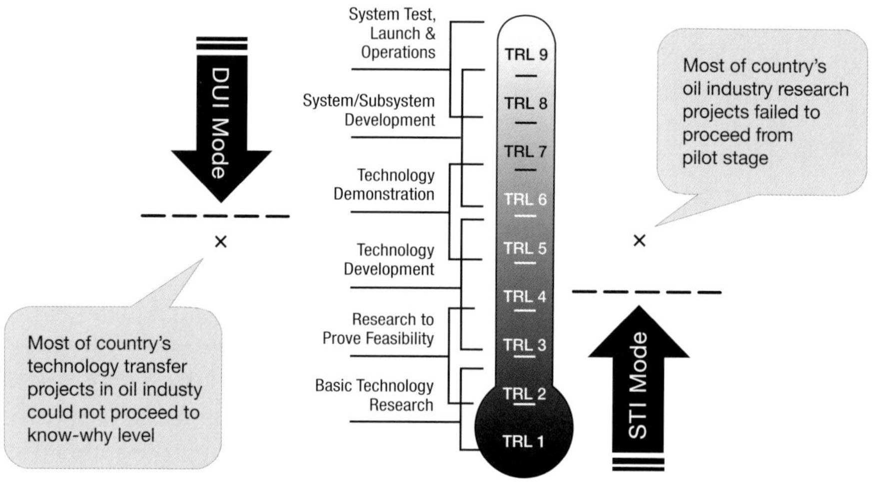

Fig. 11–1. Status of different technology approaches in Iran oil industry

The STI mode is based on the production and use of codified scientific and technical knowledge, while the DUI mode relies on informal processes of learning and experience-based know-how. As detailed in the literature of technology innovation, it is expected that the STI learning mode will have a more direct effect on product innovation, as its core activity (R&D) tends to lead to discovery and testing of new product properties, qualities, and configurations. The DUI learning mode tends to focus more directly on process innovation through exchanges and interactions among workers and engineers, managers and clients, or suppliers and service providers in order to improve the way activities are developed.

Iran's petroleum industry experience in the past 100 years of oil production reveals the domination of the DUI mode of learning. In other words, most of country's technological capability has been achieved through learning by working with or interacting with different departments. Examples include work conducted in different areas like geology and geophysics (G&G), petroleum engineering, or drilling and surface facilities in integrated reservoir studies. This method of learning tends to produce new engineering processes and organizing methods (e.g., drilling efficiency in high-pressure zones above the target reservoirs, reservoir characterization of fractured carbonate reservoirs, etc.). Additionally, as depicted in figure 11–2, in the short term, a stronger impact on process innovation (improvements in productivity) is expected. In the longer term, product innovations are also expected, as time is available to make changes in work routines (e.g., making new tools and equipment for increasing the drilling and oil recovery efficiency).

The product and process division of innovation is a widely known typology, commonly used by scholars, practitioners, and policy makers. Based on studies, it is claimed that product innovation is expected to positively affect the firm's performance, due to the technological novelty and a higher sophistication level (i.e., quality) of the new product. Similarly, process innovation yields efficiency and productivity gains attributable to the introduction of better-performing ways of producing (preexisting) products. Both, in turn, lead to superior improvements in innovation that benefit the firm's performance.

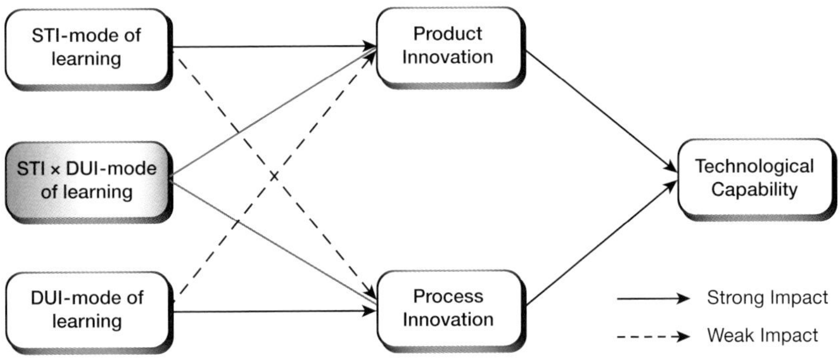

Fig. 11–2. Impact of learning mode on product and process innovation

The STI × DUI learning mode was chosen by Iran's Ministry of Petroleum to pursue both product and process innovation. The ministry's goal is to increase the technological capabilities of Iranian firms through a technological catch-up program.

The idea behind a technological catch-up effort is very simple. Countries that are technology followers can reduce their technological gap with technology leaders by imitating (duplicating) the technologies and by adopting them creatively to make new innovations.

As a result of internal studies conducted by the Ministry of Petroleum in 2014, technological catch-up strategies involving reverse product life cycle strategy and reverse value chain strategy were chosen for the upstream and downstream sectors, respectively (see fig. 11–3).

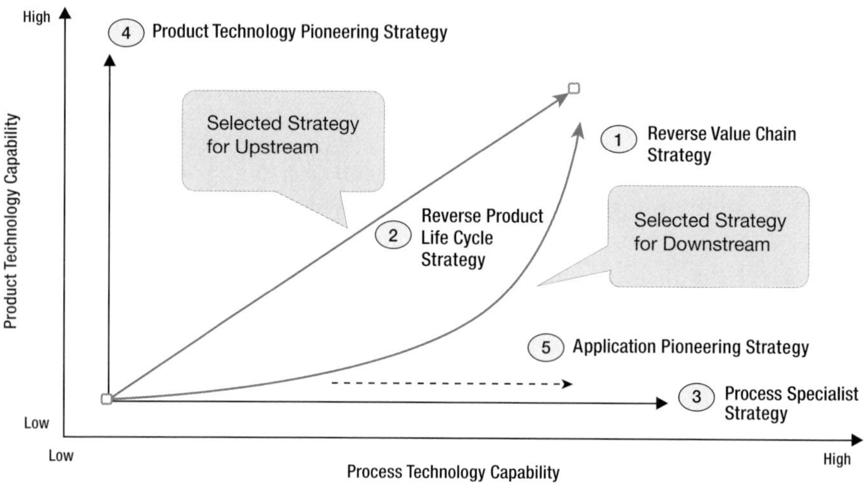

Fig. 11–3. Technological catch-up strategy for upstream and downstream sectors

The reverse product life cycle strategy (RVLC) is advised for the upstream sector of the country, but it requires deep pockets for the simultaneous focus on both product and process technology innovation. With this strategy, a latecomer company could eventually control the complete value chain of E&P activities from development to marketing of the produced oil and gas. The technology of this mature industry is not very advanced for leading international oil companies and service companies and thus is relatively easy to acquire and to imitate. The low-end market can be captured if the latecomer has a competitive advantage through lower labor costs or better knowledge of the domestic market, for example. When this mature market is captured, more sophisticated technology can be learned, and the local company may move from being a late follower to a fast follower. This strategy makes use of imitative R&D and reverse engineering for simultaneous innovation in products and processes.

For development of downstream companies of the country, the reverse value chain (RVC) strategy has been suggested. In this strategy, companies emphasize process innovation first and then move into product innovation; learning is mostly done by doing. This strategy requires lower capital projects than those in the upstream sector, and technologies can be duplicated for subsequent projects of the company. The development of gas plants in the downstream sector of phases 12 to 24 of the South Pars gas field development is one example. It mainly duplicated the phases 2 and 3 designs done by Total in early 2000.

The related technological catch-up strategy implementation road map is shown in figure 11–4.

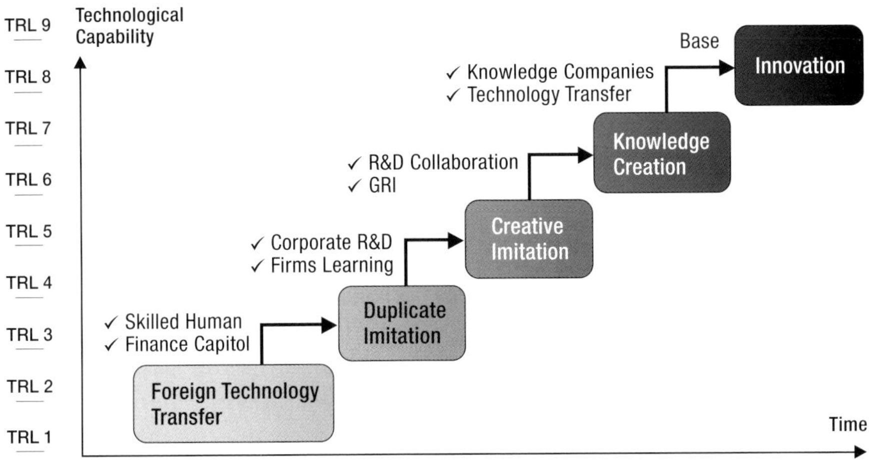

Fig. 11–4. Technological catch-up strategy implementation road map

Key Players

NIOC and subsidiaries

The National Iranian Oil Company (NIOC) was established in 1951 just after oil nationalization. It was formed to run the nationalized oil operations, acting as the executive owner of the oil resources on behalf of the government. Foreign companies are "contractors" to the NIOC, which manages and exploits an estimated 158 billion barrels of oil and 34 trillion cubic meters of gas in proven reserves. Overall, with the total volume of oil and natural gas reserves of 400 billion barrels, Iran has the world's largest hydrocarbon reserves.

Before establishment of the NIOC, the Anglo Persian Oil Company (APOC), later renamed the Anglo Iranian Oil Company (AIOC), had been conducting successful E&P efforts in Iran since 1908. Six years later, the British government acquired 50% of the voting rights. The AIOC continued production until oil nationalization in 1951. After the 1953 coup, the NIOC was forced to sign a contract with a consortium of major oil companies (40% BP, 40% five major American companies, 14% Royal Dutch Shell, and 6% French oil company CFP). Twenty years later, the consortium contract was finished, and companies agreed to work under a service contract for the NIOC.

After the revolution in 1979, the Ministry of Petroleum was established, and all activities of foreign companies in the country were concentrated under the umbrella of the NIOC, which had many subsidiaries. Currently, the NIOC has three sister organizations that are technically independently incorporated under the Oil Ministry, with responsibilities that include handling gas, petrochemicals, and refining. The NIOC is internally organized into numerous subsidiary organizations, and according to latest estimates, the total number of employees exceeds 200,000. Six organizations are charged with managing specific geographic regions or fields. The National Iranian South Oil Company (NISOC) produces more than 80% of country's oil production, and Iranian Central Oil Fields Company (ICOFC) and Pars Oil and Gas Company (POGC) are the main gas producers. Others have functional expertise in drilling, exploration, terminal management, general contracting, and engineering. Three subsidiaries are based overseas and are charged with sales and procurement. There are also some other administrative and organizational units under supervision of the CEO of the NIOC.

An organization chart of the NIOC and its main subsidiaries and affiliates is shown in figure 11–5.

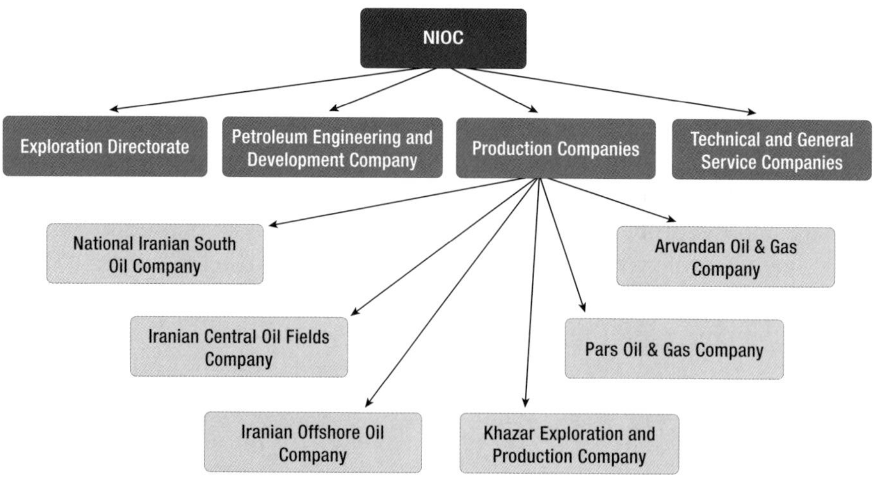

Fig. 11–5. NIOC organization chart

Recently, NIOC approved some restructuring, and all exploration and development activities and subsidiaries were placed under supervision of the newly established deputy of Engineering and Development Affairs. Similarly, the deputy of Production Affairs will supervise all production companies. With authorities given to these two deputies, the resulting NIOC organization chart is more vertical.

Ministry of Petroleum

Since its establishment, the government-owned NIOC has been responsible for all upstream oil and natural gas projects and has been directing and making policies for exploration, drilling, production, research and development, refining, distribution, and export of oil, gas, and petroleum products. The Iranian Constitution prohibits foreign or private ownership of natural resources. However, international oil companies (IOCs) can participate in the exploration and development phases through buyback contracts or the recently approved Iranian petroleum contracts (IPCs).

The energy sector is overseen by the Supreme Energy Council, which was established in July 2001 and is chaired by the president of Iran. The council is composed of the ministers of petroleum, economy, trade, agriculture, and mines and industry, among others. Under the supervision of the Ministry of Petroleum, state-owned companies dominate the activities in the oil and natural gas upstream and downstream sectors, along with Iran's petrochemical industry.

As mentioned earlier, after the revolution in 1979, the Ministry of Petroleum was formed and was given responsibility for all types of oil and gas activities. The ministry has four major subsidiary bodies as follows, which undertake implementation and operation of O&G activities:

- **National Iranian Oil Company (NIOC).** The NIOC is the national E&P company, as introduced previously.
- **National Iranian Gas Company (NIGC).** Established in 1965, the NIGC is responsible for the treatment, transmission, and delivery of natural gas to the domestic, industrial, and commercial sectors and power plants. It directs export of gas through its subsidiary, the National Iranian Gas Export Company (NIGEC).
- **National Iranian Oil Refining & Distribution Company (NIORDC).** Since its establishment in 1991, the NIORDC has been the domestic oil refiner and petroleum products distributor for Iran.

- **National Petrochemical Company (NPC).** The NPC is responsible for investing, producing, supplying, and marketing of petrochemical products. Founded in 1964, the NPC began its activities by operating a small fertilizer plant. Today, the NPC is the second largest producer and exporter of petrochemicals in the Middle East. Over the years, it has not only expanded the range and volume of its products, but it has also taken steps in areas such as research and technology to achieve more self-sufficiency.

The purpose of forming the oil ministry was to implement the major policies of Iran in the oil and gas sectors, including producing, developing, planning, and supervising all operations in both the upstream and downstream sectors of oil industry.

The organization chart of the Ministry of Petroleum is shown in figure 11–6.

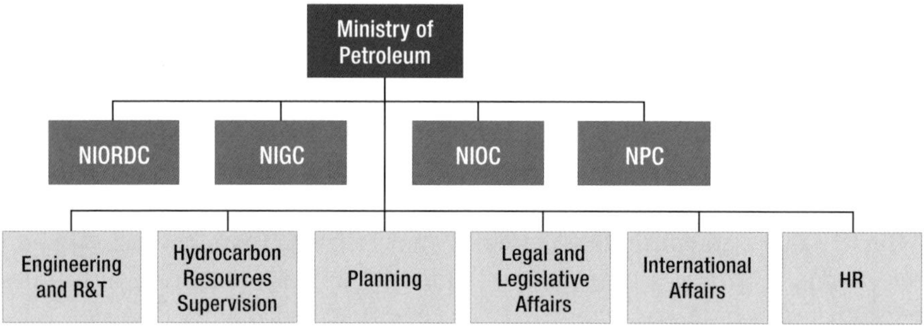

Fig. 11–6. Organization chart of the Ministry of Petroleum

Besides the main four above-mentioned companies, the ministry has several administrative and support organizational units, including HR, international affairs, legal and legislation affairs, engineering, research & technology, planning, and hydrocarbon resources supervision.

Local oilfield service companies, EPC/general contractors and manufacturers

The local oilfield service companies and engineering, procurement, and construction companies were mainly created after the revolution and the subsequent departure of the foreign companies. Iran started to renovate its old petroleum industry and develop its petrochemical industry after the Iran-Iraq War ended in 1988. The main private local downstream players during the war were a few engineering companies such as Sazeh Consultants, Enerchimi, and Nargan, and construction companies like

Jahanpars and Kayson. Most of these companies were created before the revolution and were primarily engaged in civil engineering and construction before shifting to the oil and gas industry during the war.

In the 1990s, the NPC initiated an effort to expand the potential of Iranian private engineering companies to become EPC contractors for executing EP & EPC projects. Among the more than 60 engineering companies active in oil and gas sector, 15 planned to expand their capabilities to become EP & EPC contractors. Most of these companies are members of Engineering & Construction Companies Associations (ECCA). These Iranian companies have completed the majority of the refineries, gas treating plants, and petrochemical plants in Iran, either on a stand-alone basis or in a joint venture with international companies.

At the same time, the NIOC created some subsidiaries, such as the Iranian Offshore Engineering and Construction (IOEC) company for rendering offshore EPC projects in the Persian Gulf. Oil Industries Engineering and Construction (OIEC) was founded in the late 1980s as one the biggest offshoots of the Iran Oil Pension Fund (which belongs to employees of NIOC). Petropars, as one of the main players in EPC projects as well as gas field development projects, was also established in the mid-1990s as a general contractor for development of phase 1 of the giant South Pars gas field.

In the upstream sector, the only Iranian service provider and drilling company both onshore and offshore was the NIOC-owned National Iranian Drilling Company (NIDC), founded just after the revolution. Similar to the downstream sector, few private companies were created after the end of the Iran-Iraq War. Tehran Energy Consultants was the only national upstream consultancy company. It was established after determining the need for an Iranian entity to estimate Iran oil and gas reserves during the arbitration process of a dispute between the NIOC and an American company.

Among the manufacturers, Iran Power Plant Project Management (MAPNA) and Azarab are the leading companies to construct major equipment for power plants and refineries.

A new era encouraged the establishment of local companies in both the upstream and downstream sectors as investments through buyback contracts began in the mid-1990s. The development of local companies was also aided by the concurrent approval of a law encouraging maximum utilization of Iranian content, enacted by Iran's Parliament in 1997. The number of local companies increased steadily both in the public and private sectors. According to estimates, this number now exceeds 250 companies in downstream engineering, procurement, construction, and installation. More than 50 of these companies are capable of handling big EPCI projects. Currently there are more than 50 upstream engineering and oilfield service companies active in the country. Of course, almost half of the companies created in past two decades are big public or semi-private companies such as Petroiran, Pars Energy Gostar Drilling and Exploration (PEDEX), Industrial Projects Management of Iran (IPMI), Tadbir Energy, Petro Gohar Farasahel Kish, Global Petro Tech, North Drilling Company, and MAPNA Oil and Gas. In the private sector, Dana Energy, Mehran Services, and Pasargad Energy Development (PEDC) are among the main players.

Professional associations

There are three main private associations related to the petroleum industries:

- Engineering and Construction Companies Association (ECCA); downstream engineering
- Society of Iranian Petroleum Industry Equipment Manufacturers (SIPIEM) and Iran Industrial Equipment Manufacturers Association (SATSA); procurement
- Association of Petroleum Industry Engineering & Construction Companies (APEC); construction

The main focus and general coverage of companies in these associations are presented in figure 11–7.

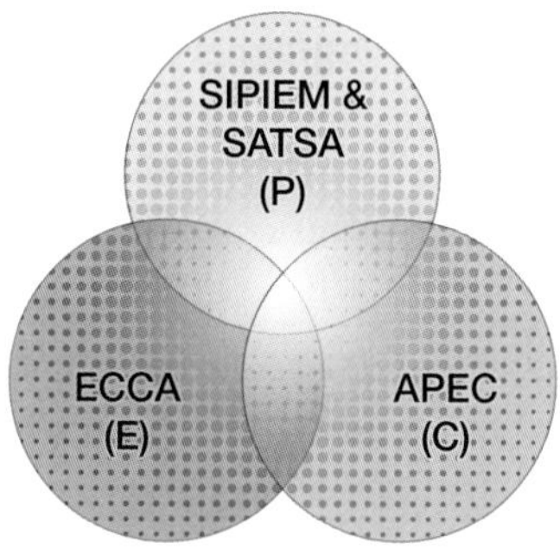

Fig. 11–7. Focus of main local engineering, procurement, and construction companies' associations

SIPIEM was established in March 2000 as the largest private association in the field of equipment manufacturing in Iran, with more than 600 members. Based on the reports provided by SIPIEM at the 2015 Tehran Summit on Iranian Petroleum Contracts (IPC), the current capacity of

Iranian manufacturers is able to meet about 70% of the equipment needed by the petroleum industry. This domestically produced equipment meets almost all international standards and requirements, such as API, ASTM, ASME, and DIN. Among the more than 600 company members, there are 50 companies with the ability to operate on a global scale.

The ECCA members have more than 9,000 employees, of which 95% are full-time. Furthermore, 75% are university graduates and have executed more than 120 projects with the value of US$15 billion. At the moment, ECCA members are able to increase their capacities to handle about US$28 billion in projects. Their capabilities in various units of gas plants, refineries, petrochemicals, and oil production units and processes can be summarized as follows:

- Feasibility and conceptual studies
- Technology evaluation and selection
- Finance and insurance arrangement
- License and PDP arrangement
- Basic design and front-end engineering design (FEED) preparation
- Detail design and engineering
- Procurement engineering services
- Procurement of material and equipment
- Engineering, procurement, and construction management (EPC)
- Precommissioning, commissioning, and start-up management
- Field engineering, technical service, and construction supervision
- Project management contracting (PMC)
- General contracting (GC)
- Joint ventures

APEC was established in 2000 and consists of 217 members who are active in major oil, gas, and petrochemical projects. Over the past half century, APEC members have shown their abilities in implementing major projects in various fields. APEC members have had a formidable and key role in cooperating with the Iranian government, and particularly the Ministry of Petroleum, in the development of South Pars oil and gas projects. In the last two decades, APEC members have been working with international contractors in joint ventures or consortia. Members of the APEC have the capability to form consortia or JVs to participate in major oil, gas, and petrochemical projects in other countries as well.

According to the latest annual report of APEC, distribution of company experience of its members and their valid grades from the Iran Management and Planning Organization are shown in figure 11–8 and figure 11–9, respectively.

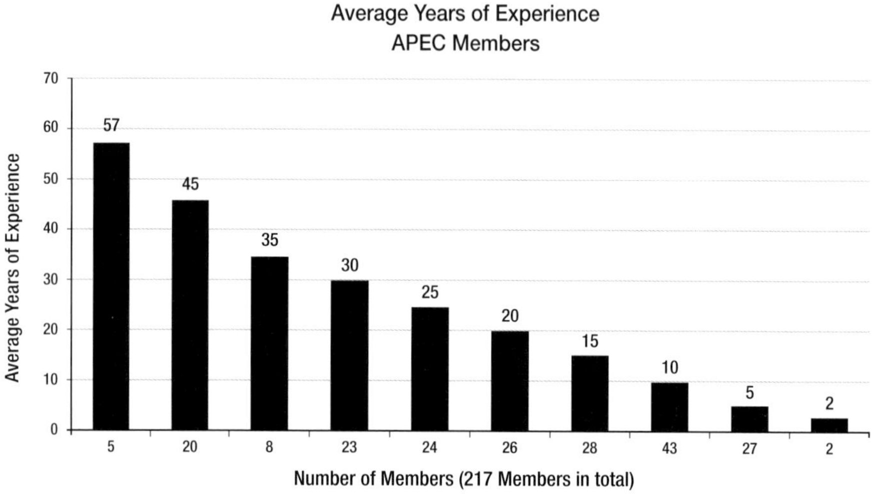

Fig. 11–8. Average years of experience of APEC members

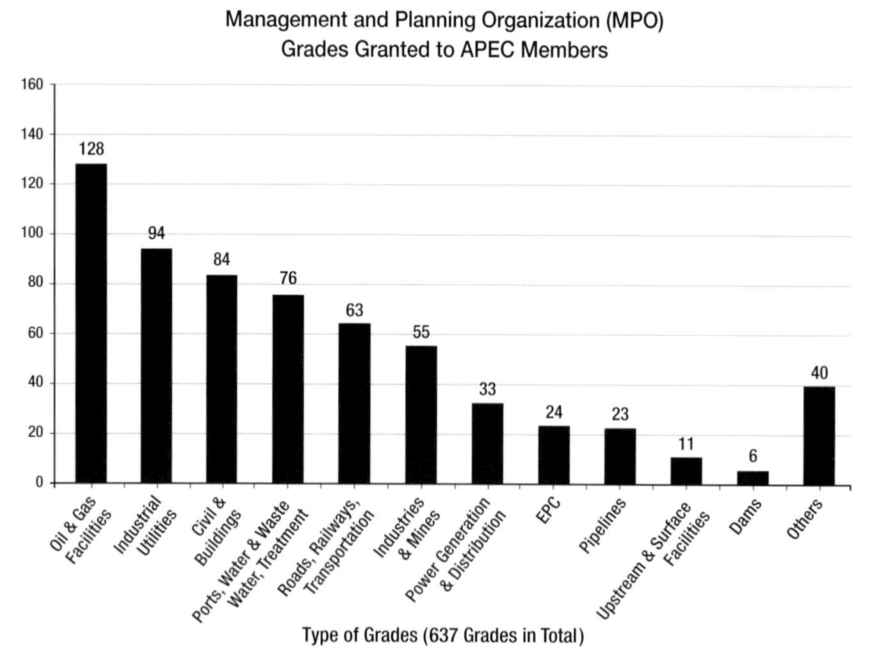

Fig. 11–9. APEC members' grades from the Iran Management and Planning Organization

Aside from the above-mentioned associations, which are mainly devoted to the midstream and downstream sectors of the oil industry, there two other professional associations in the upstream sector:

- **Association for Oil and Gas Drilling Contractors (AOGDC).** AOGDC was founded in 2014 with 18 drilling companies and service providers under the umbrella of the Iran Chamber of Commerce. Almost all key private and public drilling companies are members of the AOGDC. This association is expected to play a more significant role in coming years.
- **Association of Upstream Consultancy Companies.** This association is newly established and consists of about 15 founding companies.

Other active associations in the country are as follows:

- **SPE-Iran Section.** SPE-Iran Section is experiencing an increase in activity since sanctions have been lifted. Its main function is to promote petroleum engineering knowledge by holding SPE contests among university students and by facilitating professional training opportunities.
- **Iranian Petroleum Institute (IPI).** The IPI, with the recent reform of its bylaws and support from the Ministry of Petroleum, is going to be the main player concerning petroleum-related standards and regulations. IPI endorsed the Iranian Petroleum Standard (IPS) as the national standard. Almost all the petroleum-related associations are now members of the IPI, giving it more validity.
- **Iranian Society of Petroleum Engineers (ISPE).** The ISPE is an academic association registered under the Ministry of Science, Research, and Technology and is focused on holding academic conferences and publishing academic journals.

Industry Knowledge

Research and training centers

Research is a crucial part of the oil industry and is a vehicle for advancements in technologies that allow more oil to be located, and to locate it more cheaply. Research also permits oil to be produced and transported with increasing safety. As mentioned previously, nearly a century ago, the APOC decided to build its research center for Iranian petroleum in

London, thousands of miles away from the oil fields. Establishment of a similar national research center located in Iran was postponed until the enactment of the first national petroleum law a few years after oil nationalization and the creation of NIOC.

The Research Institute of Petroleum Industry (RIPI) was initially established as the Iran Petroleum Industry Research Development Office in 1959. According to RIPI's website, "its primary aim was carrying out research on application of petroleum materials."[5] After the Islamic Revolution of 1979, the office became the Research and Scientific Services Center, and it has continued its activities toward furthering various aspects of petroleum research. In 1989, upon the formal agreement of the Development Council of the Department of Culture and Higher Education, the center was renamed the Research Institute of Petroleum Industry. Its aim was to carry out fundamental, applied, and developmental research, and its activities toward that end have continued.

RIPI was the only research center of the NIOC before the revolution. With the maturing of the major oil fields, the need for a center focused on EOR was recognized. To this end, the EOR Research Center was created in 1980, and it became an independent research center in 2005.

The Institute for International Energy Studies (IIES) was also founded in the early 1990s as an affiliate of the Ministry of Petroleum. The Ministry of Science, Research, and Technology granted an official certification to the IIES research faculty in 1996. IIES has developed into a credible research center and reputable think-tank for the ministry, helping the ministry to assume leadership within the industry and beyond national borders. Energy issues are the focal point for the IIES studies, concentrating on oil and gas at local, regional, and international levels. The institute's contributions span a wide range of activities in the form of research, seminars, advisory services, and training courses.

As a recent trend, in the past decade larger universities and some knowledge-based companies have founded research centers in science and technology parks. These centers support research and technology projects in various technical, economic, and managerial aspects of the petroleum industry.

Courses, degrees, and specializations dealing with oil and gas

Iran's oldest university to offer degrees in the field of petroleum engineering is the Abadan Institute of Technology (AIT), located in Khuzestan Province in southwest Iran. AIT was founded to train engineers and managers for Iran's oil industry. Originally established as the Abadan Technical College in 1939, the institute was first funded by the NIOC in 1956 and then renamed to the AIT in 1962. Initially it offered bachelor's degrees in petroleum engineering, economic and administrative sciences, and accounting. At that time, the programs were still internship orientated. Considerable changes to both educational content and management have occurred since 1980. Today, AIT has expanded into the Petroleum University of Technology (PUT). The university now includes several colleges, one of which trains petroleum engineers.

In its original form, the Abadan Technical College was established under a clause in the contract with the Anglo-Iranian Oil Company (AIOC). The clause called for the establishment of a technical school to train skilled Iranian technical staff for Iran's oil industry in Abadan. Iranian high-school graduates were employed with the oil company as part of a four-year sandwich course, alternating studying and working in the Abadan refinery in four-month terms. Upon successful graduation, students were awarded a technical diploma equivalent to a bachelor of science degree in engineering. The graduates then began employment as senior staff in the Abadan refinery or in the oil fields. In 1956, a few years after the nationalization of Iran's oil industry, Lafayette College of Easton, Pennsylvania was selected to help with updating the educational program at the institute.

PUT currently offers four specializations in petroleum engineering at the undergraduate level, including petroleum exploration, reservoir engineering, drilling engineering, and production engineering. In recent years, master's degrees in other relevant disciplines have been added, including reservoir management and oil and gas project management. Currently, a doctorate in hydrocarbon reservoir engineering is also offered. Other degree programs offered by PUT related to the petroleum industry are health, safety, and environment (HSE), accounting, industrial management, and safety and inspection.

Until the end of the Iran-Iraq War in 1988, the only university offering degrees in petroleum-related disciplines was PUT. The graduates were hired by the Ministry of Petroleum or one of its four main companies (NIOC, NIGC, NIORDC, and NPC). Even before the revolution, however, many universities offered coursework related to geology and geophysics (G&G). After the war, other Iranian universities were permitted and even encouraged to offer degrees in petroleum engineering by the Ministry of Higher Education (now the Ministry of Science, Research, and Technology). One of the first private universities to offer petroleum-related degrees was a newly established local university, the Islamic Azad University (IAU). The IAU admitted students for undergraduate-level courses, mainly at its branch institutions in the southwest part of the country near the oil fields. Course work was offered for minors in production engineering and drilling engineering. Some of this group of graduates found employment with NIOC affiliates, which were growing in number because of the geographical split due to the scope of activities. Other graduates were hired by the private sector for new investment projects in upstream and downstream parts of the oil industry.

The second movement was concurrent with buyback projects. Soon after taking office in 1997, Minister Bijan Zangeneh (Ministry of Oil) recognized the lack of reservoir engineers in NIOC subsidiaries and the need for new private companies. Under his directives, and with the support of NIOC funds, a decentralization policy was created to encourage petroleum engineering training. Universities involved in the new effort included several high-ranking institutions, such as the Sharif University of Technology, the University of Tehran, and Amirkabir University of Technology (Tehran Polytechnic). These institutions were equipped with labs and required facilities, and professors with international credentials were hired so as to be able to offer doctorate degrees and related research. Later on, other universities also created petroleum engineering departments. Fortunately, Iran is now self-sufficient with respect to its national university education programs, and is well positioned educationally among oil-producing countries. In recent years, educational efforts have branched out to include energy law and other economic disciplines related to the petroleum industry, with master's and doctorate degrees offered at top universities.

Currently Iran has 34 universities with programs in oil- and gas-related disciplines; half are public universities, and the rest are mainly campuses of IAU in different cities. As can be seen in table 11–1, two-thirds of the degree programs in the public universities are master's and doctoral degrees, while the remaining one-third are bachelor's degrees. Most of IAU campuses provide bachelor's degrees.

Table 11–1. University programs in petroleum industry–related disciplines

Row	University	Year of Establishment	Faculty Name	Year of Establishment	G&G	Petroleum Engineering						
						Exploration	Reservoir Engineering	Drilling Engineering	Production Engineering	Reservoir Management	Oil & Gas Project Management	Oil and Gas Law
1	Petroleum University of Technology	1939	Petroleum Engineering	1957		BSc, MSc	BSc, MSc, PhD	BSc, MSc	BSc, MSc	MSc	MSc	
2	University of Tehran	1933	Institute for Petroleum Engineering, Industrial Engineering	1999	BSc, MSc, PhD	BSc, MSc	BSc, MSc, PhD	BSc, MSc	BSc, MSc	MSc	MSc	MSc, PhD
3	Sharif University of Technology	1966	Chemical and Petroleum Engineering	1999			BSc, MSc, PhD	BSc, MSc	BSc, MSc			
4	Amirkabir University of Technology	1958	Petroleum Engineering	2000			BSc, MSc, PhD	BSc, MSc	BSc, MSc			
5	Tarbiat Modaress University	1981	Petroleum Engineering	2012	MSc, PhD		MSc					
6	Iran Science and Technology University	1939	Chemical Engineering	2012			MSc					
7	RIPI	1959	Upstream Faculty	2014			PhD					
8	Shiraz University	1967	Chemical, Oil and Gas	2001			BSc, MSc					

Row	University	Year of Establishment	Faculty Name	Year of Establishment	G&G	Petroleum Engineering						
						Exploration	Reservoir Engineering	Drilling Engineering	Production Engineering	Reservoir Management	Oil & Gas Project Management	Oil and Gas Law
9	Sahand University of Technology	1990	Petroleum Engineering	2008			MSc, PhD	MSc				
10	Sabzevari University	1987	Petroleum Engineering	2008					BSc			
11	Shahrud University	1973	Mining, Petroleum and Geophysics	2011	BSc, MSc, PhD	MSc		MSc				
12	Garmsar University	2011	Petroleum Engineering	2011			BSc					
13	Semnan University	1974	Chemical, Oil and Gas	2008			MSc	MSc				
14	Shahid Bahonar University	1975	Faculty of Engineering	1998				MSc, PhD				
15	Persian Gulf University	1991	Oil, Gas and Petrochemical	2010			MSc	MSc				
16	Yazd University	1988	Mining Engineering	2008		MSc						
17	Islamic Azad University	1988	Petroleum Engineering	1988	BSc, MSc, PhD	BSc, MSc	BSc, MSc, PhD	BSc, MSc				
18	Shahid Beheshti University	1959	Geology, Law, Economics	1975	BSc, MSc, PhD							MSc, PhD

Row	University	Year of Establishment	Faculty Name	Year of Establishment	G&G	Petroleum Engineering						
						Exploration	Reservoir Engineering	Drilling Engineering	Production Engineering	Reservoir Management	Oil & Gas Project Management	Oil and Gas Law
19	Tabriz University	1947	Geology	1969	BSc, MSc, PhD							
20	Ferdowsi University	1949	Geology	1967	BSc, MSc, PhD							
21	Chamran University	1955	Geology	1972	BSc, MSc, PhD							
22	Isfahan University	1946	Geology	1965	BSc, MSc, PhD							
23	Allameh Tabatabaei University	1971	Law, Economics	2006								PhD
24	Khawrazmi University	1933	Geology	1974	BSc, MSc, PhD							
25	Bu Ali University	1974	Geology	1974	BSc, MSc, PhD							

Legislative Hydrocarbon Regime Concerning Local Content

Iran's first petroleum law was passed in 1957, about half a century after the first oil discovery in Masjed Soleyman in 1908. Since that time, several regulations and petroleum act revisions have been enacted, including the Petroleum Law of 1974, the Petroleum Law of 1987, the Petroleum Law Reform Act of 2011, and the Maximum Utilization Law (the "51% Local Content" Law) in 1997, along with its reform in 2011. A more recent law is the Duties and Authorities Law of the Ministry of Oil in 2012. At the same time, different contracts and fiscal regimes were introduced and implemented. These range from production sharing agreements (PSAs) to different types of service contracts, including buyback service contracts, and the more recent so-called Iranian petroleum contracts (IPCs). In this section, the laws and regulations is briefly reviewed from a local content point of view.

Local content in Iranian petroleum laws (1957, 1974, 1987, and 2011)

Before oil nationalization, there were no laws protecting or requiring local content in Iran's oil exploration and production activities. Local employees were hired or trained by the Anglo-Persian Oil Company for low-level jobs or work as technicians. These national employees, along with the agreed-upon creation of training schools such as the Abadan Technical School and a few elementary and secondary schools, were seen either as a sign of social responsibility or viewed as a means of decreasing personnel costs. The number of Iranian nationals working at higher level positions in all sections, from exploration through production, refinery, and transportation, was less than 30, based on some estimates.

After oil nationalization, control of petroleum operations was transferred to the Iranian government. Subsequent sanctions on Iranian oil production, however, resulted in economic hardships, and Dr. Mohammad Mosaddegh's cabinet was overthrown in a 1953 coup. A consortium contract was then signed with the newly created NIOC. The consortium held the rights for oil exploration and exploitation in a vast majority of the country, mainly in the south and southwest.

The momentum from oil nationalization led to enactment of the Petroleum Law of 1957 by Parliament, which formed the basis for many PSA contracts and a few service contracts. While this law was based on oil nationalization, the IOCs still held the main power and control over Iran's oil industry.

Regarding local content, almost all versions of Iran's petroleum laws emphasize employment of Iranian nationals. Gradually attention has focused on services and equipment used in petroleum operations, requiring procurement from local companies.

Article 11 of the Petroleum Law of 1957 relates to the procurement of local goods:

> The second party (IOC) is not allowed to procure anything outside the country if...similar items are manufactured in the country with [the] same quality, price and availability."

Unfortunately, this article did not help local companies grow, since they were not competitive at that time, and the percent of local goods and equipment was negligible. Nonindustrial works of the consortium like housing, warehouses, etc. were assigned to the NIOC.

From 1957 to 1974, when the second petroleum law was approved, the main achievement was employment of Iranian engineers. These engineers were mainly graduates from AIT and other major universities of the country, such as the University of Tehran, hired by the consortium company and by joint operating companies under PSA contracts.

In the Petroleum Law of 1974, two articles were related to local content. Article 20 of this law, which is related to employment of foreigners, stipulates that foreigners may only be employed if national employees with the necessary training and experience are not available. Article 20 states:

This regulation is cascaded to all sub-contracts. Such employments are subject to consultation and approval of NIOC and NIOC is obliged to set a timeframe [for] such employments and consider the training requirements for replacing [foreign staff with] Iranian nationals in the said timeframe.

Article 24 of the same law relates to use of local goods and services in petroleum operations:

> All goods and services required for petroleum operations shall be procured from local companies. NIOC is only allowed to use foreign goods and services when the quantity of locally procured goods and services is...[inadequate] for petroleum operations or their quality is lower than foreign ones or local prices are...[more] than...20 percent [higher than that] of foreign goods or services.

This law is slightly more protective than the first law with respect to local content.

It is notable that based on this law, the oil industry of Iran was once again declared national, this time in a true sense. The NIOC was permitted to attract investments, but only through service contracts.

After the revolution, employment of foreigners was prohibited. During the eight-year Iran-Iraq War, the NIOC had to replace international companies with some local engineering and construction companies. This helped local E&C companies to grow and become more proficient, preparing the way for joint venture agreements with international E&C companies a few years after the war was over.

The Petroleum Law of 1987 offered a step forward in respect to local content. According to Article 10 of this law:

> The Ministry of Petroleum shall be under obligation to exert regular and consistent efforts for recruitment and training of the necessary personnel and gaining access to advanced technologies and growth and development in various fields of the oil industry through appropriate ways and means in coordination and conformity with the policies of the Ministry of Science and Higher Education by preparation of appropriate educational programs and creation of training centers and establishment of research and laboratory complexes and shall take appropriate measures for promotion of the level of knowledge and scientific and practical information of the employees and experts and creation of an appropriate atmosphere and environment for attraction and encouragement of the competent devoted and expert individuals.

Now the ministry was positioned to consider local content in subsequent legislation. In 1995, the Local Content Committee was created within the Ministry of Petroleum (MOP). This committee was comprised of plenipotentiary representatives of the four main companies of the MOP and the managing director of Kala Naft Tehran, the procurement body of the NIOC. Duties of this committee were then assigned to Kala Naft Tehran, and its name was changed to Manufacturing Support and Procurement Kala Naft Company Tehran, a subsidiary of the NIOC.

The only changes related to local content in the Petroleum Reform Act Law of 2011 from the Petroleum Law of 1987 concerned technology transfer activities related to petroleum operations (Article 1). This allowed costs related to technology transfer to be reimbursed as petroleum costs in new upstream oil service contracts.

Local content in Law of Duties and Authorities of the Ministry of Petroleum (2012)

As stated in Article 1 of the Law of Duties and Authorities of the Ministry of Petroleum,

> Ministry of Petroleum (MOP) is established to fulfill the general policies of the Islamic Republic of Iran in the oil and gas sector, policymaking, governance, planning and monitoring of all upstream and downstream operations in the oil, gas, petrochemical and refining [sectors], and applies its sovereignty [over] and public ownership...[of] oil and gas resources on behalf of the Islamic government.

This law and the Petroleum Law Reform Act of 2011 are the main current petroleum-related laws and legislations regarding hydrocarbon resource management in the country.

Despite the clear definition of duties and authority of the Ministry of Petroleum on policy making, governance, and its supervisory role, the NIOC still plays a key role in policy making and other MOP-related affairs. However, in the recent years this has gradually shifted from the NIOC to the MOP. For instance, the MOP is now in charge of defining and qualifying all local and foreign contractors and manufacturers in the petroleum industry. In July 2016 the Ministry of Petroleum announced a list of qualified local companies for participation with the IOCs in new IPC contracts. Also, in a seminar in April 2016, the MOP elaborated the principles and executive steps of Iran's Resistance Economy implementation road map in the petroleum industry, to be followed by all its affiliated companies, including the NIOC.

The main local content–related duties of the MOP law are provided under Clause HR, Science and Technology Affairs, as follows:

- Provision for a comprehensive quality certification system for goods and services in order to promote quality of local goods and services, and exerting supportive policies of localization and design and construction of required oil, gas, and petrochemical goods and services, with coordination of all other related authorities of the country
- Provision for local content policies in order to promote local oil, gas, and petrochemical companies and increase their competitiveness, along with export of technical and engineering services aligned with the national productivity index

- Provision for a comprehensive research program, and planning and supporting knowledge-based activities for utilization of advanced technologies and technology transfer in the oil, gas, and petrochemical industry, with cooperation of other related ministries and authorities of the country
- Alignment of petroleum industry research objectives and priorities with other national research programs and policies
- Provision for research and technology infrastructure by reinforcing its experts and specialists in coordination with Ministry of Science, Research, and Technology

Responsibility for implementation of these duties was assigned to the recently merged deputy of Engineering, Research, and Technology at the Ministry of Petroleum.

Local content–related laws

Maximum Utilization of Iranian Content Law (1997 and 2011). The Maximum Use of Iranian Technical Engineering and Manufacturing Capabilities for Projects Implementation and Provision of Facilities for Export of Services Law is the most relevant law regarding local content. It was first approved on March 3, 1997 and was revised in 2011 and renamed the Law of Maximum Utilization of Local Production and Service Capacity and Reinforcement for Export and Reform of Article 104 of Direct Tax Law.

The concept dates back to early 1990s, when Iran's oil and gas projects were made available to foreign companies for development. At that time, the provisions were manifested only in the annual budgets of the country, wherein 30% of the tasks listed in the budget were allocated to local companies to create job opportunities and help transfer advanced technology into Iran as well. Later, however, it was transformed into a law that was finally enacted in March 1997.

According to this law, in connection with the execution of the work under any contract, the use of local resources, services, equipment, materials, and supplies from the Islamic Republic of Iran shall be maximized. As a minimum, 51% of the work (i.e., the lump sum price) shall be awarded to Iranian companies.

The local content also includes any subcontracting of the work by Iranian firms to foreign firms. In the acquisition of local resources, services, equipment, materials, and supplies, preference shall be given to services, including construction, provided by Iranian firms, joint ventures, or consortia, and to goods made or produced in the Islamic Republic of Iran. This applies as long as the services and goods are similar to services and goods provided by foreign firms in terms of quality, quantity, deliverability, and suitability for their intended purpose.

Under Article 5 of the same act, the ministry and its subsidiary companies are required to receive all their services, production, installation, and procurement of goods, equipment, instruments, and products needed for their projects only from Iranian qualified companies. If those services are not offered by the Iranian companies, then, with the approval of the highest executive official, they may be procured from Iranian-foreign or foreign companies.

Initially the state's Management and Planning Organization (MPO) was appointed in 2002 to supervise implementation of this law, working with all the other governmental executive organizations. Later, according to the newly revised law, the Ministry of Industry, Mine, and Trade (MIMT) was named as the supervisory body.

The MIMT is obligated to announce the list of local goods, services, equipment, and their capacities every six months, with the help of all related organizations, syndicates, associations, etc.

The general execution procedure for implementation of Article 3 of the Maximum Utilization of Iranian Content Law (2011) is as follows:

- Government executive organizations shall provide all services, construction and installation, procurement, and equipment and goods required for their projects, either individually or collectively, exclusively to Iranian companies and institutes.
- If after having made the necessary consultations and coordination with the MIMT or after having made the required inquiry or having held the tender, the executive organizations shall come to the conclusion that no competent Iranian company or institute is available, or that competent Iranian companies or institutes are not sufficient in number, such executive organizations, upon production of a well-reasoned and justified report, and after having obtained the authorization of the minister of the MIMT and the highest executive official of the executive organization, shall be authorized to assign and cede their projects to Iranian-foreign joint ventures, provided that the share or participation interest of the Iranian company or institute in the Iranian-foreign joint venture shall not measure less than 51% of the value of contract compensation of the work so assigned.

- In the tender and inquiry notices to be published, it shall be required to clearly and unequivocally reiterate that Iranian companies and institutes may declare their readiness to participate in the tender either independently or in the form of civil partnership with foreign companies, on condition that the share and participation interest of the Iranian company or institute in the civil partnership shall not measure less than 51%.
- The Iranian-foreign joint ventures established in the form of civil partnerships willing to take part in the tender shall be required to submit to the executive organization a copy of their letter of agreement or understanding, simultaneous with the production and submission of their bids.
- It shall be authorized to assign work exclusively to the bidders who shall propose to execute and implement 51% of the value of the work utilizing the equipment, accessories, and services that are already produced and rendered in Iran, or which are capable of being provided in Iran. It shall be authorized to assign work to the bidders who shall either propose or undertake to execute inside Iran less than 51% of the value of work, only after having obtained the required authorization and approval of the minister of MIMT and the highest executive official of the executive organization (mentioned in Article 4 of the Law on Maximum Utilization).
- In order to gain thorough knowledge as to the contractual value of the work to be executed and performed inside Iran, the executive organizations shall be required to clearly and unequivocally state the following items in the tender documents:
  - Concerning engineering and procurement (EP) projects or engineering, procurement, construction, and management (EPCM) projects, it shall be required to state the amount and contractual value of the engineering work to be performed inside Iran by Iranian entities.
  - Concerning the amount, value, and supplier of the commodities, raw materials, or bulk materials, it shall be required to state the standardized items that are manufactured in Iran and that will be utilized in the course of implementation of the project.
  - It shall be required to state the amount and value of the equipment the manufacture or fabrication of which shall be assigned to Iranian manufacturers or producers, together with a

list of Iranian vendors to which the manufacture or fabrication of such equipment shall potentially be assigned.

- It shall be required to state the amount and value of the work which shall be executed in Iran directly by the Iranian company being a party to the contract (either independently or in the form of an Iranian-foreign joint venture).
- It shall be required to state the amount and value of the equipment and machinery which are not currently manufactured in Iran but which the foreign partner shall arrange to have manufactured in Iran by foreign manufacturers (either independently or jointly with Iranian manufacturers).
- It shall provide a thorough list of what shall be developed and what shall eventually remain in Iran as a benefit to Iran, upon assignment of the work to the Iranian-foreign joint venture, and after implementation of the subject of the contract.

- If it shall not prove possible and practical to acquire the license and to assign the basic engineering to the nominated license holder through Iranian companies or institutes, it shall be authorized to do so through foreign companies, upon approval by the highest-ranking authority of the MIMT or the executive organization concerned.

Tender Law (2004). This law, approved in 2004, is also one of the local content promotion laws of the country. According to Clause D of Article 20 of this law,

> In international bidding, domestic bidders shall have preference over foreign bidders. The domestic preference procedure shall be specified in the bidding documents. The Economy Council will determine those biddings in which the provision of this rule is not advised.

Also, in Clause D of Article 13 of the same law, it is stated,

> In case it is necessary to organize an international bidding, or use foreign financing, relevant permissions should be obtained with due consideration for the Law on Maximum Utilization of National Technical, Engineering, Production, Industrial, and Implementation Capacity (approved in 1997), and the relevant notice should be published in one of the highly circulated newspapers and at least once in one of the domestic English newspapers and one international magazine or newspaper related to the bidding subject.

Marine Industries Development and Support Law (2008). In order to encourage local companies to build, repair, or maintain the marine industry, the Marine Industries Development and Support Law was approved in 2008. According to Article 1 of this law, the Iran Marine Fund (IMF), a state-owned enterprise of the Ministry of Industry, Mines, and Trade of the Islamic Republic of Iran, is established in order to convert the country's marine capacities into sustainable national interests.

Marine Industries Development Fund policies are as follows:

- Development of construction and repair of ships and offshore industries
- Provision of secure supply chain
- Reduction of risk of investment in maritime activities
- Coastal development and water accessibility
- Expansion of transit and reexport
- International partnerships
- Export-oriented
- Development of standardization
- Development of marine insurance
- Participation in the offshore industries
- Support of the private sector and the public demands

In Note 5 of Article 1, the IMF is authorized to award 20% of vessel contracts to qualified Iranian bidders to help them compete with international companies in country requirements. The bidding procedure shall follow tender law and maximum Iranian content utilization law.

The State Management and Planning Organization shall consider a budget of US$1 billion for supporting local companies in manufacturing or repairing vessels.

Foreign Investment Promotion and Protection Act (FIPPA, 1955 and 2002). After nearly 48 years, a new law on foreign investment in Iran, the Foreign Investment Promotion and Protection Act (FIPPA), was ratified by the Parliament in 2002. FIPPA replaced the Law for the Attraction and Protection of Foreign Investment (LAPFI), which had been in effect since 1955. FIPPA's replacement of LAPFI has further enhanced the legal framework and operational environment for foreign investors in Iran. This law encourages JVs and foreign-local companies in the country.

The 2002 law offered broader fields for involvement by foreign investors, including major infrastructure in the oil and gas sector. Based on Article 3, a broader definition was given to foreign investment, covering all types of investments from foreign direct investment (FDI) to different types of project financing methods, including civil participation, buyback arrangements, counter trade, and various build-operate-transfer (BOT) schemes.

Of course, this law sets some limitations on foreign investments, as stated in Article 2 of the same law:

> **Article 2.** Admission of Foreign Investment shall be made in accordance with the provisions of FIPPA and with due observance of other prevailing laws and regulations of the Country, for the purpose of development and promotion of producing activities in industry, mining, agriculture and services, and based on the following criteria:
>
> (a) Bring about economic growth, upgrade technology, enhance the quality of products, increase employment opportunities and exports;
>
> (b) Does not pose any threat to the national security and public interests, and cause damage to the environment; does not disrupt the Country's economy and jeopardize…production by local investments;
>
> (c) Does not entail grant of concessions by the Government to Foreign Investors. Concession means special rights which place Foreign Investors in a monopolistic position.
>
> (d) The ratio of the value of the goods and services produced by the Foreign Investments, contemplated in FIPPA, to the value of the goods and services supplied to the local market, at the time of issuance of the Investment License, shall not exceed 25 percent in each economic sector and 35 percent in each sub-sector (field). The sub-sectors and scope of investment in each sub-sector shall be determined in the by-law to be approved by the Council of Ministers. Foreign Investment for the production of goods and services for export purposes, other than crude oil, shall be exempted from the aforementioned ratios.
>
> *Note.* The "Law for the Ownership of Immovable Property by Foreign Nationals" enacted on June 6, 1921 shall remain in effect. Ownership of land of any type and to any extent in the name of Foreign Investors is not permitted within the framework of FIPPA.

Other local content–related laws, such as the Law of Removal of Barriers to Competitive Production and Improving the Country's Financial System (2015), that have not had significant impact on local content development are not described here.

Local content in Iranian upstream contracts

Utilization of local content was not explicitly mentioned in upstream contracts before oil nationalization. Since oil nationalization in 1951, the local content–related articles of petroleum contracts have mainly been devoted to employment of Iranian nationals, with lesser emphasis on Iranian procured goods and services. In the following paragraphs, the related articles are briefly reviewed.

Consortium contract. Clause F of Article 4 of the consortium contract regarding the consortium's commitments to the NIOC and the government of Iran states:

> [E]mployment of foreign nationals should be minimized and is allowed when there is no Iranian competent staff. Also, [the] consortium shall...endeavor to prepare [and train]...Iranian nationals in a way that after a short period of time, [they] can replace the foreigners and attain high rank[ing] positions in petroleum operation[s] of operating companies.

PSA contracts. These contracts generally followed the Petroleum Law of 1957, and local employment clauses were very similar from contract to contract. For example, Article 5 of Pan American Co.'s contract with the NIOC requires that the chairman of the board of the joint venture company (IPAC in this case) be Iranian, along with the managing director from the other party. Also according to Article 13 of the contract, an IPAC-like consortium contract requires employment of no foreigners unless there are no Iranian staff available with the required expertise. Furthermore, training of Iranian nationals shall be provided in such a way that after 10 years from the contract's effective date, the number of foreign staff of the company shall not exceed 2% of the total number of company staff. Finally, the number of foreign managerial and executive personnel shall not exceed 49% of the total number of managers and executives of the company.

Buyback contracts. Two articles refer to local content in buyback contracts: Article 13 and Article 24.

Article 13 relates to local employment and training and includes the following clauses:

> 13.1. Contractor shall give priority to qualified and experienced Iranian citizens in employment of personnel to carry out…Development Operations in Iran, in accordance with the laws of Iran limiting the employment of foreign personnel to only positions where qualified and experienced Iranian citizens are not available.
>
> 13.2. Contractor shall conduct and provide during the Development Phase training for Iranian employees and trainees nominated by NIOC from time to time, up to [a] maximum [of] one percent (1%) of the Capital Costs Ceiling as provided for in Clauses 18.2 and 18.3. The type of training, as well as the level, duration, time and place of training shall be agreed by NIOC and [the] Contractor in order to facilitate the transfer of technical know-how and the upgrading of the technical knowledge and skills of the nominees, and acquiring necessary transfer of appropriate technology to enable such employees and trainees to assume and perform the responsibilities and tasks for which they are being trained as required by NIOC for the purpose of, among others, the proper and successful conduct of operations on the wells and facilities drilled and installed by Contractor in the Contract Area, after the completion of the Development Operations. The training programs shall be jointly prepared by NIOC and [the] Contractor, approved by NIOC, and implemented by [the] Contractor in co-ordination with NIOC.

Article 24 refers to the utilization of Iranian content in these contracts in following three clauses:

> 24.1. Contractor shall, in compliance with the laws, rules and regulations of Iran make provisions for supplies of equipment, articles, materials and services as set out in Appendix E to the Contract with respect to the maximum utilization of Iranian content. For the purpose of the Contract, the target for maximum utilization of Iranian content as per the applicable laws, rules and regulations is agreed to be at least fifty-one percent (51%) of the Capital Costs.
>
> 24.2. Contractor shall report quarterly to NIOC on the amount of the utilization of…Iranian content.

> 24.3. Without prejudice to any other rights of NIOC under this Contract, should [the] Contractor not comply fully or partly with such provisions, and achieve actual Iranian content of less than fifty one percent (51%), NIOC shall have the right to reduce the Contractor's entitlement to Remuneration Fee (i) by twelve percent (12%) of the amount of the Iranian content not achieved in the range of fifty-one percent (51%) to thirty (30%), and (ii) by fifty percent (50%) of the amount of the Iranian content not achieved in the range of thirty percent (30%) and zero (0%).

IPC contracts. Since international sanctions were lifted in late 2015, Iran has been seeking to attract large investments in oil and gas exploration and production. Iran benefits from both an attractive geological profile and staggering oil reserves. It holds the second largest natural gas reserves globally (almost 34 trillion cubic meters) and the fourth largest oil reserves in the world (with an estimated 158 billion barrels). Petrochemicals production is the second- largest industrial sector after oil and gas.

With 20 years of experience in three generations of buyback service contracts, the Ministry of Petroleum seeks to unveil the new so-called Iran petroleum contracts (IPC), which are also service contracts. It is intended to make these contracts more attractive so the IOCs choose Iran's oil industry as their first option in the region. Several years of underinvestment created a need for new investments to enable Iran to boost its production position in OPEC. The main differences between the IPCs and previous buyback contracts can be summarized as follows:

1. Extending IOC presence to production and IOR/EOR phases of the reservoir to avoid or postpone the field production decline with utilization of state-of-the-art technologies and technology transfer. This will also help the NIOC and other local partners receive the experience and technology of more advanced EOR methods.
2. The new IPC requires the IOCs to establish joint ventures with local Iranian companies, using one of two options: (i) an incorporated joint venture company (joint operating company, JOC model) or (ii) an unincorporated joint venture (JOA model). The intention is that the joint venture will be the contractor under the IPC. Although Iranian companies have demonstrated some expertise in the supply chain side of the oil and gas sector, the number of experienced E&P companies is very limited. Thus, the MOP is preparing the way for more Iranian companies to enter the country's E&P business. (Of course, some local companies have experience in oil and gas

field development in the framework of buyback contracts.) This is the most considerable change to buyback contracts in regards to local content.

3. The local content requirements are almost the same as buyback contracts and follow the Maximum Utilization of Iranian Content Law. The main differences are explicitly cascading the local content requirements down to all subcontracts and accepting all costs related to technological upgrades as petroleum costs and subject to recovery. These requirements allow the IOCs to establish or upgrade research institutes and laboratories as a part of the annual work program and budget.
4. The remuneration fee is changed from a contractual rate of return scheme to a fee per barrel scheme, encouraging the IOCs to produce more from the fields.

Recently, the MOP completed the first round of qualification for local companies interested in E&P business in the country through partnerships with IOCs in new rounds of IPCs in late 2016 or early 2017. From a local content point of view, fostering local E&P companies together with NIOC subsidiaries promises a more productive and dynamic local petroleum industry. It is hoped that this will replicate the growth that occurred in the EPC and oilfield services sectors during the past two decades.

The need for local E&P companies was recognized first in the *Petroleum Industry Resistance Economy Implementation Report*, compiled by the deputy of Research and Technology of the MOP with the help of industry professionals. In that report, local E&P companies were introduced as the locomotive for local content development. It was hoped that its effects would cascade across EPC/GC contractors, oilfield service companies, equipment manufacturers, and materials suppliers, finally reaching research institutes and universities.

MOP initiatives for local content development in line with National Resistance Economy policy

Implementing Resistance Economy policy in the petroleum industry via E&P and general contractors. In a research study done by the deputy of research and technology of the Ministry of Petroleum in 2014, local E&P companies and general contractors were recognized and introduced as vehicles for implementation of the Resistance Economy policy in the upstream and downstream sectors, respectively. Based on this research, which was conducted with help from many national industry professionals, companies can bring improvements to the supply chain with catch-up strategies as described previously.

This supply chain can be seen as a network of key players, including the following:

- **Tier 1: IOCs/the NIOC/local E&P companies.** These companies have deep pockets with which to develop and support technology for themselves and other parts of network. They normally play the "pull" roll, acting as lead in respect to other parts.
- **Tier 2: Integrated oilfield service companies/general contractors.** These companies are main technology developers from the bottom, utilizing and aggregating smaller technology companies and manufacturers into technology packages. This group of companies normally plays a "push" strategy in technological packages in relation to first-tier companies.
- **Tier 3: Equipment manufacturers and suppliers.** This group of companies represents the biggest part of the network and is comprised of hundreds to thousands of companies. These companies choose a balanced technology strategy, as they are not usually large enough to invest in R&D and introduce new products.
- **Tier 4: Universities and research institutes.** This group represents the knowledge booster of the other tiers; depending on the tier, different disciplines and research programs are engaged.

The relations within this supply chain network are shown below in figure 11–10.

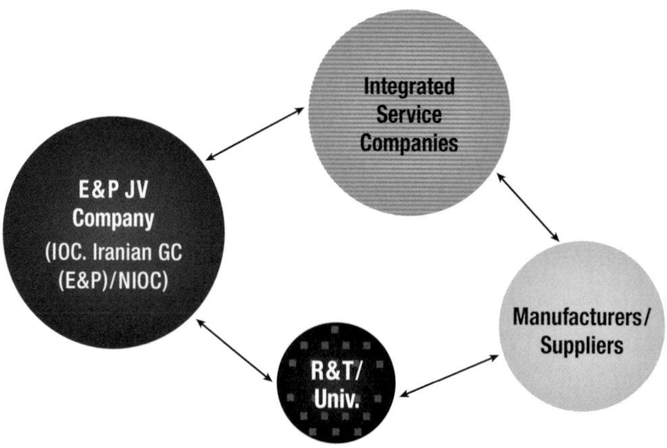

Fig. 11–10. Relation of E&P supply chain network

There are other outside entities communicating with this network as well, such as the following:

- **Regulatory and supervision.** These responsibilities are currently conducted by different entities such the MOP, the MIMT, the MPO, etc.
- **Audits.** Audits are intended to be carried out by local associations and independent organizations such the Iran Petroleum Institute (IPI), the Research Institute for Petroleum Industry (RIPI), and, mainly for quality certification, the National Standard Organization.
- **Policy making.** Policy making is primarily carried out in relation to the first and second tiers by the MOP. Policy making for the third tier is the responsibility of the MIMT, while policy making for the fourth tier is conducted by the Ministry of Science, Research, and Technology, with the vice president for Scientific and Technology Affairs.
- **Facilitation and support.** Facilitation and support mainly come from the Islamic Consultative Assembly (Parliament), the vice president for Scientific and Technology Affairs, and local financial institutions and banks.

These elements, together with the key players of the E&P supply chain, are shown in figure 11–11.

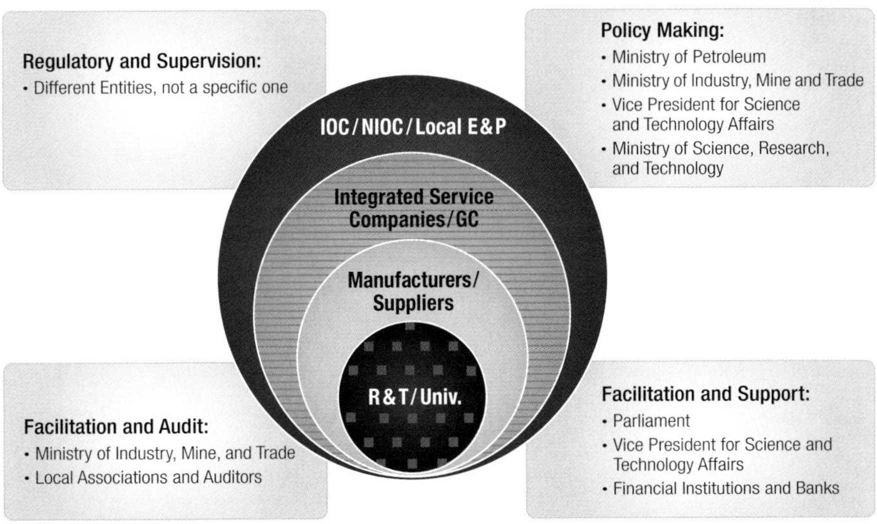

Fig. 11–11. Network of environmental entities together with key players of the E&P supply chain

According to results of a research study approved by the Ministry of Petroleum, the first tier, including the IOCs, the NIOC, and local E&Ps, is responsible for adopting and developing new technologies in the main reservoir life cycle stages from exploration to development and production. The IOCs have responsibility for transferring technology and promoting the knowledge and expertise of the NIOC and local E&P companies in the context of new petroleum contracts.

A general schematic of the conceptual model for technology transfer and development of technologies both in DUI and STI modes in three main stages of reservoir life cycle are presented in figures 11–12 to 11–14.

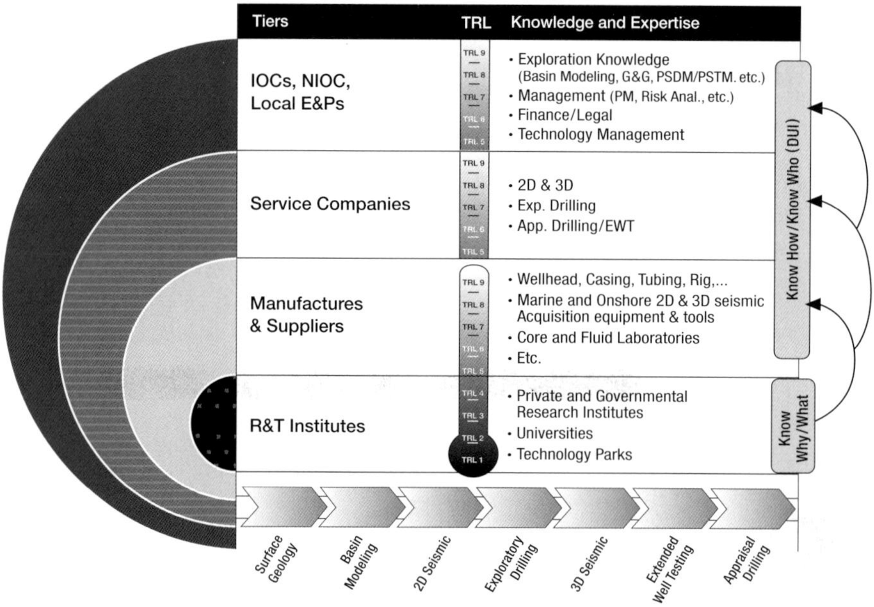

Fig. 11–12. Technology transfer and local content enhancement in exploration phase

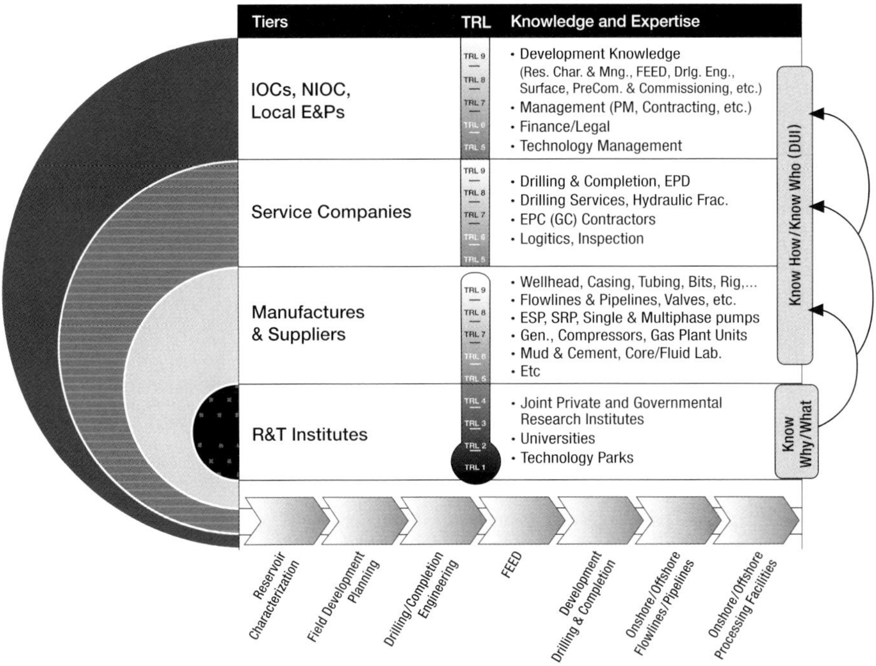

Fig. 11–13. Technology transfer and local content enhancement in development phase

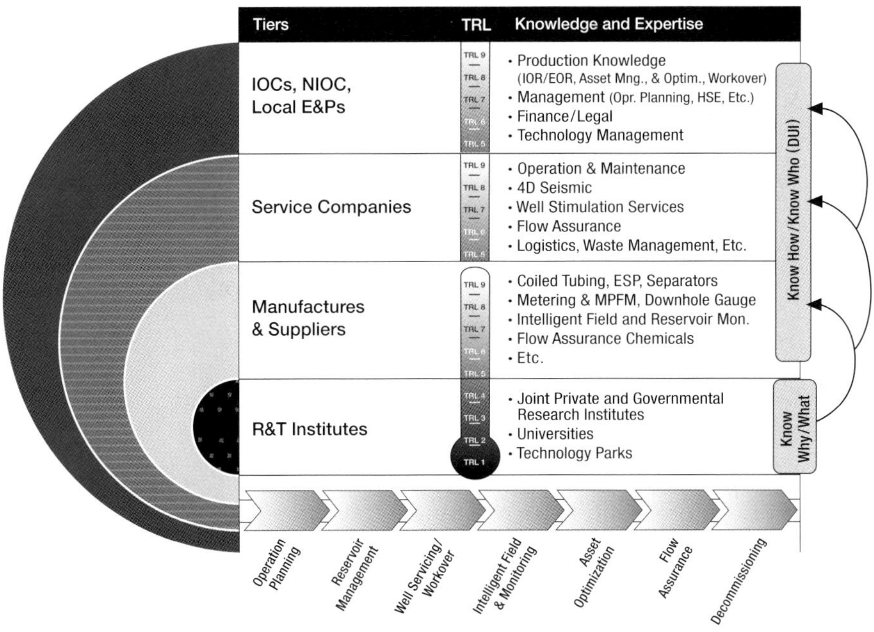

Fig. 11–14. Technology transfer and local content enhancement in production phase

The above conceptual models utilize the STI×DUI approach in the E&P supply chain. The first three tiers are focusing on the DUI approach, and the forth tier is targeted for STI mode of the other three tiers.

Localization Program of 10 Strategic Groups of Equipment of the Petroleum Industry. This three-year program was initiated in 2016 with an initial budget of US$1 billion to promote localization of 10 strategic groups of petroleum industry equipment as follows:

1. Bit (rock bits + polycrystalline diamond compact [PDC])
2. Wellhead and downhole completion
3. ESP (electric submersible pump) + SRP (sucker rod pump)
4. Downhole measuring tools and instruments
5. Seamless pipes (casing and tubing + flow lines)
6. Intelligent pig
7. Rotating equipment (API pump + turbine + compressor)
8. Variable frequency drive (VFD) + explosion-proof electro motors
9. Cryogenic steels
10. Control valves and safety valves and accessories

A roadmap related to implementation of the Resistance Economy was presented at the Ministry of Petroleum Conference held in Tehran in April 2016. It introduced the following localization process:

1. Minister's Order.
2. Steering committee is appointed, headed by ministry's deputy for Research and Technology.
3. A responsible company is appointed for each group of equipment.
4. All future requirements are collected.
5. The "MSP KALA Naft Co." goes to tender.
6. Technical, commercial, and quality audits are carried out.
7. Technology readiness level (TRL), manufacturing readiness level (MRL), and corporate reliability level (CRL) of each manufacturer are verified.
8. Joint ventures with foreign companies with a reasonable technology gap is encouraged, as well as market in exchange of technology transfer (MOTO).
9. The winners of the tenders on each group are selected and announced.

10. The equipment must be competitively priced and of high quality, with two years from production to delivery.
11. The goal is to export to neighboring countries and other customers.
12. The Ministry of Petroleum will act as the driving force behind the new equipment industry.

EOR Mega Science Project in Collaboration with Iranian Universities and Research Institutes. This 10-year program has been signed for EOR studies of 52 oil and gas reservoirs of 20 Iranian fields, with seven well-established universities and two research institutes. The program started in 2014 with an initial overall budget of US$100 million. Initial contracts with these universities and institutes were signed for 9 important fields. Universities and research institutes, together with their assigned fields, are tabulated in table 11–2. This program is the largest science program to be undertaken in Iran.

Table 11–2. Universities and research institutes together with their assigned fields

Row	University	Field Name	Client
1	Sharif University of Technology	Koupal Oil Field	NISOC
2	Amirkabir University of Technology	Darquian Oil Field	AOGC
3	Sahand University of Technology	Soroush Oil Field	IOOC
4	Petroleum University of Technology	Gachsaran Oil Field	NISOC
5	Institute for Petroleum Engineering, University of Tehran	South Azadegan Oil Field	AOGC
6	Shiraz University	Mansouri Oil Field	NISOC
7	Islamic Azad University	Karanj Oil Field	NISOC
8	Research Institute of Petroleum Industry	Ahwaz Oil Field	NISOC
9	EOR Research Center	South Pars Gas Field	POGC

Similar to the EOR mega science program, another mega science program has been defined by the MOP. The program was created with 12 well-established universities and research institutes in the downstream sector of the oil, gas, and petrochemical industries, in line with MOP's STI initiative. This program lasts for five years and started in 2015, with an initial budget of US$100 million.

Petroleum Industry Electronic Procurement Portal and Comprehensive Quality Certification System. This portal started in 2015 and currently is operational with more than 3.4 million parts in about 550 warehouses. It is the evolved material management (MM) system of the petroleum industry that was created in 1990s. Unfortunately, the procurement system was not integrated with the MM system from the beginning, and it remained unconnected until the employment of a new electronic procurement portal that integrated both systems. With this new integrated system, the updated and consolidated demands for each part can be prepared in order to aggregate all demands from various subsidiaries of the four main affiliates of the MOP (the NIOC, NIGC, NIORDC, and NPC). The consolidation of this data offers better support for local companies, with known economies of scale for each part. Together with these systems, another system had been implemented in order to monitor and issue quality certificates to local manufacturers and suppliers. A national coding system called the Iran Code is used for coding of each part, which is compatible with international standards. The qualification process of local manufacturers and suppliers was defined, and more than 1,100 companies have been evaluated with this procedure. More than 1,500 companies have completed the required questionnaire for evaluation and acceptance in the MOP's approved vendor list (AVL).

Positive and Negative Experiences, Issues, and Challenges

Implementation of local content policies and laws resulted in significant improvements in localization of petroleum operations and required goods and services. Based on some internal reports of the Ministry of Petroleum, the latest levels of localization in recent field development projects are as follows:

- Exploration and G&G: about 70%
- Drilling and completion:
  - Onshore and offshore drilling engineering: almost 100%
  - Onshore and offshore drilling and services: more than 80%
  - Goods and equipment: about 50%
  - Logistics: almost 100% for onshore and 70% for offshore
- Pipeline and flow line engineering: almost 100% for onshore and 70% for offshore

- Production units and processing facilities: more than 60%

Unfortunately, these estimates seem to be very rough, and in the writer's opinion can be viewed as educated guesses.

On the negative side, lack of a centralized monitoring or regulatory body is the main challenge for local content development, as there is no centralized oversight of local content within Iran's petroleum industry.

Also according to another internal report of the MOP, statistics of petroleum-related procurement of goods and equipment for the period of 1980 to 2010 are as follows:

- Procurement of 70,000 units of local goods and equipment, with a value of about US$5 billion
- Procurement of about 634,000 units of foreign goods and equipment, with value of more than US$21 billion

These statistics show that despite a significant growth in procurement of local goods and equipment, the ratio is about 25% local procurement, which is still very low.

According to a report provided by SIPIEM, about 52% of the capacity of local manufacturers in the petroleum industry is unused (fig. 11–15).

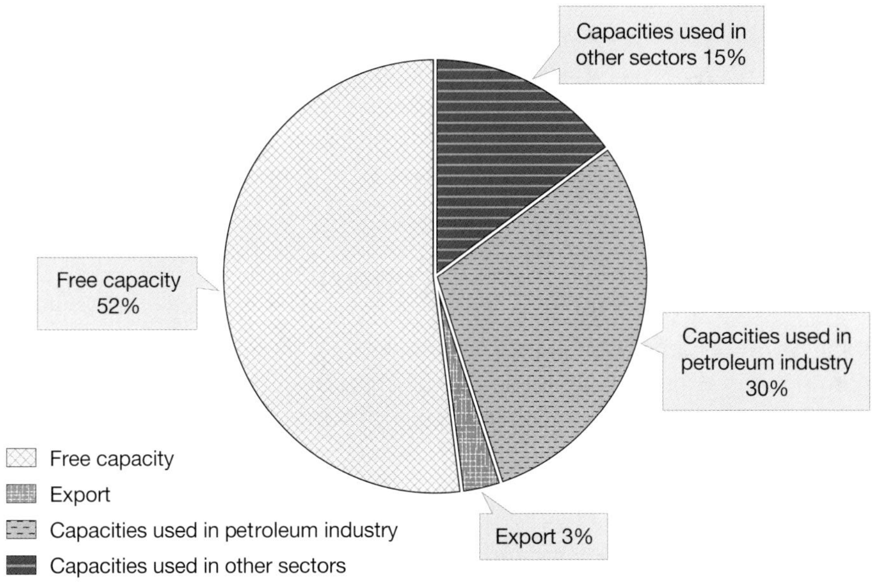

Fig. 11–15. Local petroleum industry manufacturers' capacity

Also, there several deficiencies related to local content laws. For example, there are certain issues and challenges in the tender law, mainly in its implementation. Some of these challenges are listed below:

- Lack of provision of AVL by related associations and their approval
- Noncompliance of private companies to tender law
- Inexact general and technical qualification criteria
- Unjust differences between local and international contracts in bid and contract stages, such as full compliance of local companies with bid documents versus conditional proposals by international bidders

The main challenge for international oil companies under the new petroleum contracts would be collaboration with nonexperienced or less-experienced local E&P companies in their joint venture companies. Less challenge is posed to the second and third tiers of the E&P supply chain network because of previous experiences with buyback contracts. More challenges would be expected for the fourth tier of the network for implementation of the STI mode of Iran's petroleum industry catch-up strategy with the IOCs. Local universities are not expert in handling bid research programs, and the NIOC and local E&P companies are not familiar with the STI mode and are DUI-mode dominant.

Lack of a supportive law for local E&P companies remains a challenge, along with the need for clear-cut definition and measurement of local content. Further challenges remain with changing the mindset of the NIOC and local E&P companies to support local companies by making trade-offs with project delays or lower quality. These issues and challenges will affect development of local companies in all four tiers in upcoming years.

Conclusion

Local content development before the Islamic Revolution of 1979 was mainly focused on human resource training, to allow successful transfer of the country's petroleum operations from international companies to NIOC's national employees in 1979. Nonetheless, in regards to equipment manufacturing and oilfield service companies or EPC companies, no outstanding achievements were attained. One of main reasons was lack of planning and supporting laws and policies for local content development in the mentioned areas.

During the Iran-Iraq War, some inevitable attention was directed toward local E&C companies, but not as a result of planning. After the war and during the post-war development era of country in the early and mid-1990s, many investment projects in the oil industry prepared the ground for local companies to flourish. This was facilitated and supported by policy making for local content utilization in petroleum laws and other related laws. Buyback contracts followed. Despite many efforts during the sanction era for using local companies, no comprehensive local content development plan was prepared for the petroleum industry.

With the announcement of the Resistance Economy policy, concurrent with the postsanction era, and the approval of the new Law of Duties and Authorities of the MOP in 2012, a new age of localization began. The entire petroleum industry was bolstered by implementation of several mega initiatives. These initiatives targeted a simultaneous STI × DUI mode of catch-up strategy to boost both know-how and the knowledge and technological aspects of local content development. The results of these new regulations can be evaluated when new investments in the country have been made, mainly under the new petroleum contracts or previous third-generation buy-back contracts.

Notes

1. Iranian Oil Exploration and Production Company. In April 1965, members of the Iranian Oil Consortium set up two companies called the Iranian Oil Exploration and Production Company (IOEPC) and the Iranian Oil Refinery Company (IORC), which were together named the Iranian Oil Operating Companies (IOOC).
2. Ayatollah Seyed Ali Khamenei, "Announcement of Resistant Economy Policies," 2014, http://www.leader.ir/fa /content/11480/.
3. Later in 1935, APOC was renamed to Anglo-Iranian Oil Company (AIOC), and in 1954 it became the British Petroleum Company (BP) after Iran's oil nationalization.
4. John Williamson, *In a Persian Oilfield: A Study in Scientific and Industrial Development* (London: Ernest Benn, Ltd., 1927); emphasis added.
5. "History," Research Institute of Petroleum Industry (June 7, 2014), http://www.ripi.ir/index.php/en/about/history.

12

KENYA

Berryl Claire Asiago

Introduction

At the inception of the oil industry, local content policy was not a distinct and separate objective but instead was included as a means to improve the industrial policy of a given economy. This remained the case throughout the implementation of productive development policies (PDPs) during World War II and in the years that followed.[1] From a general economic strategy, local content policies were referred to as *import substitution policies/industrialization* or ISP/I, which encouraged the development of the domestic industry and the elimination of foreign goods and services through exchange control barriers, such as tariffs and quotas, exchange rate policies, and fiscal and credit policies.[2]

Presently, the traditional concept has changed with consideration to sector-specific requirements. First, investors and states acknowledged the need to ensure coherent, steady growth in a industry. Second, there is now a global recognition that different countries and industries are at different stages of business and economic development. Today, the local content concept is argued to have two simple but powerful objectives: to create jobs at home rather than abroad, and to channel business to domestic firms rather than foreign firms.[3]

Even though exploration for oil in Kenya began in the early 1950s, only recently has that oil been declared commercially viable. As a result, Kenya is now considered a hot spot in petroleum exploration, offering some of the last frontiers in the industry.[4] With Kenya's commercial oil production set to begin in 2017,[5] interest from international oil companies and service producers has increased exponentially.

With the country's oil exploration only at the inception stages, the Kenyan government—as with many developing countries—is grappling with the problem of outdated legislation and policies governing oil exploration and production.[6] Like many other countries with exhaustible resources, the Kenyan government wishes to benefit from such resources. This is evidenced by the introduction of government policies and laws aimed at maximizing the resource benefit to the government and its people. One area subject to revisions of the law and adoption of new policies concerns the promotion of local content in the oil and gas sector through legally enforceable mechanisms.

Pertaining to local content, Kenya's focus has been aimed at the upstream subsectors. According to the definition, *local content* involves "the participation and development of national labor, goods and services, technology, and capital."[7] At the very least, a measurable quantity or percentage of materials, personnel, or financing rendered in the petroleum industry value chain must be sourced locally. The objective therefore is to encourage and promote the participation and development of domestic industries and local labor. The provisions seem to be directed at mainstreaming local participation in upstream petroleum activities in Kenya.

Local Content Policy and Baseline

The promulgation and enactment of the 2010 Constitution of Kenya substantially transformed the governance structure of the country and subsequently enhanced participation of citizens in the administrative decision-making processes.[8] This was accomplished against the background of attaining economic, social, and cultural development as envisioned under the country's Kenya Vision 2030 development program for the years 2008 through 2030.[9]

As a result of the recent discovery of oil, the assumption is that the development of the resource will accelerate and contribute enormously toward attainment of the Vision 2030 goals. These goals can only be realized if proper policies, laws, and strong institutional frameworks are put in place to protect against the ailment of the "resource curse." Therefore, challenges remain to formulate effective laws within the petroleum sector that are still coherent enough for stakeholders to understand.

Such coherence will be necessary to ensure the growth of the industry both locally and internationally.

Currently the Kenyan petroleum regime offers a broad spectrum of laws, including Kenya's Constitution and various energy policy legislations and bills. Pertinent legislation includes the Energy Bill 2014, Local Content Regulations-EB, Local Content Regulations-PB, Model PSC 2014, National Energy Petroleum Policy, National Energy Policy 2014, and the Petroleum Exploration and Production Bill 2015, among others.[10] Notably, most of the bills are at the National Assembly stage, except for the Petroleum Exploration and Production Bill, which currently is in the final stages of review in Kenya's Senate.

Policies

The 2004 energy policy paper[11] acknowledged that there is inadequacy of human resources within the energy sector. It called for concerted efforts from stakeholders toward human resources development and capacity building in the energy sector. It decried the glaring deficiency of local specialized research and consultancy services in the sector and the lack of specialized training programs in institutions of higher learning. The government proposed to address this deficit through a forward looking and dynamic capacity-building program consistent with emerging needs in the energy sector. Further, the government made commitments toward cooperating with industries, universities, and other tertiary institutions of learning to train local professionals in the sector. From the spirit and letter of the policy the government was evidently committed to building local capacity in the energy sector.

Additionally, in 2014 the Ministry of Energy and Petroleum published the Draft National Energy Policy Bill,[12] acknowledging the need to develop petroleum capacity in the country. Like its predecessor, the bill equally recognizes the enormous challenges the country faces concerning local content development. It highlights the inadequacy of skilled manpower, technical capacity, and local suppliers of services and goods required in oil exploration. It decries the inadequate mechanisms for technology and skills transfer to locals. The government is committed to promoting cooperative partnerships and to creating effective regulations to achieve local content goals in the country. Recommendations for local content provisions also included efforts to ensure that supply bottlenecks are mitigated.

Baseline reports

Final Report on Legal and Regulatory Guidance. Against the backdrop of oil discoveries, the government, in partnership with the World Bank, contracted two consultant companies to conduct a review of the existing oil and gas laws and regulations and offer recommendations on areas for improvement.[13] There was an urgent need to review and update oil and gas laws and policies to reflect the new developments in the country. The recommendations for the Kenyan legal framework, as revealed in the July 2013 consultants' report, Final Report on Legal and Regulatory Guidance, state that local content regulations must be "harmonized, gradual and forward-looking."

The report states that for local content regulations to work, Kenya must go the route of Nigeria and enact specific legislation concerning local content development and capacity building in the upstream petroleum sector. However, this recommendation came with a caution that as a country, the government must be wary of the fact that Nigeria has a long history in oil exploration and production. Nigeria has, over time, built a wealth of both institutional and human resources capacity to be able to implement the local content regulations of 2010 within the stipulated times. The report further recognizes the role of cooperation between oil industries and universities toward the enhancement of local capacities in the sector. With government efforts to promote local content in its policies and regulations, the inherent protection of local firms must be temporary. There also must be a mechanism to ensure the effective phasing out of protectionism. According to the report, this standard is in compliance with international standards in the oil and gas sector. Developing local capacities of domestic firms during the period of protectionism and later exposing domestic firms to international competition increases and promotes their ability to compete internationally in the future. This is the reason the report insists that for local content regulations to succeed, they must be forward-looking so that the national industry and employees are prepared to face international competition. Further, in the long run, the report recommends that the protection be gradually removed as dynamic industrial and technological development take center stage.

The Petroleum Technical Assistance Project (KEPTAP) in Kenya

The KEPTAP project, funded by the World Bank, aims to strengthen the capacity of the Kenyan government to manage its petroleum sector for sustainable development impacts. Kenya's Ministry of Energy and Production (MEP) sent a request for technical assistance funded by the Extractive Industries Technical Advisory Facility (EI-TAF) on June 22, 2012. The World Bank approved the project on July 24, 2014 and expects the project to close on February 28, 2021.

The project is mandated to assist Kenya to build the required capacity at various levels to enable the country to play active and leading roles in the development and management of the petroleum sector once production and utilization of the discovered oil and gas is on stream.

KEPTAP focuses on informing the government about any gaps in the legal, regulatory, and management framework for all sectors involved in exploration and production of oil. Such matters include environmental and social aspects of the upstream petroleum sector, where the Strategic Environmental and Social Assessment (SESA) was carried out by the National Environmental Management Authority (NEMA) on behalf of KEPTAP.

KEPTAP primarily concentrates on public administration, with 40% of its focus on energy and mining, 40% on oil and gas, and the remaining 20% on other mining and extractive industries. It is intended to help Kenya better manage oil and gas developments and revenues to invest in lasting growth and development of the country. Oil and gas should be able to create sustainable growth across all areas of the country's economy. Thus there is a need to strengthen the government's capacities in the sector to avoid mismanagement and the infamous resource curse.

The World Bank states:

> In order to help stimulate economic growth in Kenya, the project will promote petroleum activities to contribute to fiscal and foreign exchange revenues. It will also support entrepreneurial activities by improving the investment climate for the private sector and enhancing the oil and gas legal and institutional framework.
>
> In addition, the project will support the drafting of key policy and planning documents as well as capacity building among existing government institutions and clarification of their

roles and responsibilities. There is significant emphasis on training so government staff is well equipped to deliver the expected outcomes. The project will also increase vocational training capacity for the oil and gas sector of Kenya, therefore increasing the availability of trained staff to the private sector.[14]

Key Players

The host government and its agencies

The Ministry of Energy and Petroleum (MEP). The Constitution of Kenya (2010) under Article 69 (1)(h) states that "[t]he state shall utilize the environment and natural resources for the benefit of the people of Kenya." Thus the utilization of natural resources, which includes minerals and fossil fuels, must be for the benefit of all the citizens of the Republic of Kenya. The ministry involved must endeavor to undertake policies toward implementation of this constitutional provision.

Under the 1984 Petroleum (Exploration and Production) Act (Chap. 308), which was last revised in 2012, the minister for Energy and Petroleum is empowered to sign petroleum contracts. (*Note:* the Constitution of Kenya has changed this title from "minister" to "cabinet secretary.") *This is outlined under Section 4 (3)(b), where the cabinet secretary is empowered to sign petroleum agreements on behalf of Kenya. Further, the 1984 Petroleum Act, as revised, required the cabinet secretary to make a model agreement available to potential contractors.*

These duties of the cabinet secretary are retained under the 2015 Petroleum Bill, with additional responsibilities that include conducting any investigations, carrying out due diligence, or undertaking consultations before entering into a petroleum agreement. Further functions of the cabinet secretary under Section 10 of the 2015 Petroleum Bill include the following:

- Determining terms of production in a PSC
- Engagement with stakeholders
- Creating an environment well suited to oil and gas exploration through reference to international best practices in designing new legislation in relation to petroleum agreements, whether new or existing

- Supervising upstream petroleum operations carried out under a petroleum agreement
- Developing, publishing, and reviewing upstream petroleum policy and strategic plans
- Reviewing and approving any proposed exploration activity contained in the annual work program, appraisal program, and production forecasts submitted by a contractor
- Reviewing and approving budgets submitted by a contractor
- Upon advice of the Advisory Committee, suspending, revoking, or terminating a petroleum agreement or recalling the security therein on behalf of the government as provided for under the act, its regulations, and the petroleum agreement
- Approving transfer or assignment of any interest in a petroleum agreement in accordance with the provisions of this act and its regulations
- Taking any action or decision, or giving any permission or consent, or exercising any other control as may be necessary or desirable as provided for in the act, its regulations, and the petroleum agreement
- Approving field development plans upon advice by the Petroleum Authority
- Conducting bidding rounds and direct negotiations for granting of a petroleum agreement in accordance with the act and its regulations

The cabinet secretary's role in the granting of petroleum agreements is the most critical governance and regulation area in the upstream petroleum sector. This is because the agreements govern issues of taxation, local content, and production sharing contracts between the host government and the contractor. These issues form the backbone of government engagement with international oil companies.

The Energy Regulatory Commission (ERC). The ERC was established by the 2006 Energy Act and is responsible for regulating the energy sector in Kenya. As the regulator, the ERC approves operational permits, enforces compliance, and has wide-ranging powers to "issue, renew, modify, suspend or revoke licenses and permits for all undertakings and activities in the energy sector." This includes the upstream, midstream, and downstream petroleum sectors.

At the moment there is duplication of roles, since the cabinet secretary in the Ministry of Energy and Petroleum has regulatory powers in the upstream sector as well. This duplication of roles curtails the ERC's regulatory power and inhibits its ability to act as a check and balance on the ministry's decisions. Further, the ERC's independence is compromised because the president of the Republic of Kenya is authorized to appoint its chair, and the cabinet secretary of the Ministry of Energy and Petroleum appoints all other commissioners, including the one from the ministry.

Nevertheless, the ERC, as established under the 2006 Energy Act, has regulatory oversight over the entire energy sector. With the duplication of roles, there emerged the necessity to establish a specific body to regulate the upstream petroleum sector, especially with the recent oil discoveries in Kenya. This led to the anticipated formation of the National Upstream Petroleum Authority, as stipulated under the Petroleum (Exploration, Development and Production) Bill of 2015. The Petroleum Authority's role is limited to the upstream sector, which means that the ERC will halt its coverage of upstream operations once the Petroleum Bill is passed. This move was informed by the challenges of managing the upstream petroleum sector, which requires specialized knowledge, attracts vital international investments, and involves multifaceted regulatory systems and legislations.

The National Oil Corporation of Kenya (NOCK). The National Oil Corporation of Kenya is a state corporation founded by legislative act in 1981, with a mandate of participating in all aspects of the Kenyan petroleum industry. It is 100% owned by the Kenyan government. NOCK's initial role was to spearhead exploration on behalf of the Kenyan government. This remains a key role, but since 1997, it has also built up a retail business and today controls around 5% of the retail market for petroleum products in Kenya. It is a limited liability company incorporated by the government of Kenya for the development of E&P in the oil and gas industry. The company acts on behalf of the government in the coordination of exploration activities. It operates both upstream and downstream.

Previously, NOCK has acted as both a participant and policy maker in the upstream petroleum sector. Whether this creates conflicts of interest and favoritism is subject to debate, since all minerals belong to the government, and the government is entitled to exclusive rights of minerals extraction. NOCK represents and holds the government's interest in all petroleum operations. Previously it performed both regulatory and commercial roles, but this has been limited in recent times. It also formerly conducted policy review in the energy sector and made recommendations to the MEP, but these roles created potential conflicts of interest. Its principal role is now that of a mainstream oil company taking part in oil and gas exploration. The

Petroleum Bill of 2015 exclusively accords the MEP's cabinet secretary with a policy-making role in the upstream petroleum subsector.

Despite being in existence for decades, there is evidence of limited capacity in NOCK to harness the full potential of the hydrocarbon basins in the country. Thus, in order to fulfil its function as the commercial arm of the government in petroleum activities and to make recommendations to the ministry, NOCK's internal capacities must be upgraded. This lack of capacity is a concern highlighted in the 2014 Draft National Oil Policy.

There are fears that NOCK continues to have a strong influence on policy making, while at the same time being an industry player itself. This obviously creates an undesirable conflict of roles and interests. The consultants contracted by World Bank in their 2013 report recommended that NOCK be split to separate its regulatory affairs from upstream oil exploratory activities.[15] This has yet to be considered by the government.

Oil and gas companies operating in the country

The key roles held by the industry players in the Kenyan upstream petroleum industry include employment, skills and technology transfer to locals, social development through corporate social responsibility, and remittances of revenue to the government. Employment contributes directly to transfer of wealth to locals. It also indirectly contributes to social development through creation of an environment that fosters education and engagement of alternative means of earning livelihoods.

Skills and technology transfer to locals is gained from training, local sourcing of labor, and educational scholarships to locals provided by upstream petroleum industry players. These industry players provide the market with skilled personnel for absorption into the sector, and in so doing, they promote development of the economy, provide employment opportunities, and consequently raise the living standards of the locals. Societal development is achieved through the various corporate social responsibility activities in which the industry players engage. These activities include building of classrooms and roads, donations promoting education and clean water, creation of awareness on pertinent societal issues, and promotion of renewable energy. Remittance of revenue to the government through taxes provides more resources to the government for provision of better services for its citizens. This, in turn, enables the building of schools, roads, rail networks, hospitals, and providing other core government utilities to locals.

Tullow Oil. Tullow Oil is a leading independent oil and gas E&P group. Tullow Oil first acquired interests in the rift basins of Kenya in 2010. To date, the IOC has discovered an estimated 600 million barrels of oil resources in the South Lokichar Basin in Turkana County.[16]

Tullow Oil has continually stated that it endeavors to find ways to maximize participation of local businesses within its supply chain to ensure the oil and gas sector creates a positive impact on the economies and people in local communities in which it operates.

Tullow Oil local content policy. Tullow Oil's local content policy, titled "Shared Prosperity," focuses on provision of locals with business opportunities, local employment, and capacity building in its local content commitments. It states that the company intends to "[c]reate a positive and lasting contribution to economic and social development in the communities and countries where Tullow operates."[17]

Tullow's summary of its local content policy includes a discussion of the national employees from the countries in which it operates, such as Uganda, Kenya, and Ghana. The policy is categorical that the ultimate aim of shared prosperity is accomplished through commitment to local sourcing of goods, services, and expertise, and by building up local capability, education, and knowledge and skills transfer so that local people benefit from development of natural resources.

The impact of Tullow Oil. Tullow Oil reports that the company has been able to make the following impacts from implementation of the local content policy:

1. The policy states that 83% of staff are nationals in the African countries where Tullow operates. This creates employment and ultimately improves lives of locals and promotes economic growth of the host countries.
2. Across Ghana, Uganda, and Kenya, a total of US$225 million has been spent with local suppliers. Ultimately, this local sourcing of goods and services is one of the cornerstones of local content requirements.
3. It estimates that there has been a total economic impact of US$1.4 billion across the 22 countries of operation since its inception. This underlies its commitment toward contribution to economic and social development.

4. Tullow is committed to social investment to promote Turkana's long-term sustainable development, namely water, education, health, environment, and alternative livelihoods.
5. Partnerships with local companies have helped the local companies to gain experience and skills, which in turn enables the host country to be able to promote its products internationally.

Challenges faced by Tullow Oil. On local content issues, Tullow Oil has encountered numerous challenges, especially politically instigated demonstrations and unrest related to issues of local content and allegations of shortchanging locals. Furthermore, acute shortages of skilled staff and professionals in the field of oil and gas, especially exploration and production of oil, have presented additional challenges. One way Tullow is addressing these challenges is through its annual Tullow Group Scholarship Scheme (TGSS), which in 2015 was in its fifth year.[18]

Africa Oil Kenya (AOK). Africa Oil Kenya (AOK) is one of the leading companies in Kenya involved in oil exploration activities. It operates Block 9 in the Cretaceous rift system in northern Kenya between Marsabit in the west and Wajir in the east. AOK is involved in a JV partnership with Tullow Oil as operators in the East African Tertiary Rift trend in the northwest part of the country around Lake Turkana.[19]

Most of the AOK's local content work is conducted through its partner, Tullow Oil.

Baker Hughes. Baker Hughes is an oilfield services provider focusing on basic oilfield infrastructure, such as cement pouring and laying cables. In terms of employment, Baker Hughes employs two-thirds expatriate and one-third non-Turkana nationals. It does not report any semi-skilled positions in its structure.

Schlumberger. Schlumberger is the world's largest oilfield services company, with an almost identical employee breakdown as that of Baker Hughes. Its functions include a mix of services in directional drilling, fluids management, drilling, fishing, and well testing. They report 9 semi-skilled positions and no unskilled jobs.[20] Of Schlumberger's workforce of 119, all are Kenyans, though none are from Turkana district.

Service providers operating in the country

Twelve subcontractors reported no unskilled positions with their firms, including the following:

- KK Security
- Baker Hughes
- Citiscape
- Earthview Geoconsultants
- Freight Forwarders
- Kurrent Technologies
- New Edge
- Spica Marine Inspections
- Alpha Logistics
- Halliburton
- Schlumberger
- Transeast

These companies cover all oilfield services from providing cranes, forklifts, and catering services to inspections, telecommunications, and security services for oil-producing companies.

Of the seven companies with more than 50% Turkana employment, Turkana employees generally fall into three categories:

- Security (ROLSS Tullow, KK Security, and Newport Africa)
- Camp maintenance and catering (AFEX Group and Ardan)
- Direct petroleum-related services (Global Geophysical Services Inc. and the BGP, or the China National Petroleum Corporation).

Industry Knowledge

Research and training institutions

The Petroleum Institute of East Africa (PIEA). This is the professional body for the oil and gas industry in East Africa. The mission of the PIEA is to provide a forum for expertise and excellence in the oil industry and to promote professionalism and free enterprise in the petroleum business, supported by the highest operations and business standards.[21] The School of

Petroleum Studies was incorporated in 2007 as a wholly owned subsidiary of the PIEA, with the prime objective of offering first-class specialized training through a curriculum focused on downstream and upstream oil and gas operations.[22]

The School of Petroleum Studies offers longer diploma courses that are designed to equip persons already working in the oil sector, or those aspiring to work in the oil sector, with broad knowledge of the oil industry. Diploma courses are designed to be offered part-time to benefit working personnel. The school also offers short five-day courses geared toward meeting specific knowledge and skill needs in specialized areas.[23]

The University of Nairobi (UON). UON has a department of geology[24] with a four-year geology degree program in which the students cover a total of 52 course units. It offers a bachelor of science degree in petroleum geosciences, and master's and doctoral degrees in geology.

South Eastern Kenya University (SEKU). This is a public university located in the eastern part of Kenya. It offers a bachelor of science degree program in geology, which is a full-time, four-year undergraduate course. It teaches mineralogy, petrology, paleontology, economic geology, structural geology and tectonics, hydrogeology, exploration geophysics, exploration geochemistry, and stratigraphy, among other subjects.

Technical University of Kenya (TU-K). TU-K offers diplomas in geological science. In July 2009, "the university college had the first batch of the programme of study approved by [the] Senate at the University of Nairobi."[25]

Legislative Hydrocarbon Regime Concerning Local Content

The existing regulatory bodies

The Ministry of Energy and Petroleum. The MEP has responsibility for formulation and articulation of policies governing the sector, along with responsibility for the legal functions of the negotiating and licensing process. The MEP is also responsible for providing an environment that benefits all stakeholders and for mobilization of resources (both human and financial). Under current legislation, the cabinet secretary is the chief authorizing officer. The cabinet secretary is in charge of policy and legal functions, as well as licensing and monitoring of operations in accordance with the law and any signed petroleum agreements. The cabinet secretary is empowered to sign petroleum agreements on behalf of the government. In a nutshell, the MEP is involved in all regulations, licensing, and policy making in the Kenyan energy sector. This includes the upstream petroleum sector.

Policies and regulations made by the ministry include the following:

- The *Sessional Paper No. 4 on Energy* (2004), also called the National Energy Policy
- The Petroleum Law of 1984
- The Model Production Sharing Contract (of Kenya)

The National Fossil Fuels Advisory Committee (NAFFAC). NAFFAC is the licensing arm of the government. It was created to assist the cabinet secretary of the MEP in negotiations with would-be exploration contractors. The NAFFAC is comprised of the MEP's principal secretary as chairperson and NOCK's managing director as the secretary. The NAFFAC's other members are the attorney general, principal secretary of the National Treasury, commissioner of petroleum, the MEP's chief geologist, the director general of the National Environment Management Authority (NEMA), and the commissioner of the Kenya Revenue Authority.

However, the NAFFAC is expected to be abolished because the proposed law to govern the upstream petroleum industry does not refer to it. The 2015 Petroleum Bill places the licensing function with the MEP's cabinet secretary, without any reference to the NAFFAC. At the moment, PSCs are negotiated by the NAFFAC and signed by the cabinet secretary for the MEP.

National Oil Corporation of Kenya. As previously discussed, the National Oil Corporation of Kenya is a state corporation founded by an act of Parliament in 1981, with a mandate to participate in all aspects of the Kenyan petroleum industry. It is 100% owned by the Kenyan government. NOCK was established in the 1980s to spearhead exploration on behalf of the Kenyan government. This remains a key role, but since 1997, it has also built up a retail business, and today controls around 5% of the retail market for petroleum products in Kenya.

The proposed bodies under the Petroleum (Exploration, Development, and Production) Bill of 2015

The National Upstream Advisory Committee (NUAC). The 2015 Petroleum Bill proposes to appoint the NUAC as an advisor to the cabinet secretary of the MEP on matters of negotiation, signing, or revoking of petroleum agreements. It is also meant to advise the cabinet secretary on suspension, revocation, or termination of petroleum agreements or on the recalling of the security therein on behalf of the government, as provided for under the Petroleum Bill 2015.

This committee will take the entire role of the National Fossil Fuels Advisory Committee (NAFFAC). The primary roles of the committee are as follows:

1. Participating and advising the cabinet secretary in the negotiation of petroleum agreements
2. Granting and revoking licenses
3. Submitting a report to the cabinet secretary on the terms negotiated with contractors
4. Advising the cabinet secretary on upstream petroleum operations
5. Advising in the evaluation of the bids and applications for awarding petroleum blocks
6. Conducting all due diligence and investigating all the affairs of contractors prior to entering into petroleum agreements
7. Advising the cabinet secretary on the grant of nonexclusive exploration permits under which a person may enter an area to prospect and/or carrying out geological, geochemical, and geophysical surveys as may be provided in the permit

The National Upstream Petroleum Authority (NUPA). NUPA will be the regulator of upstream petroleum operations. The proposed functions of the Petroleum Authority include the following:

1. Regulating upstream petroleum operations in Kenya
2. Providing such information and statistics to the cabinet secretary as required from time to time
3. Collecting, maintaining, and managing upstream petroleum data

Unlike the ERC, which has regulatory oversight on the entire energy sector, the proposed NUPA's role is limited to the upstream petroleum sector. This means that the ERC will halt its regulation of upstream operations. Since the upstream sector is very challenging and requires specialized knowledge and close attention, the practice in oil-producing countries has been to establish an industry regulator to oversee its operations. The proposed body in the 2015 Petroleum Bill is therefore consistent with modern trends in petroleum regulations.

Local Content Development and Monitoring Unit. The Production (Local Content) Regulations Bill of 2014, if enacted, will establish the Local Content Development and Monitoring Unit.[26] The mandate of the unit will be to monitor, coordinate, and implement the provisions of the regulations concerning all matters on local content development. The formulation will be through the MEP.

Relevant policies, legislative framework, and key provisions

The Kenyan laws and ministerial policies focusing on the upstream petroleum sector since independence include the following:

1. The Constitution of Kenya (2014)
2. The Petroleum (Exploration and Production) Act, Chap. 308, of 1984 (revised 2012)
3. *Sessional Paper No. 4 on Energy* (2004), or the National Energy Policy
4. The 2014 Draft National Energy Policy
5. The Petroleum Regulations (Training Fund)
6. Petroleum Exploration, Development, and Production (Local Content) Regulations, 2014
7. The Petroleum (Exploration, Development, and Production) Bill of 2015, which will replace Chap. 308, Petroleum Act of 1984, as revised.

The Constitution of Kenya. Article 62 classifies all minerals and mineral oils as public land. This means that minerals and mineral oils belong to the national government and are held by the state in trust for the people of Kenya. Under customary international law, state sovereignty over natural resources is sanctioned, and the resources are regarded as the state's assets. This is completed through the process of compulsory acquisition where, under Article 40 (3), such acquisition must be for a public purpose or in the public interest and carried out in accordance with the law. Basically, this means that a just compensation must be paid.

The Petroleum (Exploration and Production) Act (Chap. 308) of 1984 (revised 2012). The Petroleum (Exploration and Production) Act (Chap. 308) is the principal statute governing petroleum exploration in Kenya.[27] It vests the ownership of hydrocarbons to the government while granting powers over the sector to the cabinet secretary of the Ministry of Energy and Petroleum (MEP). This was enacted to govern and regulate petroleum agreements relating to exploration for development, production, and transportation of petroleum. Hence it covers all aspects of the value chain from upstream to downstream. Since it was enacted at a period when the oil discoveries were not yet made, it does not exhaustively cover pertinent issues within the industry. It does, however, touch on local content development in the country by expressly requiring establishment of a training fund, preference in employment and training of Kenyan nationals, and also the use of locally available goods and services in petroleum operations.[28]

The government is the licensor to the oil and gas companies and is in charge of regulating and monitoring the sector. Revised in 2012, it is meant to regulate the negotiation and signing by the government of petroleum agreements relating to exploration, development, production, and transportation of petroleum. Under Section 5, the act empowers the cabinet secretary with regulatory and supervisory powers in all petroleum agreements in the upstream petroleum sector.

In this regard, under Section 9, the act provides that in every petroleum agreement, there will be an express term that the contractor shall give preference to the use of products, equipment, and services locally available. This is in line with the internationally accepted principle of promoting local content, as further reinforced under Section 11, which established the Training Fund to facilitate training of Kenyan nationals in petroleum operations.

The Petroleum (Exploration and Production) (Training Fund) Regulations of 2006. The regulations came into force in July 2006 and have the sole purpose of providing for the Training Fund to be utilized solely to finance the training of Kenyans in petroleum operations. Under Clause 5 of the regulations, funding will originate from the contractors' training contributions as stated in their respective petroleum agreements.

The 2006 Training Fund Regulations, only three pages, do not address how specifically the funds will be utilized in the anticipated training of locals on petroleum activities. It is not clear whether aid will be offered in the form of grants, loans, or scholarships. There is no framework or mechanism concerning how the funds are to be utilized. Consequently, it leaves huge discretionary powers to the cabinet secretary. In theory, this exposes the fund, if it even exists, to the risk of mismanagement, misuse, and misappropriation, especially with the prevalent corruption in the Kenyan public sector.

The Petroleum (Exploration, Development, and Production) Bill of 2015. The proposed Petroleum Bill of 2015 is intended to provide a framework for the contracting, exploration, development, and production of petroleum. If enacted, it will replace the 1984 Petroleum Act and its revisions. It seeks to create tangible bodies to ensure effective administration, regulation, and monitoring of the growing industry in Kenya. It creates the National Upstream Petroleum Advisory Committee (NUPAC),[29] which is an interministerial committee with the role of ensuring contractors uphold the relevant laws, regulations, and petroleum agreement terms. Further, it establishes the Upstream Petroleum Regulatory Authority (UPRA), whose functions will generally include regulating the upstream petroleum operations in Kenya.[30]

It defines local content as follows:

> [T]he added value brought to the Kenyan economy from petroleum related activities through systematic development of national capacity and capabilities and investment in developing and procuring locally available work force, services and supplies, for the sharing of accruing benefits.[31]

The definition emphasizes all aspects of local content that include preference to services and goods produced locally and using qualified and skilled local personnel.[32] Further, it requires that the contractor maintain a local content plan detailing the long-term and annual strategies to implement the local content requirements.[33] Such a plan must include strategies on employment and training of locals, areas of research and development, technology transfer to locals, industrial attachment openings, legal and financial services required, and succession plans for positions not held by Kenyan nationals. Further, it obligates the cabinet secretary to make specific local content regulations.[34]

According to Sections 77 through 79, two objectives appear pertinent and crucial in successfully conducting operations in Kenya. First, an investor must give priority to domestic goods or services as long as they meet the specifications of the petroleum industry as prescribed by the Kenya Bureau of Standards. In the absence of a Kenyan standard, any other internationally acceptable standard that the authority will approve may be used. Second, an investor must ensure that priority is given to the employment or engagement of qualified and skilled Kenyans at all levels of the value chain.

It is clear that Kenya has taken the route of national content rather than community content. This is apparent given the consideration accorded to employment of personnel, hiring of petroleum-related services, and purchasing of goods manufactured in any part of Kenya, as long as specifications of the petroleum industry are met. There is no immediate necessity for an investor to utilize services or goods from the local community, which is common in some countries with oil and gas activity. Nevertheless, the concept of social license from the local community cannot be disregarded, in that success of a project—without unwanted interruptions—solely depends on community consent. Hence investors will have to review their strategies on how to incorporate community content to any extent possible. This is particularly true given the existing devolved government that tends to concern itself with immediate community interests rather than national interests.

While reviewing Kenya's local content policies, it may be helpful to evaluate them against those of Tanzania. In Tanzania, the government considers that in order for local content requirements to be successfully met, hiring and purchasing must include both regional as well as local content in oil and gas activities. Presently, Tanzanian local content policies mandate that an investor incorporate personnel, goods, and services from both regional and local community perspectives.

Furthermore, an investor will be required to submit both short- and long-term (annual) content plans, which are subject to approval by the Petroleum Authority (who will be appointed by the ministry). These plans should address a number of issues, such as: (a) employment and training, (b) research and development, (c) technology transfer, (d) industrial attachment and apprenticeship, (e) legal services, (f) financial services, (g) insurance services, and (h) succession plans for positions not held by Kenyans.

The bill further mentions that the cost of local content hiring or purchasing should not be higher than that of any other place. This is a concept that will be very difficult to attain, since the manufacturing industry in Kenya is still at inception stages. As a result, achieving certain percentages of labor, goods, and services may not be feasible given the country's limited industrial base. Also, where the local supply issues involve equivalent goods, the price does indeed become predominant issue. In many other instances, however, it may just be secondary to the quality factor. Nevertheless, the bill requires that such specifications be in line with Kenya's standards or internationally acceptable standards. Since the country has yet to develop the industry, it is obvious that the standards being referenced are not yet in place. This creates a bottleneck that may encourage investors to circumvent the law ("creative compliance") and utilize international standards. Use of international standards is likely to favor foreign goods as opposed to local goods, making the objective of local content completely obsolete. Hopefully through its mandate, the ministry will make local content regulations that foresee the loopholes and avoid creative compliance.

The Petroleum Exploration, Development, and Production (Local Content) Regulations of 2014. The Local Content Regulations Bill of 2014, if enacted, will be the principal statute governing all local content concerns. This translates to employment of Kenyan citizens and local contracting for goods and services required in petroleum operations in the country. Further, the provisions demand short- and long-term plans to ensure skills and technology transfer through deliberate prioritizing of qualified Kenyan suppliers and employees to guarantee that they eventually are able to replace foreign companies or employees in the industry. The regulations further set the minimum local content levels for all contractors in petroleum agreements within the Republic of Kenya.

The Petroleum Exploration, Development, and Production (Local Content) Regulations, of 2014, also referred to the Local Content Regulations, were adopted by the cabinet secretary of the Ministry of Energy. They are specific and forward-looking in securing the requirement for local content in the upstream petroleum industry. The regulations define local content as "the quantum or percentage of locally produced materials, personnel, financing, goods and services rendered in the petroleum industry value chain and which can be measured in monetary terms."[35] This definition emphasizes a quantifiable aspect of local content in the upstream petroleum industry. In this aspect, the regulations promote the participation of Kenyan citizens in petroleum operations and impose a duty to hire them preferentially.[36] The preference extends to employment and training of Kenyan citizens as a matter of first priority, and to services and goods manufactured within the country.

The regulations contain conditions as to filing of both long-term and annual local content plans with the Petroleum Authority established under the 2015 Petroleum Bill. The plan must include a research and development plan, succession plan, technology transfer plan, and a legal and financial services plan.[37] A local content plan submitted to the Petroleum Authority must ensure that first consideration is given to services provided within the country and goods manufactured in the country. This applies where the goods meet the specifications of the petroleum industry as established by the Kenya Bureau of Standards or by other internationally acceptable standards. The plan must also ensure that qualified Kenyans are given first consideration with respect to employment and that adequate provision is made for training Kenyans on the job.[38]

Further, the regulations specify the minimum local content levels that are consequently annexed, as per Clause 11.[39] These minimum levels specify concerns like the acceptable percentage of positions held by Kenyans in the different categories of employment positions within the contractor's business. Conditions are set for services like insurance brokerage and legal services, which must be provided by firms whose principal office is based in the country.[40] For financial services, the service provider must be regulated by the Central Bank of Kenya.[41]

On monitoring and measuring of compliance, a contractor must submit a "Local Content Performance Report" annually.[42] This report must list the achievements of the contractor in employment, training and technology transfer, and procurement of goods and services from within the country. The report will be assessed by the proposed Petroleum Authority for purposes of measuring compliance and monitoring and addressing emerging challenges regarding local content.

The minimum acceptable levels of local content are shown in the following:

1. Goods and services
2. Fabrication and construction
3. Materials and procurement
4. Well drilling services
5. Research and development relating to in-country services
6. Exploration, subsurface, petroleum engineering, and seismic services
7. Transportation, supply, and disposal services
8. Health, safety, and environment services
9. Information systems, information technology, and communication services
10. Marine operations and logistics services

For goods and services, the contractor must achieve 50% national content, as a minimum amount, within the first five years of operation in the country. On recruitment and training, the regulations split the area into three areas that include management staff, technical core staff, and other staff. Across the three classifications, the minimum level within five years of operation is 50%. For other staff, within 10 years of operation, the contractor must have achieved at least 100% national staff. On materials and procurement, there are different levels required for specific commodities, but within five years, the contractor must have achieved at least 50% national content.

Although there is great emphasis on promotion of local content through capacity building and creation of a strong workforce, surprisingly, the 2014 Local Content Regulations do not refer to establishing a training fund, which might very well promote the development of the concept. However, the concept has only been mentioned in the proposed 2015 Petroleum Bill under Section 88. This discrepancy displays the confusion and lack of consistency in the proposed legislations and regulations in the upstream sector of the petroleum industry. It begs the question whether or not the training fund is a key ingredient of local content requirements in Kenya. There is no clarity as to whether or not a training fund, as established by the 1984 Petroleum Act, was actually created. Contributions to the proposed training fund are to be remitted by the contractors as stipulated under the petroleum agreements.

Relevant contracts

The 1982 Model Production Sharing Contract (PSC). An annex to the act is the Model PSC,[43] which contains local content provisions as well. It enlists employment and training of locals as one of the ways of implementing local content requirements.[44] Further, it requires the contractor or subcontractor in the upstream oil industry to consult with the cabinet secretary (minister) on the training program, as well as contributing to the training fund established under the 1984 Petroleum Act. The statute is currently under review, with a new statute pending adoption by the Parliament. This proposed legislation is the Petroleum (Exploration, Development, and Production) Bill of 2015.

The model PSC has been updated through the years since its adoption in 1982, and it now contains specific provisions on local content in the upstream sector of the oil industry. It also introduces additional ways to implement local content requirements by requiring preference be given to local goods and services.[45]

The Model PSC Draft Bill of 2015. The Model PSC expects a contractor to provide all capital, machinery, equipment, facilities, technology, and personnel necessary to successfully conduct upstream petroleum operations. Nevertheless, there are also provisions in Sections 19 through 24 of the 2015 Draft Bill in which the contractor must utilize a certain percentage of local personnel, goods, and services. To an investor, this utilization would be an ideal and cost-effective solution, if these indeed measured up to the industry's best standards. The Model PSC Draft Bill of 2015 has made an effort to define the expected industry best practices. These are described as practices, methods, standards, and procedures generally accepted and followed internationally by prudent, diligent, skilled, and experienced operators in upstream petroleum operations.

As mentioned earlier, definitions on local content determine its successful implementation. The Model PSC defines local content as "the use of Kenyan local expertise, goods and services, people, businesses and financing for the systematic development of national capacity and capabilities for the enhancement of the Kenyan economy." Interestingly, there seems to be a variance in definition of local content between the Petroleum Bill and the Model PSC. While the Petroleum Bill emphasizes value addition, sharing benefits, developing locals, and investing in the Kenyan economy, the PSC, on the other hand, emphasizes the value of expertise and financing in the systematic development of national capacity. Nevertheless, both follow the national content concept rather than the community content concept.

As indicated in the PSC, local content laws and regulations will apply to a contractor as amended periodically. Hence, it would be correct to state that despite having a stabilization clause that guarantees an investor waiver, in light of the changes that might occur in the life cycle of a petroleum project, local content development will not be safeguarded by the clause.

Local content is therefore guaranteed in two aspects: employment and training of Kenyans and preference given to Kenyan goods and services. In relation to employment, it will be made until the expiry or termination of the contract. In fact, the investor will be expected to also offer training courses for the employees subject to approval of the cabinet secretary. It is not clear, however, who is responsible to offer the courses. In respect to goods and services, the contractor will have to give preference to Kenyan goods and services (especially materials and supplies). It is not clear, however, if the materials or supplies referenced must be manufactured in Kenya or merely assembled in Kenya. It is easy to import products to the country and pass them off as "goods in Kenya." Also, it is not clear what constitutes a "Kenyan service," since incorporation of companies through Kenyan laws allows a company to pass for a Kenyan service company.

Finally, emphasis to ensure a successful local content policy should not be relied on from equity participation. Equity participation does not necessarily have value addition to the economy, other than to just a few political elite who benefit from the creative compliance of regulation.

Positive and negative experiences and expectations regarding the regulatory framework

Kenya is at the preliminary stages of exploration, and hence at the center of overhauling its existing legislative and regulatory framework. Most of the laws were enacted before oil discoveries were made. It is therefore very difficult to highlight experiences both positive and negative concerning local content development. Nevertheless, the Kenyan government, in its 2006 Energy Policy, has acknowledged the shortage of expertise within the sector and has promoted the training of nationals in the short term to take up jobs in the long term. Training will be undertaken (locally) through funds put aside for the training fund. As mentioned previously, it is not clear who will undertake the training (government or contractors). Also, it remains unclear how much an investor is expected to put toward the training fund; however, the option has been left open for negotiation in the petroleum agreement.

Since the government has been criticized for not encouraging inclusive participation among the stakeholders, it is not yet apparent if the local content development will stand the test of time and actually come out successful, as the case in Norway. Indeed, the government has solicited the assistance of many organizations, including the World Bank, the INTSOK, and many others.

Nevertheless, it is argued that the requirement for preference and first priority to local employees, services, and supplies raises challenges of quality, competencies, and capabilities. The upstream petroleum sector, being a highly capital intensive sector, cannot afford to compromise on quality. Contractors must not be punished for preferring quality, reliability, and competence.

Current Issues and Challenges

There are pertinent questions that arise concerning whether the country is ready for the expansion of the petroleum industry with available and capable suppliers of goods that meet international specifications. Without this, the minimum local content levels cannot be achieved.

Although there is a potential duplication of roles in relation to LC development among the regulatory bodies, there is slight difference in the roles and responsibilities played by NUPAC, UPRA, and LCDMU in the wider context of petroleum industry. Nonetheless, duplication of roles among public regulators while implementing LC is by far one of the biggest challenges of implementing LC worldwide. The lack of institutional design to avoid this challenge is yet to be at the center of discussions of LC implementation. This clearly displays inconsistencies in the proposed legislations. There is also no clear law on the method of dispute resolution for an aggrieved investor. There is lack of clarity on what transpires when the Monitoring Unit recommends to the Upstream Authority that it reject a local content plan, but the Petroleum Authority goes ahead and accepts the plan. Further issues concern independence and whether the Monitoring Unit will possess the requisite expertise and capacity to evaluate the plans effectively.

Inconsistencies and confusion are also caused by the 2014 regulations when they purport to have been adopted by the cabinet secretary in exercise of powers conferred by the 2015 Petroleum Bill. The 2015 bill has yet to be adopted, and this means that the cabinet secretary does not yet have those powers. The inconsistencies are further displayed by the absence of the training fund in the 2014 Local Content Regulations and the different definition of "local content" between the 2014 Local Content Regulations and the 2015 Petroleum Bill. Finally, there is a lack of research and training centers, specialized drilling courses, and professional examinations and degrees that pertain specifically to the oil and gas industry.

Conclusion

African countries with hydrocarbon resources, including Kenya, continue to grapple with how best to secure their energy resources in a manner that accrues the greatest benefit to their citizens. The economic and political climate environment presents enormous challenges to improvement of the existing regulatory frameworks or to addressing the lack of legislation. This creates a barrier to the changes in the energy sector envisioned by Kenya's Millennium Development Goals and other programs. Meanwhile, as projected by the *World Energy Outlook*, the energy need is set to increase in the future, with fossil fuels expected to remain the largest provider of the world's total energy supplies. These assessments affect the search for more petroleum (oil and gas), at least in the short term, which fundamentally drives and changes the decisions of stakeholders within the industry. Notably, investors such as international oil companies tend to withdraw from mature regions and compete instead for the vast underexplored resources commonly found in African countries. This search occurs against the backdrop of some pertinent issues concerning governance, transparency, political stability, and a serious human resource crisis in the energy sectors.

Successful implementation of local content requirements—those that ultimately result in substantial technology transfer, knowledge transfer, capacity building of local industry, and generation of local employment opportunities—will be difficult to achieve due to the many variables at play. These variables include political interference and lack of a workforce capable of effectively taking advantage of the required skills and technology transfer. Training and technology adoption are hindered by poor educational and infrastructural systems and poor planning by the government. Poor planning leads to restrictive local content regulations instead of forward-looking regulations that can grow local capacities in the sector in the future.

Notes

1. For an overview, see Werner Baer, "Import Substitution and Industrialization in Latin America: Experiences and Interpretations," *Latin American Research Review*, 7, no. 1 (Spring 1972), 95–122, http://www.jstor.org/stable/2502457.
2. Ibid.
3. For more discussion, see Michael E. Porter, "What Is National Competitiveness?" and "Patterns of Achieving National Competitiveness," inserts in "The Competitive Advantage of Nations," *Harvard Business Review*, March–April 1990.
4. Katrina Manson, "Kenya's Oil and Mineral Prospects Fire Global Interest," *Financial Times* (November 25, 2016), https://www.ft.com/content/20badca4-504b-11e3-9f0d-00144feabdc0.
5. Turkana County is located in the former Rift Valley province of Kenya. It is the second largest county in Kenya by area. It covers 68,680 square kilometers and is located in northwestern Kenya.
6. Most of the laws were enacted in early 1980s, when resources had not yet been discovered. The current Petroleum Law was enacted in 1982 and was amended in 1984.
7. Silvana Tordo, Michael Warner, Osmel E. Manzano, and Yahya Anouti, *Local Content Policies in the Oil and Gas Sector* (Washington, DC: International Bank for Reconstruction and Development, World Bank, 2013), 37, doi: 10.1596/978-0-8213-9931-6, or http://documents.worldbank.org/curated/en/549241468326687019/pdf/789940REVISED000Box377371B00PUBLIC0.pdf.
8. Ibid.

9. The overall national development objectives of the government of Kenya, as set out in Vision 2030, are as follows: accelerated economic growth; increasing productivity of all sectors; equitable distribution of national income; poverty alleviation through improved access to basic needs; enhanced agricultural production; industrialization; accelerated employment creation; and improved rural-urban balance. See more about Vision 2030 at http://www.vision2030.go.ke/?page_id=3247.
10. "Energizing Kenya: Publications," National Oil, http://nationaloil.co.ke/site/6.php?id=1.
11. "Human Resource Development and Capacity Building," Para. 5.9, *Sessional Paper No. 4 on Energy* (Nairobi, Kenya: Ministry of Energy, May 2004), http://www.renewableenergy.go.ke/downloads/policy-docs/sessional_paper_4_on_energy_2004.pdf.
12. "Draft National Energy Policy," Ministry of Energy and Petroleum (February 24, 2014).
13. The two consultants are Challenge Energy and Hunton & Williams for the World Bank. The report is titled "Kenya Oil and Gas Sector Development: Review and Update of the Legal, Regulatory, and Fiscal Framework; Final Report on Legal and Regulatory Guidance."
14. "Kenya: New World Bank Project Will Support Country Efforts to Better Manage Oil and Gas Developments and Revenues to Invest in Lasting Growth and Development," *WorldBank.org* (July 24, 2014), http://www.worldbank.org/en/news/press-release/2014/07/24/kenya-new-world-bank-project-will-support-country-efforts-to-better-manage-oil-and-gas-developments-and-revenues-to-invest-in-lasting-growth-and-development.
15. Hunton & Williams and Challenge Energy Consultants in their 2013 report.
16. "Operations Map: East Africa, Kenya," Tullow Oil, http://www.tullowoil.com.
17. Ibid.
18. Ibid.
19. "Project Overview," Africa Oil Corp., http://www.africaoilcorp.com/s/home.asp.
20. "Tullow Oil Sub-Contractor Employee Data Reveal Contracting Driven by Market Forces, Not Economic Policy in Turkana," *InternewsKenya.org* (March 4, 2014), http://landquest.internewskenya.org/tullow-oil-sub-contractor-employee-data-reveal-contracting-driven-by-market-forces-not-economic-policy-in-turkana/.
21. "Welcome," Petroleum Institute of East Africa (PIEA), http://www.petroleum.co.ke/index.php/petroleum-institute-of-east-africa.
22. "Welcome," School of Petroleum Studies, Petroleum Institute of East Africa (PIEA), http://www.petroleum.co.ke/index.php?option=com_content&view=article&id=27&Itemid=102.
23. Ibid.
24. "Department of Geology," University of Nairobi, http://geology.uonbi.ac.ke/.
25. "Technical University of Kenya," https://en.wikipedia.org/wiki/Technical_University_of_Kenya.
26. Clause 7 of the proposed "Petroleum Exploration, Development and Production (Local Content) Regulations, 2014."

27. "Chapter 308: Petroleum (Exploration and Production) Act," *Laws of Kenya*, rev. ed. 2012 [1984] (National Council for Law Reporting with the Authority of the Attorney-General, 2012), www.kenyalaw.org http://faolex.fao.org/docs/pdf/ken64182.pdf.
28. Ibid., Sections 9 and 12.
29. "The Petroleum (Exploration, Development and Production) Bill, 2015," Section 13.
30. Ibid., Section 15.
31. Ibid., Section 2.
32. Ibid., Section 86.
33. Ibid.
34. Ibid., Section 84.
35. "Petroleum Exploration, Development and Production (Local Content) Regulations, 2014," Clause 2.
36. Ibid., Clause 6.
37. Ibid., Clause 10.
38. Ibid.
39. Ibid., Clause 11.
40. Ibid., Clauses 29 and 30.
41. Ibid., Clause 34.
42. Ibid., Clause 35.
43. Ibid.
44. "Model Production Sharing Contract, 1982," Clause 13.
45. Ibid., Clause 31.

13

LEBANON

Ziad Obeid and Emilie Tanke[1]

Introduction

The petroleum industry in Lebanon is at a relatively nascent stage. Geophysical reconnaissance conducted in 2010 indicates that there is substantial hydrocarbon potential in the Lebanese Exclusive Economic Zone (EEZ), including both oil (an initial estimate of around 660 million barrels) and natural gas (an initial estimate of up to 850 billion cubic meters).[2]

These promising prospects prompted the government of Lebanon to enact the Offshore Petroleum Resources Law (Law No. 132 of August 24, 2010) (the OPRL) to regulate petroleum activities in Lebanon. This overarching law forms the basis of Lebanon's regulatory regime for offshore petroleum activities, alongside the Petroleum Activities Regulations (Decree 10289/2013) (the PAR), which sets out the implementation of the OPRL.

Building on this momentum, the Lebanese Petroleum Administration (LPA) was appointed in December 2012. Lebanon subsequently conducted two rounds of prequalification for licensing of offshore oil and gas exploration and production rights in 2013 and 2017.[3] There were many challenges facing implementation of a comprehensive legislative framework to govern the exploration, production, and utilization of Lebanon's recoverable oil and gas reserves. Nevertheless, the attractiveness of the emerging Lebanese oil and gas sector was apparent during this initial prequalification licensing round. Upon conclusion of the 2013 prequalification, Lebanon listed 46 international companies as having met the prequalification requirements.[4]

Exploration and production rights are set to be granted by the end of the first offshore licensing round in December 2017, following the Lebanese government's approval and ratification of crucial decrees necessary to complete the requisite components of the aforementioned legislative framework. The first of these decrees (decree 43 of January 19, 2017) issued Lebanon's Model Exploration and Production Agreement (Model EPA), which will be the template agreement for the exploration and exploitation of energy resources between the Republic of Lebanon and the right holders, as well as the Tender Protocol with evaluation criteria for license allocation. The second decree (decree 42 and of January 19, 2017) provided for the delineation of the Lebanese maritime blocks. Between 2013 and early 2017, the licensing round was temporarily stalled due to the Lebanese government's political gridlock.

It has become a common practice for established and nascent petroleum producing countries alike to adopt, or consider adopting, policies that aim to derive the greatest benefits for their economies from the extraction of their exhaustible natural resources.[5] As Lebanon's petroleum industry is in its inception stage, it presents an opportune moment to consider the adequacy, potential effectiveness, and future role of the specific local content policies (LCPs) within the legal framework governing national petroleum activities.

This chapter examines those LCPs in the context of the practice and experience of other developing nations, with a particular focus on the development, implementation, related challenges, and effects of those policies in Lebanon. First the regulatory framework governing the Lebanese oil and gas sector are examined, with particular focus on the existing LCPs. Next, an overview of the challenges to the successful implementation of LCPs are provided. This informs the concluding remarks, concerning the ways in which such challenges may be addressed, overcome, or where possible, avoided altogether.

Local Content Policy and Baseline: Current Regulatory Framework

Key features of the regulatory framework governing the Lebanese oil and gas sector

The Model EPA forms part of the overall legislative framework governing petroleum activities in the Lebanese EEZ. It accompanies the two above-mentioned laws: the Lebanese Offshore Petroleum Resources Law and the Petroleum Activities Regulations.[6] These three regulative mechanisms work together to guide the formulation of commercial incentives for prospective oil and gas production companies. They are also aimed at affording protection to the commercial, environmental, and social interests of Lebanon.

The OPRL applies to all offshore petroleum activities within the Lebanese EEZ, reserving for the Republic of Lebanon exclusive right to the petroleum resources and their management. The OPRL provides the legal and institutional framework for the exploration and exploitation of those resources.

The PAR implements the OPRL by setting specific regulations governing the conduct of petroleum activities in Lebanon, including the following:

- Right holders' legal representation
- Applicable management system
- General duties of the operator and the right holders
- Strategic environmental assessments (SEAs)
- Exploration and production rights
- Petroleum production and transportation
- Cessation of petroleum activities and decommissioning

Finally, the Model EPA is to govern the contractual relationship between the Lebanese government and the right holders in the exploration and exploitation of offshore oil and gas in the Lebanese EEZ. In particular, the Model EPA details the split of revenues between the state and the right holders (Article 23).

Local content in Lebanon's regulatory framework

The OPRL, the PAR, and the EPA each contain local content requirements, generally including the training of Lebanese nationals and granting preference to Lebanese nationals in the awarding of contracts.[7] In particular, the requirements relating to the education, training, and employment of Lebanese nationals pursuant to these regulations include the following:

1. Education and training
   - Article 67 of the OPRL provides for the right holder to organize and fund training for Lebanese nationals associated with petroleum activities.[8]
   - Article 83 of the PAR provides that the right holder shall pay to the state its respective share of the annual costs of running the training and support programs for the training of public sector personnel, as stipulated in the EPA.[9]
   - Article 155 of the PAR provides that the right holder and its contractor shall ensure that an effective training and recruitment program is in place and further, that its personnel have sufficient qualification and experience for the safe and efficient execution of its works.[10]
   - Article 156 of the PAR provides that the right holder and the operator[11] shall provide training to government employees involved in the management or monitoring of petroleum resources and related activities.[12]
   - Article 20.3 of the EPA details the fees that must be paid by the right holders for the training of Lebanese nationals in the public oil and gas sector: Required payments amount up to US$300,000 per annum during the exploration phase, yearly increasing by 5%, and up to US$500,000 in the production phase, again with a yearly increase of 5%. The costs incurred by right holders under Article 20 of the EPA are, however, recoverable.[13]
2. Employment
   - Article 67 of the OPRL provides for the employment of qualified Lebanese nationals where available.[14]
   - Article 155 of the PAR also requires the right holder to prioritize training for Lebanese nationals in order to facilitate their employment at all levels of management of the right holder's organization.[15]

- Article 20.1 of the EPA contains provisions requiring the right holder to carry out an effective and comprehensive training program for Lebanese personnel and consequently employ Lebanese personnel wherever possible. The EPA further provides that by law, right holders will be required to hire 80% of their workforce locally. In addition, a recruitment and training program for Lebanese nationals is mandatory, and plans must include a program for hiring Lebanese nationals as management and professional staff.[16]

3. Procurement
   - Article 67 of the OPRL provides that the right holder must give priority to Lebanese nationals in awarding contracts for construction or supplies and services, where the terms and conditions offered are comparable to that of foreign competitors.
   - Article 157 of the PAR provides that major contracts for the procurement of goods and services for the purposes of petroleum activities shall be subject to public tender, and further that when the quality, capacity, and availability of Lebanese goods are comparable to that of foreign competitors, preference shall be given to the Lebanese bidder.[17]
   - Article 27 of the EPA sets a Lebanese contracting requirement, whereby competitive Lebanese contractors are given preference when the price of services is less than 110% of that of foreign competitors, and when the price of goods is less than 105% of that of foreign competitors.[18]

Although experience has demonstrated that LCPs alone do not necessarily guarantee the successful maximization of benefits from oil and gas industries in the host country, they clearly form an integral part of achieving such an objective. If carefully planned, LCPs have proven to be a valuable development strategy. Careful planning must take into account the overall context of the host country, particularly with regard to local skills, technological know-how, entrepreneurship, infrastructure, and employment.[19] The following discussion therefore addresses the interests at stake in the development of petroleum activities in Lebanon and the relevant LCPs. It also examines the current community and industry capacity in Lebanon, along with other factors that may play a role in the ultimate success or failure of such policies in the implementation and achievement of both short- and long-term objectives.

Evolving Implementation Strategies

Foreign implementation strategies and varying success in the face of domestic exigencies

Given the general profitability of oil and gas projects around the world, and their potential for welfare development, it is likely that governments will continue to set ambitious LCP laws and regulations, especially in the developing world. While their objectives often converge, multiple avenues have been taken for their implementation. For an example of the different implementation methods adopted to encourage the development of local economies, one may look to Australia and Brazil. Brazil employed a firm regulatory approach, based notably on a planned gradual increase in the legally required percentage of local sourcing.[20] This enabled a progressive buildup of the required competencies and ultimately led to a significant growth in local content.[21] Australia, on the other hand, has employed a policy of incentives related to local sourcing, rather than simply imposing mandatory obligations. This policy has proven fruitful in establishing an equilibrium between financial incentives for oil and gas exploration and the diversification and support of local economies.

Through innovative and flexible approaches toward LCP implementation, developed economies such as Australia and Brazil seek to maintain the right balance between the needs of the local economy and those of the E&P companies. It must be noted, however, that Brazil has recently undertaken some changes to its LCPs. In effect, it has sought to amend those mandatory obligations that have proven too burdensome by resulting in higher costs and delays for companies and rendering state supervision and enforcement particularly onerous. One key is government willingness to reevaluate the effectiveness of LCPs and readjust them accordingly. This has been accomplished in Brazil notably by easing certain requirements to account for present industry demands or unavoidable fluctuations in price, such as the decline in oil prices. Such adjustments are therefore essential to achieve local content targets without impinging on efficiency and growth in the sector.

In the case of Australia, oil and service companies needed to import petroleum operations equipment that was not available in Australia. These companies were granted an exemption of port tariffs in exchange for the formulation and execution of an Australian Industry Participation Plan.[22] This policy illustrates Australia's innovative approach to its LCPs that is mindful of exigencies on the ground and still encourages outside investment, as well as safeguarding and developing local industries. In comparison, however, LCPs in emerging economies often remain imprecise or inadequately designed,[23] resulting in bottlenecks in the development and diversification of a sustainable local economy. Where capabilities are nonexistent or insufficient, governments must first bridge the gap before crossing it.[24]

Similarly, LCP frameworks may be flawed to the extent that the enforceability of many LCP projects is practically unattainable. This may notably be the case due to underlying flaws such as weak institutional capacity or prevalent corruption.[25] It follows that LCP implementation strategies that have proven successful in one country cannot simply be transposed into another country characterized by a fundamentally different political and socioeconomic context.

It is clear from this brief comparative assessment that there is no one-size-fits-all approach to LCPs. Existing LCPs around the world have demonstrated that such policies, to be enduring, must not only be developed with the country's specific socioeconomic context and geographic constraints in mind, but also must be reviewed and adapted over time. In effect, the wide-ranging policies implemented by Australia and Brazil would not necessarily achieve the same success in a country such as Lebanon, whose geographic, political, and socioeconomic contexts are considerably different. Moreover, LCPs that have been devised without taking into account the specific employment or capacity shortfall in a particular region are not likely to produce sustainable long-term economic development.

Industry Knowledge and Skill Gaps

Preferential treatment approach: Common pitfalls

In some cases, the traditional approach of demanding preferential treatment for local suppliers, regardless of cost, quality, or timeliness of delivery of the goods or services, has not helped to facilitate the development of local capacity. Only those companies that are already globally competitive have an opportunity to engage.[26] Instead of making the market more accessible for local suppliers, this method can hinder accessibility if no viable opportunities are provided for local suppliers to truly learn and improve so that they can become competitive with international suppliers. For this reason, local capability development must be an important part of any implementation strategy.

Similarly, in some cases, developing local content provisions can be a more costly option for the host state than sourcing supplies and services externally, particularly in developing nations.[27] Sourcing local content becomes cheaper, however, once local suppliers and the local workforce reach the competence and capacity levels required by the industry.[28] To be sure, the cost of hiring or sourcing locally, which also includes the cost of education and training in this case, can be prohibitive in countries with potentially low prospects for petroleum resources. Companies expect to recoup the extensive funds injected into training and development, as required pursuant to the applicable LCPs, through cost recovery[29] and repeated use of local content. In order for this expenditure to therefore be justified as the costs of a "license to operate," the LCP requirements must focus on the development of local capacity in goods and services that are of immediate need, or of goods and services that have "dual use applications."[30]

Moreover, LCPs designed to impose foreign/local partnerships also run the risk of facilitating the development of shell companies. This would serve only to satisfy local content requirements at a financial gain, without the benefit of local companies learning from or contributing to the operations of the project or further facilitating the development of skills that might be transferable to other sectors of the local economy.[31] However, if LCPs are designed with a focus on domestic value addition,[32] foreign companies may be encouraged to set up and operate a structure in Lebanon in partnership with local companies. Local companies and international oil companies could then participate in projects and mutually benefit from the local content regulation.

Current capacity and competence in Lebanon

Interconnected with the issue of technological and scientific knowledge gaps is the question of both availability and quality of the necessary education and training. Certain levels of education and training will be required to sufficiently develop local human resources and competently fill highly skilled positions in the petroleum industry. This is the case particularly in circumstances where the local industry is relatively new and, therefore, the incentive or opportunity to find work in the industry has only recently emerged. The expansion and advancement of local human capital is, in any event, a pivotal facet of LCP strategy across all petroleum sectors.

While many of Lebanon's universities have established academic programs to build expertise and prepare the Lebanese workforce for this sector, little attention has been given to vocational programs. The LPA has, however, been engaged in discussions with local institutions to establish programs for technicians. The most needed certificates identified include project management and financing, supply chain management, risk management, contract management and negotiation, and energy law.[33]

One must also note that Lebanon possesses all the necessary elements to be successful in its endeavor, given its multilingual population and the quality of its institutions for higher education. Moreover, one may envisage the possibility of such projects providing attractive opportunities that are likely to draw some of the substantial Lebanese diaspora back to Lebanon. This would prove to be a valuable influx of human capital capable of direct integration in the Lebanese petroleum industry.

Confronting current legislation with realities on the ground

Assuming a successful outcome to future oil and gas exploration, if and when it materializes, one must nonetheless adopt a pragmatic approach. Valuable insight must be gained from the experiences of comparable nations as well as from the challenges faced in the implementation of certain LCPs. It is clear that Lebanon has already envisaged, and indeed enacted in its framework, the necessary outline for the inclusion of national companies and human capital in the exploration operation. This legal basis must then be confronted with the exigencies of conditions on the ground.

In this regard, some concerns have been expressed about the local employment requirement contained in the Model EPA (Article 20), which provides that "as of the beginning of the Exploration Phase, no less than eighty per cent (80%) of the aggregate number of employees of the Right Holders (including the Operator) shall be Lebanese nationals." This strict local employment requirement has raised some concerns due to the fact that the threshold may not be easy to reach given the limited oil and gas expertise and shortage of qualified workers in the oil and gas industry in Lebanon.

In light of the laws governing the exploration and production of oil and gas in Lebanese waters, it is indeed apparent that, at this very early stage, there is a clear gap between policy and expected practice. Indeed, there is room for improvement in workforce training, educational development, and other capacity-building support. Nonetheless, a caveat must be made in this regard. It is true that the level of industry expertise is presently limited. The timing of oil and gas exploration and extraction, however, is such that it affords some lead time to develop needed expertise during the initial stages of prospecting and exploration, before the beginning of the production phase, at which point the requirement applies.[34]

Any attempt to bridge this gap will therefore require broad strategic conversations and timely planning across different sectors in order to facilitate knowledge transfer and informed decisions. This includes both local and international companies, as well as government institutions and other stakeholders. A cooperative effort will be needed in order to generate consistent and comprehensive policies with the appropriate vision and strategy to achieve objective outcomes.[35] This is particularly the case where policy making is generally complex, as has typically been the case in Lebanon over the past years.[36]

Moreover, timely industrial and skills development is essential not only with regards to time-sensitive industry requirements, but also with respect to unforeseeable environmental limitations. The petroleum industry is based on the extraction and production of exhaustible resources, and this crucial reality must be taken into consideration in the development of targeted education and training programs.[37] Emphasis must therefore be placed on building a flexible and transferable local workforce.

Building transferable skills

In relation to considerations of requisite capacity and competency of local human capital, an additional challenge to the implementation of successful LCPs has been identified. This challenge relates to the ability—or lack of ability—to recognize the skills that would be best utilized and developed effectively under the circumstances. The key seems to be keeping the focus predominantly on those skills that are transferable to other sectors of the local economy (as well as identifying the requisite goods, services, trades, and infrastructure).[38]

The skills required for the oil and gas industry are particularly specialized. Different phases in the process require significantly heightened human capital, particularly the construction phase, which only lasts three to five years on average. The peak in employment levels, therefore, is short and sharp and could result in severe inefficiencies. It is possible employees could be specifically educated and trained for construction roles, only to be left without employment after that short construction period reaches completion. The number of employees required is significantly reduced in the development and production phases.[39] As a result, the emphasis must be on the development of transferable skills, ensuring that the local employees are then in a position to enter into the international industry.

As we have seen, if not carefully carried out, the implementation of LCP objectives can be undermined. This is possible in circumstances where the overall situation of the host state (including the quantity and value of its potentially accessible resources), as well as the projected trends and expectations for the global industry, are overlooked. For this reason, a pragmatic and holistic approach is fundamental in designing LCPs capable of carrying out achievable targets.

As a summary, in order for LCPs to contribute to a sustainable economic development plan, the following principles should be adhered to:

1. The sectors or subsectors that should be promoted must be identified in order to assess how these, in turn, might lead to the promotion of other sectors.
2. There must be a clear understanding of the infrastructure needed to facilitate the LCP.
3. A comprehensive assessment of skills and capacity deficiencies should be carried out to ensure that the LCP is designed to function effectively and efficiently.

These principles must be tailored to Lebanon, where the oil and gas industry is still in its infancy. This will necessarily involve calibrating the LCPs in a manner that is both attractive to potential investors and beneficial for the country's economic development and the economic welfare of its citizens.

Conclusion

The overarching goal behind the development and implementation of LCPs is ultimately to extract maximum value from the natural hydrocarbon and oil resources while stimulating Lebanon's potential for broad and sustainable economic development. Trinidad and Tobago provides a striking example in this respect, as according to the OECD, it had advanced from a developing to a developed country by 2011. It now is on the list of top 66 high-income countries, primarily as a result of its involvement in the petroleum industry.

In particular, Lebanon must use the sector's outputs and the unprecedented opportunity they represent (namely raw commodities extracted, developed human capabilities, products created, and financial revenues earned) in a beneficial manner. Lebanon must seek to capture and transfer the factors of production needed for sustainability, such as capital, assets, and capabilities, to other sectors of the economy, especially technology and capital markets.[40] This appropriation can only be achieved at the price of clear commitment to transparency and ethics in the sector, leading the way toward good governance, values, and development.

Notes

1. Ziad Obeid is a partner and Emilie Tanke is a trainee with Obeid Law Firm, Beirut. With special thanks to Billie Stevens.
2. "Assessment of Undiscovered Oil and Gas Resources of the Levant Basin Province, Eastern Mediterranean," US Geological Survey (March 2010), http://pubs.usgs.gov/fs/2010/3014/pdf/FS10-3014.pdf. Such figures necessarily involve considerable uncertainty at this stage. More accurate estimates are likely to be ascertained upon commencement of drilling and exploration activities, particularly when considered together with both previous and subsequent estimates based on seismic data of offshore Lebanon, which differ substantially. Also see Bassam Fattouh and Laura El-Katiri, *Lebanon: The Next Eastern Mediterranean Gas Producer?* Foreign and Security Policy Series 2015 (Washington, DC: The German Marshall Fund of the United States, February 12, 2015), 1, 4, http://www.gmfus.org/publications/lebanon-next-eastern-mediterranean-gas-producer.
3. The LPA is a regulatory body tasked with the management of the petroleum sector in Lebanon. The LPA's principal objective is to contribute to creating the greatest possible value for the economy and society, resulting from the activities undertaken in the oil and gas industry, while also protecting the environment.
4. Fattouh and El-Katiri, *Lebanon: The Next Eastern Mediterranean Gas Producer?* 4.
5. World Bank, "Overview: Why Is Local Content Relevant in the Oil, Gas, and Mining Sectors?" (Draft conference paper, World Bank Global Conference, Vienna, Austria, September 30–October 1, 2013), http://www.worldbank.org/content/dam/Worldbank/document/SDN/localcontent/Why%20is%20Local%20Content%20Relevant%20in%20Oil,%20Gas,%20and%20Mining.pdf.
6. A complete copy of the OPRL, PAR and EPA can be viewed at the website of the Lebanese Petroleum Administration, a body formed under the ambit of the Ministry of Energy and Water Resources and responsible for the management, monitoring, and supervision of petroleum activities in Lebanese waters: "Regulations," Lebanese Petroleum Administration, http://www.lpa.gov.lb/regulations.php; and E&P Agreement, http://www.lpa.gov.lb/epa.php.
7. While this chapter will focus only on the specific LCPs contained in legislation pertaining to the oil and gas industry, it should be noted that, aside from these sector-specific LCPs, other LCPs may be inferred from general legislation. Although Lebanon is an open country, aiming principally to encourage and facilitate foreign investment (notably through limited requirements on foreign companies seeking to operate in Lebanon), its legislation does in fact contain some form of LCP. One example is the requirement for the majority of the board of directors of Lebanese joint stock companies (SAL) to be Lebanese nationals.
8. According to Article 67 of the Offshore Petroleum Resources Law (OPRL), "Local Content":

   (1) A Right Holder as well as its subcontractors shall give priority to Lebanese persons in the award of contracts for construction of a Facility and the supply of material,

goods and services related to Petroleum Activities when terms and conditions offered by Lebanese suppliers are equal to their competitors.

(2) A Right Holder as well as its subcontractors shall employ qualified personnel of Lebanese nationality whenever available. Right Holder shall also organise and fund the training of Lebanese personnel associated with Petroleum Activities.

Note: The English versions of the articles cited in this chapter are provided by the LPA as "unofficial translations."

9. According to Article 83 of the Petroleum Activities Regulations of Lebanon (PAR), "Costs Related to Training and Support Programs":

   The Right Holder shall pay to the State when stipulated in the Exploration and Production Agreement its respective share of annual costs to be spent on training and support programs for training of public sector personnel and as institutional support.

   Any commitment by a Right Holder to training or institutional support programs may be stipulated in detail in a separate agreement. Such agreement shall form an integral part of and constitute a condition for the granting of a Petroleum Right.

10. According to Article 155 of the PAR, "Qualification and Training of Personnel":

    All personnel engaged in Petroleum Activities shall possess adequate qualifications and training for the efficient and safe execution of their work. The Right Holder shall ensure that Right Holder's personnel and those of its contractors possess adequate qualification and training for the efficient and safe execution of their work. Specific criteria shall be established for the identification of relevant tasks with regard to safety and protection of the environment, as well as for the selection of personnel responsible for certifying the project....

    The Right Holder and contractor shall give priority to training of Lebanese in order to facilitate the employment of Lebanese at all level[s] of Right Holder's or contractor's organizations.

    The Right Holder shall in consultation with the Minister, propose and carry out an effective recruitment and training program for Lebanese personnel for each phase of the Petroleum Activity and at all levels of management, taking into account the safety requirements and the need to maintain reasonable standards of efficiency in the conduct of Petroleum Activities. Such employees may be trained in the Republic of Lebanon or abroad as required by the training programs prepared.

11. An "operator" is a company that is a right holder whose appointment is approved by the Council of Ministers to execute on behalf of the other right holders the day-to-day management of petroleum activities.

12. According to Article 156 of the PAR, "Training of Public Sector Personnel and Government Support":

    The Right Holder and the Operator shall provide training to Government employees involved in the management of Petroleum resources or monitoring of Petroleum Activities and related activities. The same training obligation applies to the Right Holder and Operator with regard to teachers and trainers employed or associated with Lebanese publicly funded institutions.

    Each Right Holder shall co-operate with the Minister in giving a mutually agreed number of Government civil servants and personnel associated with Government entities in charge of Petroleum resources management, Petroleum Activity monitoring and control, the opportunity to participate in training activities provided for the employees of the Right Holder or any of its Affiliated Companies.

13. According to Article 20.3 of the EPA "Recruitment and training":

 [...]

 3. The Right Holders shall pay costs related to the training of public sector personnel with functions relating to the oil and gas sector identified by the Petroleum Administration in addition to those nominated by the concerned ministries and notified to the Petroleum Administration (including awarding public sector personnel research grants), in an amount up to $/300,000/ per year (increased by 5% each year) until the beginning of the Production Phase, and thereafter $/500,000/ per year (increased by 5% each year). The Right Holders shall, upon request by the Petroleum Administration, assist the Petroleum Administration in arranging third party training opportunities for such public sector personnel, to be paid for by the Right Holders pursuant to the preceding sentence. The Right Holders shall pay all such costs directly to the provider of the training (or to the recipient of the research grant) when due and shall, promptly upon request from the Petroleum Administration, provide satisfactory evidence of payment of such costs in order to enable the Petroleum Administration to monitor and verify the Right Holders' expenditure under this Article. [...]

 5. The cost incurred by the Right Holder for recruitment and training of Lebanese personnel and the annual contribution for the training and participation of public sector personnel as provided in this Article shall be Recoverable Costs for the purpose of the Accounting and Financial Procedure that constitutes Annex D of this EPA.

14. See Article 67 of the OPRL, "Local Content."

15. See Article 155 of the PAR, "Qualification and Training of Personnel."

16. According to Article 20 of the EPA, "Recruitment and training":

 1. The Right Holders shall develop and carry out an effective recruitment and training program for Lebanese personnel in accordance with the law no 132/2010 (OPR Law) and this EPA. A proposal for a detailed recruitment and training program shall be submitted to the Petroleum Administration for approval no later than six (6) months after the Effective Date of this EPA...

 [...]

 As of the beginning of the Exploration Phase, not less than eighty per cent (80%) of the aggregate number of employees of the Right Holders (including the Operator) shall be Lebanese nationals. Prior to the beginning of the Exploration Phase, the Right Holders and their contractors and subcontractors according to article 67 of the Law no 132/2010 (OPR Law) shall achieve this eighty per cent (80%) threshold as soon as possible, to the extent that they are able to recruit a sufficient number of Lebanese nationals with the requisite qualifications, and the Right Holders and their contractors and subcontractors shall in any event maximize the proportion of Lebanese nationals employed by them, subject to their obligation to conduct Petroleum Activities in accordance with Best International Petroleum Industry Standards. In the event that as of the beginning of the Exploration Phase or at any time thereafter the aggregate number of employees of the Right Holders (including the Operator) and their contractors and subcontractors who are Lebanese nationals is less than eighty per cent (80%), the Right Holders shall submit to the Minister on their behalf and on behalf of their contractors and subcontractors in writing an explanation of the reasons why this requirement has not been met and a request for an exemption from this requirement. The Minister may, based on the opinion of the Petroleum Administration and after consulting with the concerned ministries, grant such an exemption to the Right Holders and their contractors and subcontractors. Pending a decision on the exemption request, the Right Holders and their contractors and subcontractors shall not be considered to be in breach of this requirement.

17. According to Article 157 of the PAR, "Procurement and Local Content":

Major contracts for the procurement of goods and services for the purpose of Petroleum Activities shall be subject to public tender. The quality, price, delivery and guarantees offered shall be taken into account when evaluating bids and awarding contracts. Major contracts mean any contract that materially or substantially affects the design or functionality of Facilities, the concept or timeline of Development, Production or resource management and depletion policies. Major contracts mean also contracts of substantial value, meaning that the performance or non-performance of the contract may substantially affect the economy of the project or the financial strength of the Right Holder.

The Right Holder shall ensure that the Operator gives preferential treatment to the procurement of Lebanese originating goods and services when such goods and services are internationally competitive with regard to quality, availability, price and performance. Lebanese originating goods and services are those that in substance or measured by value added are predominantly manufactured, constructed or performed in Lebanon, by Lebanese or by an entity owned and controlled by Lebanese.

18. According to Article 27 of the EPA, "Procurement and supply bases":

[...] The Right Holders shall ensure that preferential treatment is given to Lebanese goods and services that are competitive with regard to quality, availability, price and performance, in accordance with Article 67 of the law no 132/2010 (OPR Law) and Article 157 of the decree no 10289/2013 (PAR). Without limiting the foregoing, the Right Holder shall procure that:

a) in selecting Contractors for the provision of services in relation to Petroleum Activities, the Operator shall select, when possible, individuals who are Lebanese or Entities which are majority-owned or controlled by Lebanese nationals and that provide such services through permanent establishments in Lebanon, provided that:

(i) the Operator, in its judgment acting reasonably, is satisfied (on the basis of demonstrated ability, experience, quality, timely performance, workmanship and other relevant criteria) with their ability to properly perform the work entrusted to them;

(ii) the cost of such services does not exceed one hundred and ten per cent (110%) of the lowest acceptable cost of equivalent or similar services offered by or otherwise available from foreign individuals or Entities that are not majority-owned or controlled by Lebanese nationals; and

(iii) the other terms and conditions applicable to such services are otherwise competitive with those available from foreign individuals or Entities that are not majority-owned or controlled by Lebanese nationals; and

b) in purchasing goods, the Operator shall give preference to goods which originate from Lebanon, or are manufactured or constructed in Lebanon, in each case by Entities that are majority-owned or controlled by Lebanese nationals, provided that:

[...]

(ii) the goods price does not exceed one hundred and five per cent (105%) of the lowest acceptable all inclusive cost offered by or otherwise available for equivalent or similar goods that do not originate from Lebanon, are not manufactured or constructed in Lebanon or are not supplied by Entities that are not majority-owned or... [...]

19. "Local Content Strategies in the Oil and Gas Sector: How to Maximize Benefits to Host Communities" (theme paper, 2012–2015 Triennium Work Report, IFRI, World Gas Conference, Paris, 2015), 5, https://www.ifri.org/sites/default/files/atoms/files/local_content_tf3_igu_final_may_2015.pdf.

20. Isabelle Ramdoo, "Local Content Policies in Mineral-Rich Countries—an Overview," (Discussion Paper No. 193, European Centre for Development Policy Management, May 2016), 8, http://ecdpm.org/wp-content/uploads/ECDPM-Discussion-Paper-193-Local-Content-Policies-Mineral-Rich-Countries-2016.pdf.

21. Silvana Tordo, Michael Warner, Osmel E. Manzano, and Yahya Anouti, *Local Content Policies in the Oil and Gas Sector* (Washington, DC: International Bank for Reconstruction and Development, World Bank, 2013), doi:10.1596/978-0-8213-9931-6, or http://documents.worldbank.org/curated/en/549241468326687019/pdf/789940REVISED000Box377371B00PUBLIC0.pdf.

22. Ibid.

23. This is for example the case in Uganda, where the specific local content legislation, namely the Petroleum (Exploration, Development, and Production) Act No. 3 of 2013, Section 125, provides that "[t]he licensee, its contractors and subcontractors shall give preference to goods which are produced or available in Uganda and services which are rendered by Ugandan citizens and companies." However, it provides no definition of local content, no defining criterion for "Ugandan companies" as understood under the act, and no penalties for noncompliance. See Jesse Salah Ovadia, "The Role of Local Content Policies in Natural Resource-Based Development," Austrian Foundation for Development Research (2015), http://www.oefse.at/fileadmin/content/Downloads/Publikationen/Oepol/Artikel2015/Teil1_03_Ovadia.pdf.

24. Ramdoo, "Local Content Policies in Mineral-Rich Countries—an Overview," 8.

25. To this effect, Balouga concludes that "the factors which have created the chasm between policy substance and implementation are mainly inadequate think-through, weak institutional capacity, absence of the required political will to carry through change, lack of support from relevant stakeholders and corruption." Jean Balouga, "Nigerian Local Content: Challenges and Prospects," *International Association for Energy Economics* (Third Quarter 2012), https://www.iaee.org/en/publications/newsletterdl.aspx?id=176.

26. Anthony E. Paul and Atiba Phillips, "Management of Energy Resources for National Development: The Trinidad and Tobago Model," in *Key Determinants of National Development: Historical Perspectives and Implications for Developing Economies* (London and New York: Routledge, 2015), 159.

27. Valérie Marcel, *Guidance for Good Governance in Emerging Oil and Gas Producers* (London: Chatham House, Royal Institute of International Affairs, June 2015), 24, https://www.chathamhouse.org/sites/files/chathamhouse/field/field_document/20150624GuidelinesGoodGovernanceMarcel.pdf.

28. Ibid.

29. See Article 76 of the PAR, "Cost Recovery":

    The Right Holder shall bear and pay all costs incurred in carrying out the Petroleum Activities in which the Right Holder participates. The Right Holder shall exclusively be reimbursed for Recoverable Costs incurred by means of the Right Holder's entitlement to quantities of Petroleum referred to as Cost Petroleum as stipulated by the Exploration and Production Agreement.

30. Marcel, *Guidance for Good Governance in Emerging Oil and Gas Producers*, 24–5.
31. Ibid.
32. Ibid, 24.
33. Jeremy Arbid, "Fostering Environments: Recommendations for Increased Participation in Lebanon's Petroleum Policy Discourse" (manuscript for publication by Basil Fuleihan Institute, 2014).
34. The OPRL provides that the Reconnaissance Licence may be awarded for a period of up to 3 years (Article 11), and that the EPA shall include two phases (Article 12): an exploration phase not exceeding 10 years and a production phase not exceeding 30 years.
35. Arbid, "Fostering Environments: Recommendations for Increased Participation in Lebanon's Petroleum Policy Discourse."
36. John Kingdom, "Multiple Streams Framework" concept (2003), as cited in Arbid, "Fostering Environments: Recommendations for Increased Participation in Lebanon's Petroleum Policy Discourse."
37. Marcel, *Guidance for Good Governance in Emerging Oil and Gas Producers*, 22.
38. Ibid, 23.
39. "Local Content Strategies in the Oil and Gas Sector: How to Maximize Benefits to Host Communities," 7.
40. Paul and Phillips, "Management of Energy Resources for National Development: The Trinidad and Tobago Model," 141.

14

LIBYA

Waniss A. Otman

Introduction

Libya is situated on the southern Mediterranean coast and is part of the African continent. It has a coastline of around 2,000 km and a land area of nearly1.76 million km^2, making it the fourth largest country in Africa by area. Libya is bordered by Egypt to the east, Sudan to the southeast, Chad and Niger to the south, Algeria to the west, and Tunisia to the northwest. Despite its vast area, its population in 2013 stood at a little over 6 million, with its inhabitants mainly concentrated in the Mediterranean coastal belt.

Following the political unrest of 2011, Libya has attempted to establish a new system of governance. Elections were held on July 7, 2012 for the General National Congress (GNC), which installed a new government on October 31, 2012. However, well-armed, autonomous militias continue to hold major power, causing serious internal security problems. At the time of this writing, the country has two rival governments and parliaments vying for power, based in Tripoli and Tubruq, as well as militia groups battling for control of the country's oil wealth. As stated by the IMF in June 2014, "Security conditions continue to deteriorate as political and sectarian tensions deepen and the legitimacy of the state is eroded."[1]

In Resolution 2144 of March 14, 2014, Article 6, the UN Security Council tasked the UN Support Mission in Libya (UNSMIL)[2] with a mandate covering four key areas:

1. Ensure the transition to democracy
2. Promote the rule of law and monitor and protect human rights

3. Control unsecured arms and related matériel in Libya and counter their proliferation
4. Build governance capacity

Currently talks in Geneva are ongoing, aimed at uniting disparate factions, based on UN-backed proposals on a comprehensive settlement and a unified government.

Macroeconomic Issues

As the political situation continues unresolved, it is impossible to produce meaningful macroeconomic data on Libya. Table 14–1 provides projections published by the IMF in its *Arab Countries in Transition* report of 2014.[3] At that point, the collapse in the international crude oil price had not yet occurred, and therefore this data must be seen as highly conjectural.

Table 14–1. Libya: Selected economic indicators, 2011–2015 (percent of GDP, unless otherwise indicated)

	2011	2012	2013 Estimated	2014 Projections	2015 Projections
GDP Growth, percent	–62.1	104.5	–13.6	–19.8	15.0
CPI inflation, period average, percent	15.9	6.1	2.6	4.8	6.3
Current government balance, excluding grants	16.1	27.8	–4.0	–52.1	–30.2
Current account balance, excluding grants	8.4	30.1	13.7	–27.0	–20.8
Fiscal financing needs, (excluding grants), US$ billion (1)	-----	-----	-----	-----	-----
External financing needs, (excluding grants), US$ billion (2)	-----	-----	-----	-----	-----
Public debt	-----	-----	-----	-----	-----
External debt	16.1	6.8	8.5	11.3	8.8
Reserves in months of imports	41.1	44.2	48.3	34.3	25.8

Sources: National authorities; and IMF staff calculations

(1) Budget deficit, excluding official grants, plus public external amortization.

(2) Current account deficit, excluding grants, plus amortization.

Industrial Knowledge and Manufacturing Base

Starting in the early 1970s, the Libyan government, in line with its socialist ideology, embarked on a program of state-sponsored economic development and control of the economy. Throughout this period the government expanded its role, effectively taking control of the entire Libyan economy. As well as nationalizing the key hydrocarbons industry, the Libyan state established controlling interests in the supply and operation of electricity, water, transportation, communications, and the seaports and airports. Public corporations also entered the banking and retail markets. A publicly owned import company, the National Supply Corporation (NASCO), was also awarded a monopoly in the early 1970s for the import and sale of many basic consumer items.

In the decades since this massive nationalization program, it has become apparent to all that the system in Libya had failed to deliver its economic goals. As Colonel Gaddafi stated publicly at the General People's Congress in Sirte in January 2000, "[T]he system is finished. I have to step in today to stop this wheel from spinning in a rut and wasting fuel." He accused members of the GPC of deliberately wasting the country's resources, saying, "You're holding onto obsolete methods in order to justify wasting oil."[4]

Following this, in June 2003, he called for the privatization of the oil, banking, public sector companies, and the aviation sector. On November 24, 2003 the new Libyan prime minister, Dr. Shukri Ghanem, announced a comprehensive plan for privatizing state-owned factories and companies, stating that the plan would cover "privatisation of heavy industries, especially iron, steel and chemical industries, trucks and buses assembly factories, textile, and shoe factories, and state owned farms." He added that the number of companies in the privatization plan would be more than 360, while the plan itself was to be applied in three phases until the year 2008, stating that "the Central Bank will sell shares of these companies and factories until a stock market is established." He stressed that this plan was aimed at "improving the national economic performance and the living standards of individuals through expanding the base of property owners."[5]

The General People's Committee (the Libyan Cabinet) issued Regulation No. 198 in 2000, regarding the establishment the General Board of Ownership Transfer (GBOT) as a sovereign and financially independent state body similar to the other secretariats (ministries) and under the direct supervision of the General People's Committee.

It had a wide remit:

- To perform preliminary studies relating to the companies and economic units targeted for transfer of ownership (privatization), recommend plans for their transfer, and present these to the secretary of the General People's Committee or the People's Committee of the Relevant Communes (Sha'abiyat) for approval to proceed
- To determine the most suitable way to effect transfer of ownership for each public company or economic unit in line with their levels of technical capability, and to decide on the necessary procedure to restructure them in preparation for a fast and successful transfer of ownership
- To list the surplus assets of the company or economical unit targeted for transfer of ownership, estimate their value, and set regulations for action in accordance with existing legislation and suggest the appropriate method or mechanism for each unit to be privatized
- To evaluate the target company and its assets in a way that reflects its present financial position according to normal financial practices
- To establish participation companies or shareholding companies and partnerships to hold the assets of the target companies and underwrite them
- To present the findings of the evaluation of the target units ready for ownership transfer to be approved by the secretary of the General People's Committee.
- To follow up on the outcome of transfer of ownership and to arrange for payment for privatization to a special account opened for that purpose
- To coordinate with the authorities concerned to effect a transfer of ownership that will assist in the successful implementation of the transfer of ownership program
- To suggest investment opportunities suitable for foreign investment[6]

Despite such initiatives, it must be concluded that, due largely to vested interests, the negative attitude and fear of redundancy of existing employees, and the uncooperative attitude of the Libyan General People's Committee, the privatization initiative was not successful. Inefficiency in the state-owned industrial sector continued to characterize most industries prior to the 2011 political unrest.

Infrastructure

Since the early 1960s, when Libyan oil wealth began to significantly increase, Libyan infrastructure, represented mainly by the transportation, aviation, and telecommunications sector, absorbed a large share of the annual development budgets. For example, between the years 1962 and 2004, an average of 13.37% of the government's annual development budget was spent on infrastructure. This expenditure was designed to access every geographically remote area in the country, with the Libyan government demonstrating, as in the housing, education, and health sectors, a high level of commitment to the principle that every Libyan had the right to equal treatment in terms of amenities. Very significant expenditure was therefore allocated to ensure that even the most remote settlement in this vast country was provided with road access and the basic necessities of a healthy life, such as clean drinking water and efficient waste disposal and sewer systems.[7]

In terms of the present, however, there are two factors that must be understood in attempting to gauge the overall situation regarding Libyan infrastructure, and the huge infrastructure deficit that undoubtedly currently exists in the country.

The first is the vast amount of capital expended on providing Libya's present inventory of infrastructure in the 1970s and 1980s, which means that much of Libyan infrastructure is now 30 to 40 years old. Because of the centralized system of the Libyan government, the government bears sole responsibility for maintaining infrastructure. In fact, the Libyan government virtually owns every business, public building, road, seaport, airport, university, school, community hospital, social services, library, and park and recreational facility—virtually everything in Libya. Government ownership also extends to a further range of infrastructural facilities, including water distribution and supply, sewage treatment, desalinization plants, sanitary and combined sewerage systems, storm sewers, and solid waste and hazardous waste disposal.[8]

The second factor is of course the massive war damage in the country, inflicted first during the NATO bombing raids in 2011 and subsequently during internecine militia fighting.

Looking at the situation preceding the civil war, even by the year 1999, Libya had around 100,000 km of roads. The main highway remains the 1,822-kilometer road that runs from the border with Tunisian to the Egyptian border, passing through Tripoli and Benghazi.[9]

The country is also well endowed with seaports and airports, as shown in table 14–2 and table 14–3.

Table 14–2. Libya: Principal airports

Libyan Airports ("I" signifies International)		
Airport	**Airport Code**	**Location: City**
Tripoli International (I)	TIP	Tripoli
Sabah (I)	MRA	Sabah
Benin (I)	BEN	Benghazi
Sirte (I)	SRX	Sirte
Ghat	GHT	Ghat
Ghadames	LTD	Ghadames
Misuratah	MRA	Misuratah
Ubari	QUB	Ubari
Kufrah	AKF	Kufrah
Martuba	DNF	Derna
Labraq	LAQ	Beida
El-Brega	LMQ	Marsa El-Brega
Houn	HUQ	Houn
Tubruq	TOB	Tubruq

Source: Otman and Karlberg, *The Libyan Economy*, 157.

Table 14–3. Libya: Seaports and harbors

Seaport	Location: City	Seaport	Location: City
Tripoli	Tripoli	Derna	Derna
Misuratah	Misuratah	Zuwarah	Zuwarah
Ras Lanuf	Ras Lanuf	Marsa El-Hariga	Tobruq
El Sidra	Ras Lanuf	Al-Zawia	Al-Zawia
El Brega	El Brega	Zuetina	Zuetina
Ghadames	Ghadames	Sirte	Sirte
Benghazi	Benghazi		

Source: Otman and Karlberg, *The Libyan Economy*, 159.

In terms of telecommunications, the country can be rated as sophisticated. As early as 2004, Alcatel and Nokia were selected to install a countrywide mobile phone network in Libya.

In terms of water provision, reference should be made to the so-called Great Man-Made River Project (GMRP). This project was initiated in 1983 to provide water to the water-deficient, predominantly coastal urban population, based on the extraction of gigantic water reserves underlying the Libyan Desert. The GMRP can be said to have emerged from discoveries made by the oil and gas industry. In the early 1950s, two laws pertinent to the oil industry were enacted: the Mineral Law of 1953 and Petroleum Law No. 25 of 1955. However, during the exploration for new oil fields in the deserts of southern Libya, not only were major oil reserves discovered, but also vast quantities of fresh water trapped in the underlying strata. The majority of this fresh water collected between 38,000 and 14,000 years ago, though some pockets are 7,000 years old.

In practical terms, apart from the very real benefits to the Libyan urban municipalities of the country's Mediterranean littoral area, there have been major impacts in Libya's agriculture sector in terms of irrigation projects in the coastal belt.

In terms of power provision throughout Libya, the General Electric Company of Libya (GECOL) is the state-owned vertically structured power utility responsible for generation, transmission, and distribution of electric power throughout the country. Currently the ultra-high voltage (400 kV) countrywide network has a capacity of generating more than 5,000 MW of electricity with mostly gas-fired and combined-cycle power plants in Zawiya, West Mountain, Tripoli West, Tripoli South, Al Khums, Zueitina, Benghazi North, Derna, and Tubruq. Transmission lines from the coast to Gradates, Sabah, and Al-Sarir guarantee countrywide distribution.

To get some idea of the excessive war damage caused during the Libyan political unrest to the power sector, it is useful to scrutinize the photographs in a recent GECOL presentation.[10]

Human Resources

During the Gaddafi era, the Libyan educational philosophy was essentially "education for all." School is mandatory through completion of basic education, which lasts up to the age of 15. Education was the state's responsibility and was financed by the Libyan government all the way to completion of university. However, the increasing number of school-aged children and the desire of a large number of them to pursue a college education led to a huge increase in the government's expenditure on education.

However, it can be rightly said that in terms of literacy, gender equality in education, student/teacher ratios, teacher training, and tertiary education, Libya ranks at or near the top in the Arab world.

This means that fundamentally Libyans are well equipped to participate at all levels in the oil and gas industry. This was demonstrated, for example, during the period of the United States and United Nations sanctions,[11] when Libyan technocrats and engineers were more than able to maintain the flow of produced oil and gas, despite the inability of the country to import vital spare parts for the largely US and European plants and equipment.

Local Content Policy in the Libyan Oil and Gas Sector

For the purposes of this chapter, the broad definition of local content provided by Tordo et al. will be used: "Local content can refer to jobs or value-added that is created anywhere in the domestic economy as a result of the actions of an oil and gas company."[12]

The first Libyan petroleum legalization was issued in 1953, two years after the country attained independence. The 1953 Mineral Law granted international oil companies (IOCs) undertaking preliminary geological reconnaissance the right to gather information relating to potential exploitation of the natural resources existing in the country, without any legal right of exploration or production. In view of the high level of interest from the IOCs, the first comprehensive Libyan petroleum law, Petroleum Law No. 25, was issued two years later in April 1955, replacing the Mineral Law.

Libyan Petroleum Law No. 25 of 1955

The Libyan Petroleum Law No. 25 of 1955, amended in 1961, 1965, and in the early 1970s, remains the most comprehensive piece of legislation regulating Libya's oil and gas sector.

In terms of ownership of oil and gas, Law No. 25 states in Article 1 that hydrocarbons found in their natural state in the subsoil layers of Libya are regarded as state-owned property.

In policy terms, the primary objective of Law No. 25 was to attract and encourage the IOCs to invest in Libya, diverting their historical interest from the Middle East. It was under this law that the granting of oil concessions began. Concessionaires were granted rights that covered vast areas, and the contracts were signed for long periods, typically 50 years. Fiscally, they can be described as a tax-royalty system, within which, effectively, the role of the Libyan government was reduced to that of a tax collector. The first concession was granted to Esso Standard Libya in mid-1955. One decade later, more than 20 IOCs were working in Libya in 137 concessions covering the most prolific areas in the country.[13]

According to Otman,

> [T]he classical concession agreements applied in Libya under the Petroleum Law No. 25 during the early stages of petroleum development activities were in many ways less favourable to government than were those obtained by other producing countries in terms of the economic and financial benefits. However, the raison d'être of these terms was their attractiveness to the IOCs. This turned out to be one of the key features that contributed to rapid growth of the Libyan upstream sector during the period from 1955 to 1971.[14]

As a result, Libyan production by the late 1960s had reached 3.3 MMb/d.[15]

Law No. 25 and local content—labor. Despite the rather pro-IOC tenor of the original version of Law No. 25, it is interesting to note that the drafters had at least begun to think about the future role of Libyans in the oil and gas (O&G) sectors.

Clause 15 of the Second Schedule, "Concession," entitled "Company's Labor," states explicitly that

> subject to the Immigration Laws in force at the time, the Company is hereby allowed to bring into the country such employees as may be necessary for its operations and the competent authorities shall facilitate the entry and exit to and from Libya and movement within Libya of such employees and their dependants while such employees are engaged on the Company's operations hereunder. The company shall have the right so to arrange its labour shifts that all operations may proceed by day and by night and during public holidays.[16]

On the other hand, Clause 18, "Employment and Training of Libyan Subjects," is rather interesting. It demonstrates that even from an early date, the issue of indigenous labor and the growth of Libya's domestic oil industry had been contemplated. It states:

1. Provided that the requisite number having adequate skill and ability is available, the minimum number of Libyan subjects employed by the company in Libya after 10 years from the date of commencement of operations shall have reached at least 75% of the total number of persons employed by the company in Libya.
2. The Company shall as from the date of commencement of regular exports from Libya of petroleum in commercial quantities produced in the concession area make an annual payment to the Libyan government of not less than LD 2500 and not more than LD 5000 which payment shall be applied towards giving Libyan subjects such technical training as may be agreed by the Director and the Company, in order to fit them for employment in the petroleum industry or in related undertakings, provided however that the Company may each year reduce such payments by the amount which during that year it expended for the training and education of Libyan subjects for such purpose in Libya and abroad.[17]

Law No. 25 and local content—equipment. Clause 10 of the 1955 law, "Exemption from Certain Import and Export Duties," is the clause that effectively undermined any rationale for local manufacturing. It stated:

(1) A permit or concession holder or any contractor employed by him may import free of duty:

(a) Plant, tools, machinery, equipment, supplies and materials; and

(b) Such other goods as may be specified from time to time by Regulations issued under the Customs Law.

Provided that all the goods mentioned in paragraph (a) or (b) above are used for petroleum exploration, prospecting, mining, transporting, refining operations and activities connected therewith. Provided however, that such exemption shall not apply to any goods included in this paragraph which are available in Libya of suitable type and reasonably comparable quality and at no higher price provided that in comparing price to the price of imported goods shall be added customs duties and other expenses incurred up to the time the imported goods reach Libya.

(2) Other goods which are dutiable under the Customs Law shall be subject to the payment of the appropriate duty.

(3) Any person intending to sell or transfer any goods which have been imported free of duty under paragraph (1) hereof shall before such sale or transfer make a declaration to the Customs Department and shall, unless such goods are sold or transferred to another permit or concession holder or contractor entitled to the same exemptions, pay on demand such import duty as may be assessed by the Director General of Customs in accordance with the Customs Law.

(4) Petroleum or any of its derivatives produced in the Libya and any goods imported free of duty under paragraph (1) hereof may be exported free of Customs duty and without an export license subject to the policy of the Government regarding exports in general and to such restrictions as the State may impose by Law on production and exports during a state of war or emergency.[18]

Unfortunately, because of the longstanding policy of the IOCs regarding the use in Libya of their own tried and tested equipment and contractors (e.g., Schlumberger), there are very few oilfield goods manufactured in Libya. Those that are manufactured domestically tend to be very low tech items such as catering equipment, food, or protective clothing, which have a relatively low impact on local content.

Evolution of the Libyan Exploration and Production Sharing Agreement (EPSA)

In the period after the emergence of the 1955 Petroleum Law until the emergence of the first iteration of the Libyan Exploration and Production Sharing Agreements (EPSA 1) in 1974, the Libyan fiscal legislation for the upstream sector can be described as erratic and uneven. The unsatisfactory joint venture agreements (JVAs) of the late 1960s and the nationalization/participation agreements of the early 1970s alienated foreign investors and led to a long decline in Libyan production and reserves. In an effort to counter this, a new form of petroleum exploration and development contract, known as the EPSA, emerged in 1974.

As stated by a leading researcher on the Libyan O&G industry,

> Two main factors led to the Libyan government's introduction of the EPSA. Firstly, the Libyan petroleum authorities, after the completion of the nationalization or participation process of its oil industry in the first half of the seventies, felt reasonably confident about their ability to exercise control over petroleum operations in the country.... Secondly, although global oil prices had remained low until the early 1970s, they literally took on another dimension between 1973–74, increasing by approximately four hundred percent in less than half a year. In these circumstances the oil companies were seen to be making excessive profits under the classical concession-type agreements.[19]

In its four iterations (EPSA-I, 1974–1979; EPSA-II, 1980–1988; EPSA-III, 1989–2004; and EPSA-IV, 2004–present), the Libyan EPSA gradually evolved into a successful commercial E&P arrangement. After 2004, the EPSAs began to attract major foreign investment again to the Libyan oil and gas industry. The first three EPSAs failed to attract the desired investment, due to a combination of harsh and unattractive fiscal terms and the geopolitical situation in Libya with US and international sanctions in the middle and late 1980s. The results of the first postsanctions bidding round of 2005, however, were deemed to be highly satisfactory both to the investors and the Libyan government.

EPSA-IV: Key sections on local content—labor

An analysis of the Libyan Model Exploration and Production Sharing Agreement (EPSA-IV) of 1974, a document 50 pages long comprising 27 articles, reveals the following sections relevant to local content.

Major relevant articles include the following:

5.1 Libyan Personnel and Training

5.1.1 Operator shall hire Libyan nationals to carry out Petroleum Operations in the Contract Area. Except that in cases where specialized technical personnel or key management positions are required and not available among Libya Personnel, Operator may hire non- Libyan nationals to carry out such Petroleum Operations.

5.1.2 Operator shall prepare and submit to the Management Committee for approval an annual training program. The objective of which is to enable Libyan personnel to fill posts of higher responsibility and to replace any non-Libyan personnel in the areas of specialization for which such non-Libyan personnel were hired. The annual budget allocated for such training program shall be sufficient to cover all costs associated with twenty four (24) man- months of the training per Calendar year during the Exploration period and shall not be less than ten percent (10%) of the manpower cost during each year of the exploitation period. For the purpose of this Article, manpower costs means the total direct and indirect costs associated of the Operator's employees.

5.2 Priority to Local Work

Operator shall at all times give priority for use of local supplies, equipment and services, provided that they are competitive in terms of performance, price and availability.

EPSA-IV: Key sections on local content—equipment

Section 5.2, Priority to Local Work. As presented above, this section states that the operator "shall at all times give priority for use of local supplies, equipment and services, provided that they are competitive in terms of performance, price and availability."

Section 19.2, Customs Duties. This sections states that the "second Party (i.e., IOC) shall be exempted from customs duties according to the Petroleum Law."

New Draft Petroleum Law 2004, General People's Congress

In Libya today, more than 60 years after its enactment in April 1955, Libyan Petroleum Law No. 25 is still the basic legislation that covers investment in the petroleum sector, including the exploration and production sharing agreements (currently EPSA IV). No new oil concessions have been granted in the previous five decades, and the majority of the concessions expired in 2005, their maximum duration being 60 years after a 10-year extension.

Since the early 2000s, the previous Libyan government under Gaddafi had been drafting a new petroleum law. In fact, a comprehensive draft of the law had been circulated in 2004 to senior Libyan policymakers, including the author, at that time. A detailed examination of the New Draft of the Petroleum Law 2004 reveals that the following clauses dealt with issues related to local content concerning labor:

> Article (5)
>
> Secretariat shall supervise the implementation of oil policy of the state, and is working to follow the best ways to preserve oil resources and ensure the proper utilization and achieve greater returns from, and the Secretariat shall represent the State in all international forums and organizations related to the oil and gas sector. The Secretariat shall review the technical and economic contracts and agreements involving the exploitation of oil and the Secretariat shall compose of number of qualified consultants and a number of specialized departments in the areas of inspection and measurement, environmental protection and oil accounting and international cooperation.
>
> Article (12)
>
> The Contractor shall comply to establish in the State the infrastructure capable of providing technology transfer and resettlement required for the oil operations.
>
> Article (30)
>
> The contractor for the purpose of implementing its obligations under oil exploitation contract shall exercise the following dependency rights in the contract area, according to the legislation in force:
>
> . . .

4. Using of foreigners in the fields that require specialized expertise that is not available among the Libyans or need high-technology as defined by the rules and regulations in force. The expatriates recruited from abroad may transfer their salaries and wages and any benefits or rewards given to them according to the terms and conditions specified in the rules and regulations in force.

Chapter X of the New Draft of the Petroleum Law 2004 also addressed the employment of Libyans and the use of local materials and equipment:

Article (63)

The Contractor shall comply to employ the qualified Libyans at the level of executive positions and skilled labour necessary for the oil operations and related work.

And shall prepare annual training programs to increase the number of Libyans and rehabilitating them for technical works, high positions and senior executive responsibilities for the operations described, and to replace them gradually with the foreign expatriates. The contractor is prohibited to bring in a regular employment which is available locally. And the Corporation determines the rates, stages, controls, and areas of employment and training listed and the obligations of the contractor to take account of the needs of the oil industry and expertise where required.

Article (64)

The Contractor shall comply with using the qualified Libyan contractors and give priority to the local materials, manufacturers, equipment, services and the means of crude oil transport and its products that are locally available when their prices and quality were competitive compared with those abroad.

The New Draft Petroleum Law 2004 also included key articles related to local content requirements concerning equipment:

Article (40)

The material, equipment and devices used in oil operations, including drilling rigs types, offshore materials, equipment and vessels equipped for the geophysical survey and which are imported for petroleum operations shall be exempted of the customs duties and production tax and other fees under any guise, whether the importation is temporary or permanent.

> In the case of sale or transfer of ownership of things covered by the exemption provided for in this Article for the benefit of a Non-Party who does not enjoy the same exemption, the contractor or the buyer shall pay before the sale or transfer of property as prescribed by law and taxes and fees.

The draft New Petroleum Law (NPL) evolved from a committee established by the Libyan National Oil Corporation (LNOC) in the late 1990s. Under the LNOC chairmanship of Dr. S. Ghanem, a final draft of the NPL was reviewed and sent for approval to the General People's Congress (the Libyan Parliament) in 2010. A crucial issue stalled the adoption of this law, however, since agreement could not be reached on who actually had oversight and control over the petroleum sector—the LNOC or the Ministry of Energy.

Political events in 2011 of course undermined any further attempts to agree on this issue and officially approve a definitive document.

Although the NPL attempted to modernize important aspects of both the upstream and downstream sectors, in its current form it must be rated as a missed opportunity. Key issues such as shale gas development, decommissioning, and liberalization and deregulation of the upstream and downstream sectors were not addressed. Also left unaddressed were seismic options for the high-risk basins and development of the domestic oil and gas industry. It is for these reasons, among others, that a radical overhaul of the current NPL draft needs to be undertaken as soon as civil government and security in Libya are restored.

Libyan National Oil Corporation (LNOC) and Local Content

In 1970 the new Libyan government issued Law No. 24, establishing the Libyan National Oil Corporation (LNOC), 100% owned and controlled by the Libyan government. Conceived by the new government as a key, if not the key, Libyan state agency, it assumed immediate responsibility for the development, administration, and exploitation of oil and gas resources at all stages, with a mandate to shape and structure all of these into an integrated industry. In line with Law No. 24, the remit of the LNOC was comprehensive, as intended, and it was tasked with achieving ambitious objectives for the national economy.

Its formation eliminated any duplication in control over the oil companies, and its form and structure as an independent body operating under the Oil Ministry gave it broad powers to regulate as well as participate in the Libyan hydrocarbons industry. It had sole responsibility for the sale, distribution, and transport of hydrocarbons both domestically and overseas. It also had authority to construct pipelines, harbors, and oil and gas terminals, as well as downstream facilities such as refineries and petrochemical plants.

From the beginning, the law gave prominence to the LNOC's financial independence from the state, providing it with an independent budget tailored to a purely commercial model. It created financial resources for the LNOC to keep it free from any financial constraints that might lead to rigidity and incapacity to achieve its national goals.

This farsighted policy of the new government, in setting up LNOC on such a strong foundation, enabled it to grow into the major pillar supporting the national economy. By its establishment, the role of the Libyan government changed overnight from a dormant collaborator and tax collector to that of an active player. The LNOC held the power and responsibility to formulate all policies for utilizing the vital national oil resources.

In the upstream sector, the LNOC approves bidders in the licensing rounds. It issues, and is the government's sole signatory to, new EPSAs. It is also a player in its own right and owns 100% of many of the key producing companies in Libya.

The LNOC also carries out E&P operations through its own affiliated companies or in participation with other companies under service contracts and other types of petroleum investment agreements. This is in addition to marketing oil and gas locally and abroad. To conduct business in these different sectors, the LNOC has its own fully owned companies to carry out exploration, development, and production operations, in addition to local and international marketing companies. The LNOC also has participation agreements with specialized international companies. Such agreements have developed into E&P sharing agreements, in accordance with the development of the international oil and gas industry and international petroleum marketing.

The LNOC owns refining and oil and gas processing companies. It operates the Zawiya and Ras Lanuf refineries; various ammonia, urea, and methanol plants; the Ras Lanuf petrochemical complex; and the gas processing plant. To establish petrochemical industries, anther stage of development of the ethylene plant has been completed, as well as the low- and high-density linear polyethylene plants.

The LNOC also owns national service companies that carry out oil well drilling and work over operations, provide all drilling material and equipment, and lay and maintain oil and gas pipelines. The LNOC also builds and maintains oil and gas storage tanks and carries out related technical and economic studies. It provides the sector with other services as well, such as catering, procurement of materials and equipment, and training and employment of expatriates seeking foreign jobs (see table 14–4).

Table 14–4. LNOC E&P companies

Fully Owned by LNOC	LNOC Joint Ventures with IOCs
Arabian Gulf Oil Company (AGOCo)	WAHA (Oasis) Oil Company
Sirte Oil Company	Akakus (Repsol) Oil Operation Company
National Oil Wells Drilling and Workover Company	Mellita (Eni) Oil & Gas Company
North Africa Geophysical Exploration Company	Harouge (former Petro Canada) Oil Operation Company
Taknia Libya Engineering Company	Mabruk (Total) Oil Operation Company
Zawia Oil Refining Company	Zueitina (Oxy) Oil Company
Brega Petroleum Marketing Company	Nafusah Oil Operation Company
Jowfe Oil Technology Company	RasLanuf Oil and Gas Processing Company
National Oil Fields and Terminals Catering Company	
Petro Air Company	

Source: LNOC

As can be surmised, the LNOC since its inception has expanded into every section of the Libyan economy and dominates the Libyan petrochemical and downstream distribution sector.

While it might be argued that at the time of its establishment, the LNOC was a successful effort at developing the domestic petroleum industry in Libya, completely staffed by Libyans, it has since evolved into an inefficient behemoth. Serious consideration must now be given to its radical overhaul, to bring it in line with modern concepts of corporate governance and effective management (see table 14–5).

Table 14–5. Oil and gas companies operating in Libya

IOCs Operating Under EPSA	IOCs Operating Under EPSA
Eni North Africa Company	Petrobras Company
Amerada Hess Company	Chevron Libya LTD Company
India oil Company	Shell Company
Total E&P Company	RWE Company
Petro Canada Company	Sonatrach Company
Polish Oil & Gas Company	Turkish Petroleum Corporation
OMV Company	Medco Energy Company
OXY Company	Exxon Mobil Company
BP Exploration Libya Limited Company	ONGC Limited Company
STATOIL Company	Tatneft Company
Gazprom Company	Wintershall AG Company
RepsolMurzuq Company	

Source: LNOC

Research and training centers

The section of the current LNOC website regarding personnel and local content states the following:

> In the area of manpower development, NOC provides the oil industry with qualified nationals within a well-planned scheme founded on a base of educational and training institutions, for training and developing qualified manpower in such professions as engineering, accounting and administration.
>
> Training outside Libya is limited to those technical specializations which are not available locally, to cope with the rapid development of the industry. Technical training is being carried out at the training centers and institutions belonging to NOC, to develop specialized technicians for the operation and maintenance of industrial facilities and plants.[20]

In line with the above, prior to the 2011 political unrest, research and training institutes for Libyan nationals, ranging from hands-on skills to tertiary education in the O&G sector, had been long established. Affiliated with the LNOC is a center that carries out research and technical studies related to the oil industry. It conducts technical analysis and tests for the various stages of exploration and production of oil and petroleum products, performs quality control tests, and issues certificates in this respect. It also evaluates patents and licenses of exploitation and the fees and forms related to oil operations and petroleum products. It publishes the studies carried out by the center in its in-house publication, as well as in local and international scientific publications.

Petroleum Training and Qualifying Institute

The Petroleum Training and Qualifying Institute in Tripoli was established on August 8, 1970 as the Libyan Institute of Petroleum Affairs. It is a nonprofit educational institution that has been supplying the national oil industry with well-trained, qualified technicians since its inception.

The Specific Training Center of Petroleum Industrial (Zawiya)

This organization trains technicians to meet the requirements of the oil refining and chemical and petrochemical industries.

Libyan Petroleum Institute

Libyan Petroleum Institute (LPI), formerly known as Petroleum Research Center (PRC), was established in 1977 as the technical arm of the NOC. The vision and mission of the LPI is to be the leading standard setter and preferred national laboratory services provider for the oil and gas industry. It is tasked with providing reliable laboratory testing services and assisting efforts to increase the productivity of the petroleum industry in the region.

The LPI consists of more than 40 laboratories specializing in different branches of the petroleum sector (upstream and downstream activities). It employs more than 400 competent personnel, of which 25% have professional and academic degrees, in order to provide high-quality laboratory testing services and technical studies. Fields of expertise include exploration, reservoir exploitation, oil and gas processing, geological and geophysical data management, and a wide range of services in environmental monitoring.

In addition to the technical departments, the LPI consists of four supporting departments working together to achieve LPI's goals.

In order to provide high-quality services to its customers, the LPI was granted ISO17025 accreditation, which guarantees and increases the accuracy and reliability of the tests and calibration. The LPI is also committed to documenting, implementing, and maintaining a quality management system based on ISO 9001:2000.

As well as the above, the Bright Star University of Technology in Mersa El-Brega and the University of Tripoli provide degrees in petroleum engineering and related disciplines.

Appraising the Libyan Local Content Approach

An appraisal of the legal architecture of the Libyan oil and gas industry relating to local content, as described previously, demonstrates that since the promulgation of the Libyan Petroleum Law in 1955, issues related to local content have not developed in any meaningful way. In fact it might even be argued that while the 1955 law was, for its era, farsighted in this regard, subsequent iterations of the Libyan EPSA did not attempt to build on this foundation in any structured manner.

Although, as discussed earlier, the preeminent position of the LNOC provided it with access to potentially the best E&P opportunities in Libya, it did not have the technical or managerial capacity to fully capitalize on its powerful position. Thus despite the existence of local institutions such as the LPI, the educational system did not produce either the quality or quantity of technicians/graduates required to drive an industry renowned for its constant search for technological innovation. This effectively meant that expatriates filled all the top- and medium-level positions, and accordingly commanded the top salaries.

This lack of economic empowerment among young people not only in Libya but elsewhere in the region has, in fact, been cited by some authors as a significant contributory factor to the so-called Arab Spring.[21] This is a far cry from the situation in Norway, where very early on in their experience with oil resources, the Norwegian public and policy makers grasped the crucial importance that a dynamic oil industry could have as a driver for local skills, employment, and indeed technical innovation by Norwegian companies and individuals.

Even though the EPSAs are not explicit enough in themselves regarding local content, still, the EPSA is a powerful document assigning significant leverage to the LNOC. The LNOC by default will hold at least 51% in each EPSA.

Certain articles in the current EPSA-IV Model contract bear this out, as seen in the following:

Article (4) Management Committee

4.1 Appointment of Management Committee

Petroleum Operations in the Contract Area shall be conducted under the control and supervision of a management Committee composed of four (4) members, two (2), including the chairman, to be appointed by First Party, and two (2) to be appointed by Second Party. The Management Committee shall be formed not later than one (1) month after the Effective Date. Each Party shall notify the other of the members of the management Committee it appoints.

4.2 Decisions

The Management Committee shall make all important decisions relating to Petroleum Operations, including, without limitation, the approval of Work Programs and Budgets prepared by Operator for Petroleum Operations. Decisions of the Management Committee shall be made by the unanimous vote of its members. Decisions of the Management Committee have indicated their approval of a decision in such correspondence.

Invariably by virtue of holding 51% and by convention, the chairman of the Management Committee will be an LNOC nominee. This in itself represents a clear opportunity for the LNOC, if it so desires, to influence decisions relating to approval of work programs and the AFE (Annual Financial Expenditure). The latter covers, among other things, both hiring and purchase of equipment, decisions about payment and staffing, and provisions for training of Libyan citizens. Here, a strong policy of in-country purchase and maximizing the use of Libyan citizens could easily have been implemented had the political will existed.

A similar opportunity to enhance the development of the domestic oil and gas industry in Libya is apparent from Article 5.6:

Contractors

Operator may engage contractors to carry out any part of the Petroleum Operations authorized under this Agreement, Provided, however, that

(a) Operator has obtained the approval of the Management Committee for standard terms and conditions of contracting.

(b) Operator has obtained the specific prior written consent of the Management committee in any case where the total contract price for the petroleum operations to be carried out by a contractor exceeds one million U.S. Dollars (U.S. $ 1,000,000.00).

(c) Operator shall at all-time use Libyan contractors, provided that they are Competitive in terms of Performance, price and availability.

Of course the foreign investor would argue with regard to Clause C that such Libyan contractors did not exist. But in view of the huge advantages to the Libyan economy and to Libyan citizens that would result from the existence of such contractors, this was a massive failure. Libyan government policy should have tried to ensure through "joined up thinking" that the path for such contractors to develop was clearly signposted. Financial incentives should have been made available to such local contractors through the Libyan banking system, which the government controlled in any case.

Similarly, Article 5.7, "Libyan Personnel and Training," represents another powerful tool that the LNOC had at its disposal:

> 5.7.1 Operator shall hire Libyan nationals to carry out Petroleum Operations in the Contract Area. Except that in cases where specialized technical personnel or key management positions are required and not available among Libya Personnel, Operator may hire non- Libyan nationals to carry out such Petroleum Operations.
>
> 5.7.2 Operator shall prepare and submit to the Management Committee for approval an annual training program. The objective of which is to enable Libyan personnel to fill posts of higher responsibility and to replace any non-Libyan personnel in the areas of specialization for which such non- Libyan personnel were hired. The annual budget allocated for such training program shall be sufficient to cover all costs associated with twenty four (24) man- months of the training per Calendar year during the Exploration period and shall not be less than ten percent (10%) of the manpower cost during each year of the exploitation period. For the purpose of this Article, manpower costs means the total direct and indirect costs associated of the Operator's employees.

Again, Article 5.8, Priority to Local Work, demonstrates missed opportunities:

> Operator shall at all times give priority for use of local supplies, equipment and services, provided that they are competitive in terms of performance, price and availability.

Had the government wished this to happen, there were already, in international best practices, clear and well-understood methods for this to be put in place, even if gradually.

Within the Norwegian context, as early as 1972, the Norwegian Ministry of Industry had already established the Goods and Services Office. This office performed the functions of a watchdog agency to control the oil companies' contracting and procurement activities. Prior to tender invitations, the operator had to announce the tender schedule and companies to be invited. The ministry's role was to ensure that qualified Norwegian companies were included on the list of bidders.

The strong emphasis placed on the Norwegian content (local content) made it a crucial factor for all bidders; it was a key criterion when evaluating companies competing for new acreage. The ministry's policy was to be transparent in its enforcement of the procurement policy. Moreover, with an observer present in all licensed groups, the ministry secured insight into all the operators' contracting activities. A high level of cooperation between the authorities and the operators, together with the contracting and supply industry, was essential in developing the high local content in Norway, which at times came to exceed 70%. Under the Norwegian license terms for the IOCs, it was mandatory to transfer skills and competence to Norwegian companies. Initially personnel from Statoil, Norsk Hydro, and Saga would participate in the oil majors' training courses, receiving on-the-job training in their overseas operations. The majors were forced to recruit young Norwegian engineers for overseas training for a significant period, before they returned home to help increase the involvement of the domestic Norwegian industry in their various organizations.[22]

Finally, a local content-related issue that the Libyan government failed to grasp was the importance of Libyan students mastering the English language. Like it or not, virtually all technical manuals relating to oilfield equipment and operations are written in English. The failure of the Libyan educational system to produce English language speakers and to execute a compulsory English stream in the curricula can be cited as a very grave error.

While it is clear that there was no serious and sustained effort by the Libyan authorities to initiate a sustainable local content policy, it must be mentioned, in all fairness, that certain foreign contractors initiated their own Libyan local content efforts. For example Wintershall, part of the BASF group, and operator of two concessions in southeast Libya, commenced its in-house Libyan Integration and Development Program in 2001, resulting in the construction and operation of its Occupational Training Center in Jakhira in Libya in 2007. This center was established to provide in-house vocational training to its Libyan employees. According to one case study, by 2013, Wintershall had successfully reached its initial target ratio of 75% nationals within the overall staffing of the company. Administrative and financial functions were run 100% by Libyan nationals, while technical functions were gradually moving toward that target.[23]

Current Issues and Challenges for O&G Companies

As stated previously, the key challenges to oil and gas companies, from an operational point of view, are presently related to physical internal security. This applies not only to their personnel and equipment, but also to the final product itself, crude oil. The international media has also reported countrywide protests on issues ranging from illegal immigration (Libya has become a major port of immigration to Europe) to ISIS expansion and lack of internal security. Because of these concerns, virtually all foreign oil and gas companies have completely ceased operations in Libya until the security situation is stabilized.

Again, recent media reports suggest that ISIS is making massive military inroads in many parts of Libya, for example in the battle for Sirte in August 2015.[24] This has led to many international analysts concluding that Libya is now a failed state.

Another crucial issue facing oil and gas companies is the status of Libya's National Oil Corporation (LNOC), which is at the center of a tug of war between the official internationally recognized government, based in Tubruq, and the rival government based in Tripoli, controlled by the Libya Dawn movement and backed by Qatar, Turkey, and Sudan.

The LNOC has been based in Tripoli since its inception in 1970, and all foreign contractors have signed their current EPSAs with the organization there. In Benghazi, the LNOC is now known as NOC East, but since oil and gas companies have traditionally dealt with the LNOC in Tripoli, this stand-off has posed potentially huge legal questions for oil companies. Challenging issues concern existing contract validity and the ownership and sale of produced oil. Essentially, it means that the oil and gas industry is in a state of paralysis.

Conclusion

Taking an optimistic view of Libya, at some future point in time the country will emerge from the existing chaos. At that point there will clearly be an opportunity for a radical new approach to seriously enhancing local content in Libya's vitally important oil and gas industry.

Without doubt, the past approach has failed. Nevertheless, there is currently emerging a global consensus, after many years of trial and error, on some key principles that underpin a successful approach to local content, i.e., one that can work for both the host country and the investor. Nigeria, through the Nigerian Oil and Gas Industry Content Development Act 2010 and the establishment of the Nigerian Content Development and Monitoring Board (NCDMB) soon thereafter, is a good example of this new approach.[25]

International best practices, as defined for example in such recent works as the 2013 World Bank Study, *Local Content Policies in the Oil and Gas Sector*,[26] need to be based on a new paradigm. They should encompass a range of new basic principles, including the following:

1. The relationship between the host government and the IOC needs to be more collaborative than adversarial.
2. Once the collaborative relationship is established between the host government and the IOC, both need to assess jointly existing in-country capacities with the HR and technical needs of the O&G companies.
3. The host government needs to establish and encourage strong partnerships between local businesses and the IOCs.
4. The host government and the IOCs need to agree on measurement criteria, i.e., how governments/NOCs and the IOCs measure and verify local content, for example by using an independent third party.
5. The government needs to take a long-term view. One important aspect of this is setting local content targets that are realistic and achievable in order to gradually and sustainably build up local expertise and content.
6. The government must ensure that legislative requirements relating to local content are gradually phased out so that local suppliers,

technicians, and locally manufactured equipment can emerge as globally competitive, as for example the Norwegian subsea sector, which is now an international player.

7. The government must create an environment that supports the domestic petroleum industries so that the economic linkages can evolve organically, not forcibly.
8. Through schools, technical colleges, and universities, the government must ensure that the right skills are being developed in the domestic workforce. A good example of this occurred in Denmark, where a two-year master's of science program was established in wind energy at the Technical University of Denmark. Governments should consider establishing domestic campuses that match the quality of academic providers in relevant sectors from European or US universities.
9. The government should support local entrepreneurs. This can be done, for example, through a government entrepreneurial fund or by mandating the banking sector to ensure that funding is available for new ventures related to local content in the O&G industry.
10. Finally, and very obviously, the government should ensure that the domestic infrastructure is established and maintained so that local entrepreneurs and industry can flourish. Such infrastructure includes roads, power supply, harbors, dry docks, and perhaps special economic zones.

Several of these initiatives would be very effective, and O&G policymakers in a postconflict Libya should seriously think about their implementation. For example, principle 3, that the Libyan government should establish and encourage strong partnerships between local businesses and the IOCs, would be hugely important in creating a group of techno-savvy Libyan entrepreneurs. Such businesses could operate at all levels in the oil and gas chain, ranging from seismic operations and drilling technology and equipment to pipeline design and installation. As a result, value would be created in the Libyan downstream petrochemical chain, such as refineries and fertilizer, polyethylene, polyvinyl chloride (PVC), and polypropylene (PP) plants, all of which have massive potential for Libya as a regional exporter.

Furthermore, principle 7 has resonance for Libya. Although O&G-focused educational establishments exist, such as the Libyan Petroleum Institute, as mentioned earlier, the depth and breadth of the fields of study offered need to be expanded and modernized. This is vital to the ability of Libyan nationals to fill key senior engineering roles in the industry.

Principle 9 is another key initiative that is within the Libyan government's power to implement, i.e., to support and develop a body of local entrepreneurs. This can be achieved by establishing a government entrepreneurial fund, such as the United Kingdom's Energy Entrepreneurs Fund,[27] or by mandating the banking sector to make funding available, at preferential rates, for new ventures related to local content in the O&G industry.

Finally, a crucial requirement to enhance local content, delineated in principle 10, is one that the author has explored exhaustively and in detail elsewhere.[28] This is the provision of key infrastructure such as roads, power, harbors, dry docks, special tax-free zones, etc., to ensure that local entrepreneurs and industry can evolve expeditiously. In a postconflict Libya, farsighted government planning can achieve this most important of goals.

Notes

1. International Monetary Fund, *Arab Countries in Transition: An Update on Economic Outlook and Key Challenges* (Washington, DC: IMF, April 9, 2014), http://www.imf.org/external/np/pp/eng/2014/040914.pdf.
2. "Home Page," United Nations Support Mission in Libya, http://unsmil.unmissions.org/.
3. IMF, *Arab Countries in Transition: An Update on Economic Outlook and Key Challenges*, 14.
4. Waniss Otman and Erling Karlberg, *The Libyan Economy: Economic Diversification and International Repositioning* (Berlin: Springer, 2010), 217.
5. Ibid.
6. Ibid., 222.
7. Ibid., 153.
8. Ibid., 154.
9. Ibid., 156.
10. Presentation, General Electric Company of Libya (GECOL) (2012), http://www.energyafrica.de/fileadmin/user_upload/Energy_Africa_13/Presentation_GECOL_Ashaibi_PAnel%204a_7th%20German-African%20Energy%20Forum.pdf.
11. US sanctions extended from 1981 to 2004; UN sanctions extended from 1992 to 2003.
12. Silvana Tordo, Michael Warner, Osmel E. Manzano, and Yahya Anouti, *Local Content Policies in the Oil and Gas Sector* (Washington, DC: International Bank for Reconstruction and Development, World Bank, 2013), 3, doi: 10.1596/978-0-8213-9931-6, or http://documents.worldbank.org/curated/en/549241468326687019/pdf/789940REVISED000Box377371B00PUBLIC0.pdf.
13. Waniss Otman, *Economic and Legal Aspects of Investment in the Libyan Oil and Gas Industry* (Aberdeen: Shell Oil Company, University of Aberdeen, 2004).

14. Ibid., 6.
15. Waniss Otman, "Appraising the Libyan Petroleum Agreements: A Comparative Analysis of Risk Factors & An Evaluation of their Effectiveness," *Oil, Gas & Energy Law*, 5, no. 4 (November 2007), www.ogel.org.
16. Petroleum Law No. 25 of 1955, as amended through February 12, 1970, Libyan Government, undated, 43.
17. Ibid., 44.
18. Ibid., 39.
19. Waniss Otman, "Evaluation of Investment in the Libyan Upstream Oil Sector" (PhD thesis, University of Aberdeen, 2006).
20. "About Us: NOC in Brief," National Oil Corporation, http://noc.ly/index.php/en/about-us-2.
21. Naoual Belakhdar, Ilka Eickhof, Abla el Khawaga, Ola el Khawaga, Amal Hamada, and Cilja Harders, *Arab Revolutions and Beyond: Change and Persistence*, ed. Serena Sandri (Working Paper No. 11, Center for Middle Eastern and North African Politics, Berlin, August 2014), http://www.polsoz.fu-berlin.de/polwiss/forschung/international/vorderer-orient/publikation/WP_serie/WP11_Tunis_Conference_FINAL_web.pdf.
22. Øystein Noreng, "The Norwegian Experience of Economic Diversification in Relation to Petroleum Industry," *Oil, Gas & Energy Law*, 2, no. 4 (October 2004).
23. Phil Andrews and Jim Playfoot, *Education and Training for the Oil and Gas Industry: Building a Technically Competent Workforce* (Amsterdam: Elsevier, 2015)
24. "What Comes After Sirte's Uprising?" *Libya-Analylis.com* (August 20, 2015), http://www.libya-analysis.com/?s=sirte+uprising.
25. "About NCDMB," Nigerian Content Development and Monitoring Board, http://www.ncdmb.gov.ng/index.php/about-ncdmb/what-we-do.
26. Tordo et al., *Local Content Policies in the Oil and Gas Sector.*
27. *The Energy Entrepreneurs Fund: Fourth Phase Guidance Notes* (London: Department of Energy and Climate Change, December 2014), https://www.gov.uk/government/uploads/system/uploads/attachment_data/file/378464/Phase_4_Guidance_notes.pdf.
28. Waniss Otman and Erling J. Karlberg, *The Libyan Economy: Economic Diversification and International Repositioning* (Berlin: Springer Science & Business Media, 2007).

15

MOZAMBIQUE

Paula Duarte Rocha and Tiago Arouca Mendes

Introduction

Mozambique is a country rich in natural resources, especially those that can be transformed into energy. The main sources of energy are coal, petroleum (which includes oil and gas), and renewable energies, which together provide for the primary energy sources in Mozambique, estimated in 2014 at 0.45 quadrillion Btus.[1]

It is important to note that in Mozambique, all petroleum resources located in the soil or subsoil or under inland waters, territorial seas, the continental shelf, and in the exclusive economic zone (EEZ) are the property of the state. As a result, private enterprises in the oil and gas industry will have to deal with a sovereign power.

Despite massive potential for electricity and fuel generation, energy consumption in Mozambique remains heavily dependent on firewood and other wood fuels (such as charcoal). This reliance on wood fuels is due to a lack of socioeconomic development, as well as limited access to electricity and natural gas networks.

Compared to the country's increasing natural gas sector, the oil sector in Mozambique is in an early stage. Nevertheless, international players have been lining up and prepping the field to ramp up E&P efforts. Oil production was due to commence in an oil field discovered next to the Temane natural gas field (both operated by Sasol). Unfortunately, market economic fluctuations have led to its delay.[2]

The development of the petroleum sector is an asset for the Mozambican economy. Recognizing this, the government has been actively preparing the legal framework necessary for both the industry and the nation of Mozambique to prosper in the sector.

Local Content Policy and Baseline

Policy background, including short- and long-term goals

In 2009, the government of Mozambique approved the Strategy for the Concession of the Areas for Petroleum Operations. This policy was directed toward prospective areas in the sedimentary basins of the country and other regional or international areas, including its territorial sea, EEZ, and continental shelf. The 2009 Strategy contained mandates related to pursuit of oil and gas (O&G) activities to ensure the proper and efficient management of the areas and the resources.

The 2009 Strategy was implemented to provide guidelines concerning how the state would govern the process of granting the rights for petroleum E&P in domestic, territorial, or international areas. Under international law, Mozambique has sovereign rights over these areas for the purposes of research, exploration, and extraction of natural resources.

The 2009 Strategy aims to ensure the continuation of the systematic search for hydrocarbons in the basins of Mozambique and Rovuma and to promote investment in research and exploration of petroleum in the national territory. It also defines the priority prospective areas, while complying with the necessary transparency and internationally accepted good practices of the industry.

Regarding the potential of the national sedimentary basins, these offer areas with strong potential for the occurrence of oil, and there are natural gas discoveries dating back to the 1960s, with the most recent in 2008 and 2010.[3] Even with these discoveries, Mozambique is still considered a country with few basins studied, given the low density of search boreholes per unit area (square kilometers).

The sedimentary basins of Mozambique are relatively unexplored. The Mozambique Basin, with an area of about 300,000 km^2, has a density of approximately 1 borehole per 8,000 km^2 on land and 1 borehole per 17,000 km^2 offshore. The Rovuma Basin, with 60,000 km^2, has a borehole density of 1 per 17,000 km^2 on land, and there are no offshore boreholes. The remaining basins lack any significant research activities.

To date, 202 boreholes have been drilled in Mozambique, including 108 research boreholes, 23 production boreholes, and 63 assessment boreholes.[4]

In the gas sector, the potential gas reserves amount to 180 Tcf (trillion cubic feet) of gas. Production contracts were signed for the use of the reserves and for the construction of the infrastructure necessary for processing and transport. For example, a pipeline of 865 km was constructed to connect the gas processing center in Temane to the South African market (in Secunda) and to the Maputo industrial park.[5] In addition, five interconnection points were installed for domestic natural gas use.

A major liquefied natural gas plant is currently under development in northern Mozambique. Its operators expect it to be one of the biggest operations in the world in terms of gas processing volume when it commences operation.[6]

In order to carry out petroleum operations, Mozambique has conducted five international tenders for petroleum E&P concession areas. The government has conducted successful bidding tenders since 2005, with the latest being in 2014.[7]

Through the Strategy for the Concession of the Areas for Petroleum Operations, the government defined the steps considered necessary to successful development of the petroleum sector. Considerations included the definition of the areas to be put to tender, as well as the areas that are subject to direct negotiation or other modalities of attribution of concessions. The consolidation of national business participation in the petroleum operations sector was not overlooked, and it remains a major focus of the government of Mozambique. One way the strategy to consolidate national business participation has been carried out is through strengthening the role of the state-owned company Empresa Nacional de Hidrocarbonetos (ENH). Recently enacted petroleum legislation established that the ENH must always be a party on the concession agreements granted to foreign or domestic companies. Additionally, domestic companies are encouraged to associate themselves with foreign companies that can add technical and financial value to the projects. This should help the Mozambican government to attain its goal of national companies participating during the development and production phases of the oil industry.

Overview of key prevailing local conditions

Overall, Mozambique's transport infrastructure and links lag far behind, decades after the civil war ended in 1992. Transport projects worth US$17 billion are in the pipeline, including increased rail links to ports and expanding port capacities to allow for greater exports. It should be noted, however, that these projects are being developed primarily to serve the petroleum and other energy industries. As a result, uneven infrastructure build-out will be inevitable. Transport infrastructure is mostly being developed transversally, west to east, connecting mining and agricultural clusters in Mozambique and neighboring countries to exit ports. The central transport infrastructure extends from the Port of Beira to Zimbabwe, and marginally to Malawi and Zambia.[8]

Also, the concentration of Mozambique's urban population in Maputo means that this will continue to be the hub for infrastructure development projects over the medium term.[9]

There are seven main seaports in Mozambique, of which Maputo, Beira, and Nacala are the most important. All of these have functioning rail linkages to the interior and rural areas and neighboring countries via the rail network, as well as a few adequate road linkages.[10] Despite important progress in the modernization of Mozambique's port systems, there is still a significant lag between the increase in demand and the development of infrastructure projects to meet that demand.

Aside from the Maputo International Airport, airports in Mozambique tend to service either tourist or economic destinations, such as mining hubs.

In general, Mozambique has one of the least developed road transport infrastructures in the southern African region. Not taking into account the major arteries, overall road conditions in Mozambique do not meet the expectations of the international investors who wish to cross the country to develop their projects.

Mozambique is considered to be a welfare state, as it plays a key role in the protection and promotion of the economic and social well-being of its citizens. The government upholds the principles of equality of opportunity, equitable distribution of wealth, and public responsibility for those unable to avail themselves of the minimal provisions for a good life.[11] The development of the petroleum sector has become a major focus of the Mozambican government, since the income it may generate could be channelled to other sectors, allowing for gradual and balanced development of the country.

Through the National Social Security system, the government protects those citizens who, due to illnesses, age, or other conditions, are prevented from pursuing means for self-support. As with many other countries, in Mozambique this responsibility is shifted to others who are actively producing through wage deductions by their employers.

Regarding the national workforce, one must take into consideration that Mozambique has a reasonably small population, estimated at around 27 million people in 2017.[12] The population growth rate is high at 2.5%, and reasonable estimates suggest there are 300,000 new entrants into the labor market every year. However, the poor training of Mozambican professionals is a hindrance to Mozambique's economy. The government supports the development of a talented local workforce, specifically those prepared to work in the petroleum industry. The growth opportunities that lie ahead will occur only if Mozambique continues to build and support the growth of its domestic skilled workforce. The government of Mozambique aims to develop a national workforce capable of providing the services expected by international operators. This is supported through local content requirements.

The Ministry of Mineral Resources and Energy, as the entity in charge of the development of the petroleum sector, is invested in promoting essential skills for all those who wish to work in this area in Mozambique. A training strategy was adopted in 2010 that was designed to train Mozambican citizens to work in the oil and gas industry.

Key Players

Host government and its agencies

As previously mentioned, the Constitution of Mozambique establishes that all natural resources, regardless of their location within the purview of the government of Mozambique, are property of the state. As such, the Ministry of Natural Resources and Energy (MIREME, or sometimes MIREM) is the governmental body responsible for managing and executing the governmental natural resource policies. It is also the entity responsible for the supervision of the National Petroleum Institute (Instituto Nacional de Petróleo, or INP), which is the regulatory authority responsible for the administration and promotion of petroleum operations.

The recently enacted Petroleum Law (Law No. 21/2014 of August 18, 2014) created a new entity called the High Authority of the Extractive Industry (Alta Autoridade da Industria Extractiva) to work alongside the

MIREME and the INP. The composition, status, powers, skills, and organizational structure of the new entity have yet to be defined by the Council of Ministers. It is expected to have mainly a supervisory role related to petroleum operations carried out in Mozambique.

National hydrocarbons company, ENH

As referred to previously, the state-owned company ENH[13] represents the government in petroleum operations.

It is the ENH's responsibility to participate in all stages of petroleum activity, from prospecting, E&P, refining, transport, and storage to commercialization of oil and gas and their derivatives, including LNG and GTL (gas to liquids), domestically or abroad. It is also the ENH's responsibility to manage the oil and gas quotas destined for the development of the national market and the country's industrialization.[14]

Any investor interested in the exploration of petroleum resources in Mozambique must enter into a partnership with the ENH.[15] The government guarantees the financing of the ENH as its exclusive representative for investments in the improvement and stabilization of the government's participation in the petroleum business.

Oil and gas companies operating in the country

Mozambique has been one of the African countries expected to see major development and growth of its economy due to the expected increase of foreign investment from the newly discovered major reserves of natural resources. However, the investment projects in Mozambique are facing a delay due to falling oil prices and the global economic slowdown, according to the Economist Intelligence Unit (EIU).

Eni. Even with the economic downturn, Mozambique is entering into agreements for construction and operation of a world-class LNG plant. Both Eni, the Italian multinational, and Anadarko Petroleum, from the United States, agreed on the use of gas reserves. The project was delayed, however, and LNG production is no longer expected to commence in 2018. A more realistic time line expects production in 2020, which is now the goal of Eni and the government of Mozambique.[16]

Anadarko. Anadarko also appears to be one of the major players in Mozambique. It will participate in the LNG project, taking part in the Area 1 consortium (offshore area), which is sitting on gas reserves of 75 Tcf. Once it starts producing, analysts expect that Mozambique will become the largest exporter of LNG in the world, overtaking Qatar and Australia over the next decade.

Anadarko is partnering with Eni in the Area 1 consortium, as well as the Area 4 consortium (offshore area).

Sasol. A third major player is Sasol, the South African company, which continues to explore for natural gas from the Temane and Pande fields (onshore) in southern Mozambique. The Sasol operation began in March 2004, and production has grown about 50% since then.

Other key players. Besides the key players already mentioned, according to the INP, other international oil companies involved in Mozambique's upstream offshore oil and gas sector include Galp Energia, Kogas, PPT Exploration and Production, Petronas, and Statoil.

Service providers operating in the country

Alongside the major players that will carry out oil and gas E&P, other companies have taken an interest in Mozambique. The oil industry is opening doors for many other services that are usually subcontracted by the concessionaires or even the contracted companies.

Service companies such as Bolloré Africa, Baker Hughes, Technip, Halliburton, and Schlumberger, among others, either have already established a presence in Mozambique or are being watchful for opportunities there. The latter companies are prepared to enter the market when there's a need to assist the major players in the projects that are being planned or are already underway.

Industry Knowledge

Research and training centers

Currently in Mozambique there are no research and training centers specializing in the oil and gas industry or providing training in the adjacent areas.

Courses, degrees, and specializations dealing with the oil and gas industry

The Ministry of Natural Resources and Energy, which is the entity supervising the INP, has recently established the necessity of expanding the range of courses for students from Mozambique. Plans call for new courses of study related to the gas industry and the specifics of drilling and refining, among others.[17]

The government of Mozambique is also aware that internship programs need to be created so that national students can learn from companies operating in the oil business, whether in private or public sectors. This should allow students to return to Mozambique with some experience in the oil and gas sector.

Mozambique falls way behind Angola in terms of domestic workforce development. Nonetheless, public universities have been offering specialization courses (usually at the graduate level) that explore the oil and gas framework. They focus on specific know-how in an attempt to provide students with the necessary skills to actively participate in this growing industry.[18]

Legislative Hydrocarbon Regime Concerning Local Content

Regulatory bodies

These entities are discussed in the "Key Players" section above.

Relevant policies

The principal laws and regulations applicable to upstream activities in Mozambique are as follows:

- The Petroleum Law (Law 21/2014 of August 18, 2014) establishes the rules for the granting of rights to carry out petroleum operations in the Republic of Mozambique and beyond its borders in accordance with international laws. The law applies to petroleum operations and to any infrastructure belonging to or held by the holder of rights or third parties and used in connection with oil operations, subject to the laws of Mozambique and including mobile infrastructure under a foreign flag located in Mozambique with the purpose of conducting or assisting in petroleum operations in a concession contract area, unless otherwise established by law.

- The Regulations on Petroleum Operations (Decree 34/2015 of December 31, 2015) regulate operations under the Petroleum Law, setting out the rules for awarding rights to conduct such activities in order to ensure that petroleum operations are performed in a systematic manner and on such terms that allow for its comprehensive and coordinated supervision.
- The Petroleum Tax Law (Law 27/2014 of September 23, 2014) establishes the specific tax regime for petroleum operations, applying to corporate entities incorporated and registered in the territory of Mozambique, as well as to national or foreign individuals who carry out petroleum operations under a concessions contract.
- The Special Regime for Natural Gas Liquefaction Projects in Areas 1 and 4 of the Rovuma Basin (Decree-Law 2/2014 of December 2, 2014) applies to concessionaires under existing E&P concessions contracts (EPCCs). It also applies to any special purpose vehicles established by such concessionaires, and any persons entering into contracts with concessionaires or special purpose vehicles (contractors, financiers, and employees), as well as their subcontractors. This pertains to activities relating to the development and operation of offshore Areas 1 or 4 and which are undertaken under existing EPCCs or any other contracts with the government of Mozambique.
- The Regulation of Employment of the Foreign Citizens in the Petroleum and Mining sector (Decree 63/2011 of December 7, 2011) establishes the legal regime applicable to national and foreign employers from the petroleum and mining industries and to the foreign citizens who intend to work in these industries.
- The Environmental Regulation for Petroleum Operations (Decree 56/2010 of November 22, 2010) establishes the requirements to be satisfied in order to perform oil operations. The regulation specifies environmental impact assessment procedures and protection and control measures in order to prevent environmental disasters.
- The Strategy for Concession of Areas for Petroleum Operations (Resolution 27/2009 of June 8, 2009) establishes the legal regime that will guide the concession of appraisal and production of oil rights offshore and onshore, contributing to the development of the extractive industry in Mozambique.

- The Mega-Projects Law (Law 15/2011 of August 10, 2011) establishes the legal framework that governs public/private partnerships (PPPs) and large-scale projects and business concessions. On one hand, it is conducive to a greater involvement of private partners and investors. On the other hand, it supports greater efficiency, effectiveness, and quality in the operation of resources and other national property assets, as well as the efficient provision of goods and services to society.
- The Mega-Projects Law Regulation (Decree 16/2012 of July 4, 2012) establishes the applicable procedures for contracting, implementing, and monitoring public/private partnerships, large-scale projects, and business concessions ventures.

Pursuant to the approval of the new Petroleum Law of 2014, the Petroleum Operations Regulation had to be reviewed and adjusted accordingly. This review was conducted by the INP and culminated with the recently approved Oil and Gas Operations Regulations that entered into force on December 31, 2015.

From the list indicated above, it is evident that multiple legislative decrees regulate local content requirements. Ultimately, however, the main law to be considered, and the law that is the cornerstone of local content and petroleum operations, is the Petroleum Law.

Concerning local content requirements arising from the Petroleum Law, it is important to recall that besides promoting the involvement of the Mozambican business community in the petroleum sector, the government also ensures that no less than 25% of the oil and gas produced in the national territory is destined for the national market.[19]

Furthermore, where the national interest[20] so requires, holders of rights must give preference to the government regarding the acquisition of petroleum produced in the concession contract area. This relates to all rights holders, whether related to reconnaissance E&P, infrastructure, or construction and operation of an oil or gas pipeline system. The government may also request the petroleum product at negotiable prices for use in local industry whenever Mozambique's commercial interests demand it.

Also, foreign entities that provide services to the petroleum operations must collaborate with Mozambican individuals or corporate entities and should give preference to goods and services purchased or obtained from them.[21] This regulation applies when the goods or services are comparable in terms of quality to the products, materials, and services sold abroad, and are available within the required time frame and in sufficient quantities. The final stipulation is that local content rules apply when the price, including taxes, is not more than 10% higher than the prices of the imported goods available. This decree also establishes the requirement to ensure the employment and training of Mozambican nationals, as well as their participation in the management of petroleum operations, among other requirements that were set out by the applicable legislation.

Following the government's plan to allow for economic growth based on the development of this sector, the oil and gas companies must now be listed on the Mozambican Stock Exchange. This is expected to boost the Mozambican economy and allow for a distribution of the investment across private and corporate entities, whether foreign or national.

Relevant contracts

It is important to distinguish between upstream activities and downstream activities when it comes to the relevant contracts, as each sector has specific instruments.

Upstream activities. Private investment in upstream interests for conducting petroleum operations are granted through concession contracts and are generally attributed by a public tender process. Such rights may also be attributed by simultaneous or direct negotiations in relation to areas that had already been declared available when no concession was granted pursuant to previous public tender, or from rescission, relinquishment and abandonment, or the need to join adjacent areas to a concession, where justified for technical and economic reasons.

Concession contracts are administrative contracts, subject to the authorization of and supervision by the Administrative Court, and thus their main clauses are subject to publication.

The following rights may be conferred under concession contracts:

- Reconnaissance
- Exploration and production (resulting in EPCCs)
- Pipeline construction and operation
- Infrastructure construction and operation

From the list referred above, the EPCC contract is the key contract applicable to upstream activities, as it grants an exclusive right to carry out petroleum exploration and production.

In June 2016, MIREME approved a draft for a model contract that was published by the INP on its website.[22] The approval of such a draft or model follows the recent enactment of legislation regulating this sector (the Petroleum Law and its regulations). Therefore, the new guidelines were incorporated in the draft and will serve as a basis for negotiation between MIREME, the ENH, and the operators.

The exclusive right to petroleum exploration under an EPCC cannot exceed eight years. In the event of a discovery of crude oil or a discovery of nonassociated natural gas, an extension may be granted of up to two or eight years, respectively, depending on the complexity of the work needed to conduct an appraisal program or a commercial evaluation of the discovery.

Local content clauses that are specific for these types of contracts often require contact details concerning the director or appointed representative from the operator. MIREME must be notified of this information, along with contact information for an alternate company representative. Regarding the employment of foreign workers, it is important to reiterate that the basic principle is that Mozambican workers must be given preference. In addition, extensive and very well-detailed training programs must be put in place by the operators in order to develop the local workforce.

Positive and negative experiences and expectations regarding the regulatory framework

With respect to legislation relating to the petroleum industry in Mozambique, laws and regulations are already in force and provide guidelines on how the sector will and must develop. The petroleum sector in Mozambique is still developing, however, and has not yet reached its peak. Thus in practice, the projects that are currently underway are all very recent, and the effects of the local content polices have yet to be seen.

The expectations are very high, and that is the main reason why the Mozambican government has been approving new legislation for this sector, aimed at adjusting it to international standards. The government's goal has been to create broad legislation that functions effectively, preventing the need to review it in critical situations when the projects are fully operating and immediate action is needed.

Current issues and challenges

The trade and investment policies in Mozambique are undercut by many factors, including the following:

Government interference in the economy. Mozambique is a country that relies on external financial assistance for approximately half of its government budget.[23] However, it is up to the government of Mozambique to come up with the macroeconomic policies that can create an environment capable of making the economy grow. Unfortunately, corruption and bureaucratic inefficiency are key hindrances. A very recent example is the increase of the public debt due to actions of former governors that have had (and continue to have) a negative impact on the country's economy. These actions served to throw further off balance what was never balanced in the first place, and the economy has suffered from a lack of good governance by the leaders who should have been promoting it.

Arbitrary tax policies. Among legislation that has been enacted in order to allow for more investment in the oil and gas and mining sector, one law is important to mention as it triggered a high-level unpredictable event. This law relates to the payment of taxes from the sale of assets or stakes in mining projects, sometimes worth tens of millions of dollars. The taxation of capital gains has been inconsistent, and a recent major sale in the mining industry brought that to light, since the amount subject to tax deduction was several millions in US dollars. This attracted the attention of the tax authorities. It is worth mentioning that before this major operation in 2011, there had been no significant discussion of the taxation of capital gains in the extractive industry companies and no evidence of the collection of the capital gains tax from any international company. Such companies are subject to careful regulation through the recently enacted legislation on the taxation and benefits determined for mining activities.

Minimal enforcement of property rights. According to the Mozambican legal system, the land is property of the state and cannot be sold or in any way transferred or mortgaged. Only a minor right over the land allows for its use (the DUAT, or Direito de Uso e Aproveitamento da Terra).

Given the above, foreign investors have to come up with different options for securing loans, since the plot of land attributed to the concession is off limits, i.e., it cannot be encumbered since it is public domain. Third parties, whether domestic or foreign, are only entitled to a minor right of use of that land.

Poor rule of law. Foreign and national investors often are faced with the fact that the Mozambican officials they liaise with lack a working knowledge of the relevant laws for petroleum projects. Mozambique has a well-implemented judicial system, and sometimes the application of the appropriate laws and regulations does not become apparent until a dispute is submitted to judicial review in the courts. It becomes very difficult, therefore, in some situations to predict if a project will be developed without any major setbacks.

However, the Petroleum Law specifically provides for arbitration between the government of Mozambique and foreign investors. Arbitration is to be conducted in accordance with the following:

1. The Mozambican Arbitration Law
2. The rules of the International Centre for the Settlement of Investment Disputes (ICSID), under the 1965 Washington Convention on the Settlement of Investment Disputes between States and Nationals of Other States
3. The rules of such other international conventions of recognized standing, as agreed by the parties in the concession agreement, provided that the parties have expressly defined in the agreement the conditions of implementation, including the method for designation of the arbitrators and the time limit within which the decision must be made

On July 18, 1998, Mozambique became a signatory to the Convention on the Recognition and Enforcement of Foreign Arbitral Awards (the New York Convention), which is fully applicable in national territory. However, as permitted by Article I(3) of the New York Convention, when it acceded thereto, Mozambique declared that it would apply the convention to the recognition and enforcement of awards made only in the territory of another contracting state on the basis of reciprocity. This mechanism is far more trustworthy for foreign investors than the national courts and is a common provision between parties involved in concession agreements.

The tax regime is not too restrictive to business development and growth. However, the level of inefficiency created by the Mozambican tax bureaucracy heightens the cost of operating in the country. Inefficiencies of the tax bureaucracy lead to many irregularities in application of the tax laws. Authorities often need to consult with parallel entities, but their work is hindered by the lack of a centralized information system and outdated data and communications technologies. The World Bank's Doing Business Project estimates it takes 230 hours, or 29 8-hour days, to become tax compliant in Mozambique.[24]

Conclusion

The new oil and gas legal regime is a legal manifestation of the government of Mozambique's desire to improve the business environment in the hydrocarbons sector. There have been steady improvements to this legal framework, along with increased compliance. Progress has been enhanced by the willingness of public and private initiatives to develop a safe, efficient, and sustainable partnership model. The regulatory and private efforts could provide the platform to dramatically boost the Mozambican economy and to consolidate expectations for the discoveries of natural resources in Mozambique.

Notes

1. "World Atlas Data: Mozambique," *Knoema.com*, https://knoema.com/atlas/Mozambique/Total-primary-energy-consumption.
2. "Sasol to Start Drilling in New Mozambique Oil and Gas Fields," *Rigzone* (February 1, 2016), http://www.rigzone.com/news/oil_gas/a/142797/Sasol_to_Start_Drilling_in_New_Mozambique_Oil_and_Gas_Fields.
3. "30 Anos Pesquisando Petróleo e Gás," Empresa Nacional de Hidrocarbonetos (ENH), http://www.enh.co.mz/Imprensa/Publicacoes.
4. "Furos," Instituto Nacional de Petróleo, http://www.inp.gov.mz/pt/Pesquisa-Producao/Furos.
5. "Mozambique Gas Pipeline Loop Line 1 Completed," Sasol, http://www.sasol.com/media-centre/media-releases/mozambique-gas-pipeline-loop-line-1-completed.
6. "FLNG Sera Porta de Entrada de Investimentos para Moçambique," ENH, http://www.enh.co.mz/Imprensa/Noticias/FLNG-sera-porta-de-entrada-de-investimentos-para-Mocambique.
7. "5º Concurso—2014," INP, http://www.inp.gov.mz/pt/Concursos/51-Concurso-2014.
8. "Mozambique's Infrastructure Investment Potential," PricewaterhouseCoopers (PwC), http://www.pwc.com/gx/en/industries/transportation-logistics/publications/africa-infrastructure-investment/mozambique.html.
9. Ibid.
10. Ibid.
11. All principles protected by the Constitution of the Republic of Mozambique, approved in 2004.
12. "Informação Rápida," Instituto de Nacional Estatística, http://www.ine.gov.mz/.
13. "Quem Somos," ENH, http://www.enh.co.mz/Quem-Somos.
14. As established by Decree 29/2015 of December 28, 2015, which approved ENH's bylaws.

15. As per Article 24 of the Petroleum Law, approved by Law 21/2014 of August 18, 2014.
16. "Eni CEO: First Mozambique LNG Shipment in 2020," *LNG World News*, http://www.lngworldnews.com/eni-ceo-first-mozambique-lng-shipment-in-2020/.
17. "MIREM Envia Estudantes para Formação em Geologia de Petróleo," *Jornal O País* (May 31, 2011), http://opais.sapo.mz/index.php/economia/38-economia/14380-mirem-envia-estudantes-para-formacao-em-geologia-de-petroleo.html.
18. "UEM Anuncia Três Novos Cursos Ligados a Exploração Mineira e Petrolífera," *Jornal O País* (March 9, 2012), http://opais.sapo.mz/index.php/sociedade/45-sociedade/19377-uem-anuncia-tres-novos-cursos-ligados-a-exploracao-mineira-e-petrolifera.html.
19. As per Article 35 of the Petroleum Law, approved by Law 21/2014 of August 18, 2014.
20. "National interest" is a vague concept, also referred to in the Mozambican Constitution as it links to interests that are protected by the Constitution that relate to the nation. For instance, the Mozambican state has as its main objectives (Constitution for the Republic of Mozambique [2004, rev. 2007], Article 11, "Fundamental Objectives," http://www.africanconstitution.org/home/mozambique):
    (a) the defense of independence and sovereignty;
    (b) the consolidation of national unity;
    (c) the building of a society of social justice and the achievement of material and spiritual well-being and quality of life for its citizens;
    (d) the promotion of balanced economic, social, and regional development in the country.
21. Article 41 of the Petroleum Law, approved by Law 21/2014 of August 18, 2014.
22. "Modelo de Contratos de Pesquisa e Produção," *Boletim da República*, I Series, no. 118 (October 3, 2016), http://www.inp.gov.mz/pt/Politicas-Regime-Legal/Modelo-de-Contratos-de-Pesquisa-e-Producao.
23. "Aid and Budget Transparency in Mozambique: Constraints for Civil Society, the Parliament, and the Government" (Informal Governance Group and Alliance 2015, May 2010), https://www.trocaire.org/sites/default/files/resources/policy/aid-budget-transparency-mozambique.pdf.
24. Data are for 2015, from *Doing Business 2016: Economy Profile 2016, Mozambique*, 13th ed. (Washington, DC: International Bank for Reconstruction and Development/World Bank, 2016), https://openknowledge.worldbank.org/bitstream/handle/10986/23312/Doing0business0ficiency00Mozambique.pdf?sequence=1&isAllowed=y.

16

NIGERIA

Bunmi Obembe

Introduction

Oil was discovered in Nigeria in 1956 at Oloibiri by Shell-BP after half a century of exploration. First production occurred two years later in 1958.[1] The industry has experienced tremendous growth over the years, with the nation boasting 37.062 billion barrels of proven crude oil reserves in 2016.[2] Crude oil production in Nigeria averaged 1,905 bbl/d from 1973 until April 2016, reaching an all-time high of 2,475 bbl/d in November 2005.[3]

Nigeria has earned close to US$600 billion from petroleum activities since the inception of oil production.[4] The sector constitutes 14% of the nation's GDP[5]—an indication that the economy is fairly well diversified, especially when compared with other oil producing countries, such as Kuwait (61%), Venezuela (24%), and Russia (11%).[6] In spite of the many years of exploitation, Nigeria's petroleum industry still has enormous potential for growth, especially in natural gas. Nigeria has the largest natural gas reserves in Africa, the ninth largest in the world, and supplies 7% of global LNG.[7] Gas reserves are estimated at 179.4 trillion standard cubic feet[8] and are split about 50-50 between associated and nonassociated gas.[9] Nonetheless, there is no doubt that the petroleum sector will continue to be the cash cow of the economy. The petroleum sector provides about 95% of export revenues and roughly 75% of federally collected revenues.[10]

But it is not all good news. Like other resource-driven economies, Nigeria suffers from the well-known "Dutch disease" or "resource curse," terms popular among development economists. Among the numerous downsides related to Nigeria's petroleum resources, the following are particularly worth mentioning: corruption, suboptimal performance, and value destruction.

Corruption

The rampant corruption of the elite and the flagrant displays of conspicuous consumption are perhaps the most visible signs of malaise in resource-driven economies. Nigeria is no exception. This has had dire consequences for the nation, with Nigeria ranked 136th in the Corruption Perceptions Index 2015, out of 167 nations ranked.[11] The propensity for rent collection and consumption tends to stifle initiative among the political elite. It also frustrates the emergence of purposeful leadership, the kind needed to transform a nation from a predominantly subsistence farming community to an industrialized nation. This double jeopardy—cheap money and corrupt leadership—has had serious consequences for the political economies of many resource-driven nations, of which Nigeria is one.

Suboptimal performance

The suboptimal performance of the Nigerian O&G industry over the years is a fact that has been widely reported in the media, both domestically and internationally. Nowhere is this more evident than in the near absence, after half a century of oil E&P, of viable midstream and downstream sectors. The continuing importation of fuels, anchored on a poorly administered subsidy regime, has had a crippling effect on the national economy. According to a World Bank report, the Nigerian subsidy regime consumed some US$35 billion from 2010 to 2014. Subsidies represented as much as 18% of government expenditures in 2015 and will likely increase to 30% by 2018.[12] Even with the subsidies support, the four government-owned Nigerian National Petroleum Corporation (NNPC) refineries operate at less than 30% of their nameplate capacities.[13] Any industry that has sustained an economy for this number of years should be self-sufficient and not reliant on government funds or intervention.

The country's only petrochemical plant came onstream after 18 years, but it was sold to an Indian company to prevent it from closing. The LPG sector is another disaster. Nigeria's per capita consumption of LPG is 0.5 kg, compared to 10 kg in other countries at the same latitude. Meanwhile, Nigeria's annual gas flare is believed to be capable of generating about 27 GW of electricity.[14]

The inability of the national oil company to get the downstream sector up and running, as well as its inability to explore linkages outside the O&G

value chain, constitute a compelling indictment of the corporation and its affiliate companies.

Value destruction

The inability of the industry, particularly the upstream actors, to generate added value in Nigeria is perhaps the most debilitating consequence of a badly run economic sector. In the early 2000s, the sector was earning more than US$60 billion and spending some US$20 billion annually.[15]

But the local content was estimated at less than 5%, as most of the industry's construction-related activities were literally being exported to Europe, Asia, and the United States. This period also coincided with the industry's major transition of E&P activities from onshore to offshore, with various construction work transitions. It was also a time when expectations soared following the election of a retired general, Olusegun Obasanjo, as president after Nigeria had experienced 16 years of military rule. For the first time, the massive unemployment among the younger workforce became a focus of national discourse.

The government of Nigeria was pressured to look to the petroleum sector for solutions to unemployment, but the doors were effectively closed. This generated open resentment at the highest levels of government. Government leaders could not understand that the petroleum industry does not employ many workers once development projects are over and the fields are up and running. The industry had many convenient excuses, such as lack of industrial infrastructure and capacity, deficiencies of local service companies, inadequate skill sets among the domestic workforce, and low insurance capacity. The excuses, however, began to meet with cold rebuff from the politicians, who were looking for a way to provide employment for the teeming younger workforce at any cost. Although there are many other factors, this issue of unemployment was the trigger that awakened the political leadership to the need to focus on how the oil industry does business in Nigeria. The government was determined to change the trend and get the industry to do more within Nigeria, as a way of promoting employment.

Given the issues of corruption, suboptimal performance, and value destruction at the forefront, and with a determination to address these issues, the government enacted the Nigerian Local Content Act of 2010, as discussed more fully later in the text.

Local Content Policy and Baseline

Although some believe that concerns about Nigerian local content policies are relatively new, this is far from true. A brief overview of the history of the O&G industry shows that local content efforts date back to the late 1960s.[16]

The Petroleum Act (1968)

In the Petroleum Act (1968) and its regulations, there are far-reaching provisions for the employment, training, and utilization of Nigerian manpower in the oil industry. It was an obligation imposed upon oil mining lessees and oil prospecting licensees to submit a detailed program for the recruitment and training of Nigerians as subsumed under Sections 1(2)(h), 26, and 27 of the Petroleum (Drilling and Production) Regulations, a subsidiary legislation to the 1968 Petroleum Act.

Joint operating agreements/production sharing contracts

The joint operating agreements (JOAs) and production sharing contracts (PSCs) have Nigerian content provisions. The JOAs require that preference be given to Nigerian companies in tenders, provided there is no significant difference in price and quality between them and foreign companies. In the 2005 PSC, the contractor and the NNPC are required to work toward maximizing local content in all areas of petroleum operations.

There was nothing unusual in either the Petroleum Act or its regulations when benchmarked against standard practices around the world, like those of the United Kingdom, Norway, Brazil, and Venezuela. Nation states have always pursued national interests by protecting local businesses and seeking ways to maximize opportunities for them.

Chibudom Nwuche's bill of 2001

Two noteworthy attempts to update Nigeria's local content policies in the petroleum laws did not succeed. The first was Chibudom Nwuche's bill of 2001, and the second was Chike Ofofili's bill of 2004.

Chibudom Nwuche's bill was highly prescriptive, requiring certain percentages of contracts to be awarded to indigenous contractors. With the experience of Nigeria on the repealed Indigenization Decree, many saw this bill as counterproductive. After failing to clear hurdles in both houses of the National Assembly, it was eventually discarded in 2003. The Indigenization Decree, also known as the Nigerian Enterprises Promotion Decree of 1972, was a law that gave Nigerians "the exclusive right to the ownership of some enterprises and greater participation than seen in the equity ownership of others."[17]

Chike Ofodile's bill of 2004 and other legislative developments

Ofodili's bill was less prescriptive and set high-level objectives, while leaving the initiative for the implementation details to the regulations. The argument was that the pursuit of the policy would be better accomplished through collaboration among all stakeholders—the IOCs, the international and local service companies, the industry regulator (DPR), and NAPIMS (the investment arm of the NNPC).

While Ofodili's bill was undergoing due legislative process at the National Assembly, some interesting developments were also taking place in the industry, as follows:

1. The IOCs started discussions on the setup of an industry-sponsored joint qualification system (JQS). It was to be modeled after a similar system called FPAL in the United Kingdom. This eventually gave birth to the NIPEX (Nigerian Petroleum Exchange), of which the JQS became an integral part. The original vision of the IOCs was that the JQS would be managed by a separate company, but it eventually ended up as a division of NAPIMS.
2. In 2005, the NNPC, under pressure from the political leadership, issued a 10-point directive for local content implementation. This was to increase later to 23 points. It set a very ambitious goal of 45% local content by 2006 (from about 5% in 2000), and 70% by 2010, without details on metrics. It also issued an implementation procedure that introduced certifications for local content in different stages of the contracting process of the IOCs. The contract procedures had been notorious over the years for delays as long as two years. With the superposition of a parallel Nigerian content certification requirement, the tendering process became nearly stalled.
3. There were two Nigerian content-related legislations that were also passed during this period: (a) the National Insurance Commission Act, also called the Nigerian Insurance Act (2003)[18] and (b) the Coastal and Inland Shipping Act, also called the Cabotage Act (2003).[19] Though the laws were not specifically called Nigerian

content laws, their main focus was the use of Nigerian service providers in the insurance and marine sectors, respectively.

It must be said that these efforts by the NNPC and the National Assembly substantially radicalized the conversation around the Nigerian Content bill at the National Assembly, and prevented the smooth passage of Ofodili's bill into law. This was rather unfortunate because the bill was highly collaborative. If passed, it would have resulted in a more balanced law for the petroleum industry. Unfortunately, by 2007, the tenure of that session of the National Assembly ended, and the bill lapsed.

The subsequent session of the National Assembly resumed in June 2007. Lee Maeba sponsored another Nigerian Content development bill. The bill seemed to have been inspired by the preceding bills (i.e., Chibudom's bill and Ofodili's bill), the NNPC 23-point directive, the new NAICOM and Cabotage legislations, and the general mood in the country. This was the "super prescriptive" [20] piece of legislation that was subsequently enacted as the Nigerian Oil and Gas Industry Content Development (NOGICD) Act in 2010, discussed in more detail later in the text.

Key Players

There are several key players involved in different aspects of regulation of the oil industry in Nigeria.

The Ministry of Petroleum Resources

The Ministry of Petroleum Resources "has primary supervisory oversight over the oil and gas industry. As such, the Ministry is responsible for the formulation, implementation and co-ordination of government policy for the industry. The Ministry exercises its regulatory functions through the Department of Petroleum Resources (DPR)."[21]

The Department of Petroleum Resources (DPR)

The DPR is a department within the Ministry of Petroleum Resources with the "statutory responsibility of ensuring compliance to petroleum laws, regulations and guidelines" in the O&G industry. The DPR's purview includes the following:

1. Supervising all Petroleum Industry operations being carried out under licenses and leases in the country.

2. Monitoring the Petroleum Industry operations to ensure that are in line with national goals and aspirations including those relating to Flare down and Domestic Gas Supply Obligations.
3. Ensuring that Health Safety& Environment regulations conform with national and international best oil field practice.[22]

The Nigerian Petroleum Investment Management Services (NAPIMS)

NAPIMS functions as a partner in joint ventures and as the concessionaire in PSCs. NAPIMS "is neither an operator nor a regulatory body of the industry, contrary to the notion held by some sectors of the public; rather, NAPIMS manages the Federal Government of Nigeria (FGN) interests in the oil and gas industry."[23] Its responsibilities include managing Nigeria's assets, promoting the government's plans, and promoting use of new technologies in Nigeria's O&G industry. NAPIMS also has certain funding, operational, and financial control tasks, among other roles.

Nigerian Content Development and Monitoring Board (NCDMB)

The NCDMB was created through the enactment of the Nigerian Content Law in 2010. It is tasked with the following:

- Increase indigenous participation in the oil and gas industry,
- Build local capacity and competencies,
- Create linkages to other sectors of the national economy. and
- Boost industry contributions to the growth of…[Nigeria's] National Gross Domestic Product.[24]

The Nigeria National Petroleum Corporation (NNPC)

The NNPC is Nigeria's national oil corporation, created in 1977. The NNPC conducts exploration activities and holds "operational interests in refining, petrochemicals and products transportation as well as marketing. Between 1978 and 1989, NNPC constructed refineries in Warri, Kaduna and Port Harcourt and took over the 35,000-barrel Shell Refinery…in Port Harcourt.[25]

Legislative Hydrocarbon Regime Concerning Local Content

The Nigerian Oil and Gas Industry Content Development Act

The Nigerian Oil and Gas Industry Content Development Act, also known as the Nigerian Content Act, became law on April 22, 2010.

The NOGICD Act established a regulatory agency called the Nigerian Content Development and Monitoring Board (NCDMB).[26] This is in addition to the existing industry regulators, DPR (Department of Petroleum Resources) and NAPIMS—two government agencies already active in the petroleum industry.

The main functions of the board as contained in Section 70, and other relevant sections of the NOGICD Act, are as follows:

- To promote and implement capacity development initiatives for indigenous service companies
- To manage the Nigerian Content Development Fund (NCDF) as a source of affordable funding for contracts won by indigenous contractors in the industry
- To establish and operate a joint qualification system (JQS)
- To monitor Nigerian content performance of operators and contractors
- To manage activities of the Nigerian Content Consultative Forum.

The following are the other major highlights:

- The act defines "Nigerian company" as a company incorporated in Nigeria, with 51% of its equity held by Nigerians. It also requires that international or multinational companies working through their Nigerian subsidiaries demonstrate that a minimum of 50% of the equipment deployed for execution of work is owned by the Nigerian subsidiaries.
- The act requires that first consideration[27] be given to indigenous Nigerian goods and services.
- Each operator is required to submit a local content plan to the board that includes commitments ranging from use of Nigerian goods and

services to employment and training of Nigerians and research and development.

- All contract processes above US$1 million are subject to approval of the board at every stage of the tendering process (i.e., advertisement, prequalification, technical, and commercial evaluations). A premium of 10% is accorded in commercial evaluations to support Nigerian content. (If a bid is up to 10% more expensive than the lowest bid, it must get the contract provided it is at least 5% higher in its Nigerian content score.)
- The act requires that operators establish offices in the project's local community and employ Nigerian nationals within that community.
- The act allows operators to retain up to 5% of management positions in the company for expatriate staff. Even at that, the board must approve all applications for expatriates before applications are submitted to the Immigration Service.
- Operators are also required to establish a technology transfer plan, detailing the company's technology transfer initiatives. The law empowers the board to direct operators to set up factories for manufacturing equipment and spare parts necessary for its operations.
- The act places more stringent requirements on contracts in the areas of insurance, legal services, and financial services, for which operators are also required to submit plans to the board every six months.
- All fabrication works requiring welding are required to be carried out in Nigeria.
- A levy of 1% of every contract awarded is to be deducted at the source and paid to the account of Nigerian Content Development Fund.
- A penalty for noncompliance of 5% of the project sum; or cancellation of the entire project is also stipulated.

The law has been widely criticized by various segments of the Nigerian business community. It is worthwhile looking at some of these:

- The act is considered too prescriptive and overbearing. It is also considered draconian in its administration, a situation that has in practice led to parallel approvals in the tendering process by two different government agencies, each acting independently.

- Many consider the act rather discriminatory in respect to international service companies (like Schlumberger) that are incorporated in Nigeria but owned 100% by foreign entities. These companies, they claim, not only employ Nigerians, they also send Nigerians on international assignments. Some of these companies already have almost 100% Nigerian staffing. These companies think (and rightly so) that better consideration should have been accorded them.
- Addition of another regulator into the tendering process has further slowed down the already long process in the industry.
- Building capacity on the back of EPCI projects has proven to be expensive, with consequences for project viability and profitability. Some cite as much as 20% extra project cost to fulfill Nigerian content obligations.
- The targets set in the Act, many claim, were set without any field surveys to ascertain existing capacities and capabilities in Nigeria. Many industry actors, including indigenous companies, feel that this should have been left out of the law. There are others who feel that a simple phrase in the law requiring the regulator (i.e., the board) to work with industry stakeholders to fix targets would have been sufficient.
- The 1% levy on contracts is considered another extra cost in addition to the high cost of doing business in Nigeria. The intention of the law was to create a pool of cheap funds that could support indigenous players. Unfortunately, according to the indigenous players, it has been late in implementation. Many national companies still have to go to local banks for financing at double-digit interest rates.
- Many believe that the provisions of the NOGICD Act conferred powers already being exercised by some agencies of the government. The result is a double jeopardy situation for companies who have to be answerable to different regulators on the same issue. Apart from the approval powers in tendering processes, the issue of expatriate quota approval by the board is considered an unnecessary interference in the duties of the Immigration Services.

The above criticisms and others notwithstanding, the law has received a lot of support, particularly from local Nigerian contractors. Many believe the only way to redress 50 years of exclusion by the IOCs is to “clamp down on them.” They maintain that this act, even if it is deficient, offers a far better solution than the previous situation in which there was no law specifically relating to local content in the oil and gas industry. It is argued that it is better to start with a draconian law that can be reviewed as the sector evolves. This viewpoint is characteristic of legislators who participated in the bill's enactment.

In the years that have passed since the NOGICD Act came into effect, and with the benefit of hindsight, one can say the law has not brought doomsday as many initially feared. It is worthwhile to note that the board has been quite pragmatic in its approach, working tirelessly with the industry to fulfill the spirit of the law, while nudging them to do better than before. It is to the credit of all that when the law came up for review in 2014, no substantial amendments were made to its provisions.

Industry Knowledge

There is no doubt that the NOGICD Act has changed the game in terms of local content. These achievements are evident in many sectors, as discussed in the following.[28]

Asset ownership

Nigeria has seen more investment in ownership of deepwater rigs in the last few years than in the 50 preceding years of O&G activities in the country. Thanks to the offshore rig acquisition strategy (ORAS) of NCDMB, Oando, Seadrill, and Deroteck are Nigerian companies that now have rigs that meet classification as Nigerian rigs. Also, more than 80% of land and swamp rigs are now Nigerian owned. In addition, there has been a rapid increase in Nigerian ownership of marine platforms, marine vessels, and rig maintenance capabilities.

Fabrication

Fabrication capacity of indigenous manufacturing yards has also soared, thanks to the efforts of the Nigerian Content Development and Monitoring Board. From a meager fabrication capacity of about 10,000 tonnes/year in 2005, Nigerian yards now produce 200,000 tonnes/year. It is estimated that some US$5 billion in investments has been made by different companies to achieve this increase in capacity. At the forefront of this achievement is the Samsung/Ladol US$300 million investment in an FPSO integration facility in Lagos, as part of the Egina field development project of TUPNI.[29] It is hoped that the yard could provide up to 100,000 direct and indirect jobs on future FPSO integration activities.

A new pipe mill, the Polaku pipe mill, will produce some 250,000 mT/year of pipes when completed.[30] Although this is still far short of estimated 1 million mT/year pipe consumption by the industry during the boom cycle, it is nevertheless a welcome development.

Human capital development

There is no doubt the industry offers more focused training to young Nigerians than before the NOGICD Act. This is directly related to the directive that most jobs be filled by national personnel and companies, thereby opening doors for training and employment opportunities. It is reported by the board in 2015 that more than 3 million man-hours of training have been completed, while more than 2.5 million man-hours of engineering (basic and detailed) have been carried out in the country. Various technical training events have also been conducted in collaboration with contractors.

Training, research, and development

The Nigerian Content Act stipulates that one of the functions the NCDMB is charged with is capacity building. Training is integral to any plan submitted under the act, and the NCDMB has committed to oil and gas Training centers of excellence (TCE). These centers will focus on demand and capability, providing a coordinating body and a well-trained workforce.[31] Although the NCDMB has a number of programs and activities that include training, there is limited evidence to suggest that training centers have been set up or partnerships have been formed with independent research institutions.[32] Further, the private sector has stepped in to fill the void, as seen with the programs offered by Lonadek and Delton Shareholding Company.[33]

Research and development (R&D) is considered a key component to content development. The Nigerian Content Law, under Sections 36 to 39, states that all persons submitting a proposal should have a research and development plan. In furtherance of the board's mandate under the act, the research and development guidelines were released in 2012.[34] There are a number of factors required in order to drive R&D: better monitoring and evaluation models, organizing or aggregating local businesses, creation of business hubs, and access to financing (or microfinancing).[35]

The Nigerian Content Development Fund (NCDF)

An NCDF operating model has been created. Today, 70% of funds accumulating from the 1% levy on all industry contracts is directed toward providing partial risk guarantees for indigenous service companies that obtain loans under the NCDF framework.[36] This is an area where much remains to be accomplished, however, since contractors still complain about their inability to access this funding.

Capacity development initiatives (CDIs) by operators

The operators, in response to identified gaps and to meet the requirements of obtaining waivers where there is inadequate availability of relevant goods or services, have embarked on various CDIs. These CDIs are aimed at filling identified gaps in the following areas:

- Manpower development
- Facilities upgrade
- Establishment of new plants and facilities
- Increased participation of Nigerian service companies in the industry projects
- Increased equipment ownership by Nigerian companies

Many initiatives are on the table, but it remains to be seen what level of impact they will produce in the areas they have targeted.

Current Issues and Challenges

Implementation challenges

The positive outcomes of the NOGICD Act are readily apparent. Nonetheless, there have been some challenges in its implementation. Some of these are discussed in the following sections.

Increased EPCI project costs. The scope of the Nigerian Content Act translates to increased overall costs of EPCI projects. Where there are inadequate capacities locally to meet the project needs, the act requires that such capacities be built, leading not only to higher costs, but also to delays in project delivery. Although no official numbers are available, project managers claim figures as high as 20% extra cost implications for meeting the Nigerian content requirements. A lot of money goes into on-the-job trainings

(OJTs) during the projects, to the credit of the regulators. But many industry observers claim the bulk of these trainees still end up unemployed because the projects could not absorb them all. A similar observation has been made regarding some of the extra capacities built to meet the demands of the project. On one of the major offshore projects, for example, a whole facility was brought into the country for pipe coating. It was idled after the project and had to be dismantled and shipped back overseas after three years. Experts question the economic justification for a policy that translates to such increased costs without a sustainable economic impact. This goes back to the very root of the local content philosophy. There is certainly a need for a cost-benefit analysis of these initiatives in order to ascertain whether these OJTs and capacity-building projects as presently conceived are truly beneficial. It appears the regulators will have to determine in the future what level of extra capacities the national economy can absorb beyond the needs of projects, and then use this knowledge to moderate the pressure on industry projects managers.

Lack of capacity in required skills. Inadequacy of required skills continues to be a major challenge. The act and its regulation have presumed that by locating project activities in Nigeria, indigenous employment would result. There is no doubt that it has helped increase the amount of indigenous man-hours harnessed for the projects. But in terms of the high-level technical competencies required, a lot of expertise still has to come from outside, at additional cost to the projects. It must be mentioned, however, that some projects have not made sufficient effort to reach out to highly skilled Nigerians, especially the expatriate Nigerians living outside the country. Some Nigerians are already employed internationally in subsea engineering, project management, and engineering construction. Regulators have a role to play in creating a platform whereby project sponsors and potential Nigerian experts can meet and interact. A database of such experts would be helpful for recruiters looking for employees with specific skills.

Promotion of local manufacturing/lack of infrastructure. One of the objectives of the act is to promote local manufacturing of equipment and spare parts used in the industry, a provision that many believe to be unnecessary. Some deem this provision to be excessive, regarding it as a means to force operators to step out of their core business. In addition, the lack of industrial infrastructure (ports, waterways, roads, railways, and electricity), a problem that has affected the nation's manufacturing sector, remains unsolved. As long as this is the case, the objectives of the act in this regard remain unobtainable.

Learning Points

It took Nigeria close to 50 years to aggressively embark on value extraction initiatives in the O&G value chain. As is normal in such cases, the tendency is to become overly prescriptive in a bid to recover lost time. This has been the case in Nigeria.

Local content can mean different things in different countries. For Nigeria, it has come to mean domestication of industry activities. All efforts before and after enactment of the NOGICD Act have been geared toward this single overriding objective, but indigenization eventually became one of the fallouts. It must reiterated, however, that indigenization was not the primary objective. Political expediency dictated that development of all aspects of a domestic oil industry would not be achieved unless Nigerian citizens were encouraged to participate. As a result, ownership of assets became a major point in the NOGICD Act.

Apart from its employment focus, the Nigerian content policy also has been driven by the Nigerian contractors' need for access to industry contracts. This is perfectly legitimate. But extracting value does include more than this. For new countries entering into the O&G industry, it will serve their purpose for national development to be more value-chain focused. Rather than focus on a limited section of the value-chain (e.g., upstream), attention should be given to policies that encourage the transformation of the resource into consumer products. This could occur locally at first, but then internationally as capacity grows.

For the petroleum sector, this will require enacting policies that will support local sectors such as refining, gas-to-power, CNG for transportation, LPG for cooking, and mini-LNG for industrial plants. These policies will not only reduce the amount of crude oil that is exported, and gas that is flared, but also will create a robust multiplier effect, in terms of increased economic activities and employment generation. With this kind of thinking, it is evident that local content policies as presently conceived in economies driven by natural resources should be prescribing the very opposite of what they are presently churning out.

Nigeria has a population of 170 million, a growing educated middle class, and increasing urbanization. Thus it would certainly be far more beneficial for Nigeria to enact policies to support local refining and petrochemicals, gas for electricity, LPG for cooking, and CNG for transportation, just to name a few. Each of these subsectors represents more than US$5 billion annually, with substantial potential for employment generation. For this kind of policy to succeed, foreign and local investors are needed.

Developing nations, in particular, must be mindful of the fact that a good local content policy must be underpinned by a robust industrial base. Domestication may not necessarily translate to sustainable employment. We have seen this in many resource-driven countries, including Nigeria. As mentioned previously, the policy is piggybacked on major oil industry projects, such as Akpo, Agbami, Bonga, and Egina. Putting so much industry money into building capacities and capabilities, whether manpower or industrial, for the delivery of a project, only to create unemployment or shutter the facility at the end of the project, may not be the smartest way to allocate a nation's scarce resources. The point is that there are justifiable economic reasons why sometimes activities are better located outside a nation's borders. Where the quantum of long-term industrial activity cannot justify domestication of a particular activity, it is perfectly rational not to invest scarce resources in such an effort. In the final analysis, local content should be seen as an investment, and the returns have to be justifiable to be worth it.

Ownership of assets is a subject in which the Nigerian Content Law had many implementation challenges. Perhaps one area worth exploring by asset owners is the stock exchange. Existing JV and PSC arrangements between IOCs and the government exclude this option altogether. This is rather unfortunate. One would like to see a situation in which fiscal incentives are provided by the government for companies that are listed on the Nigerian Stock Exchange. For independents, and especially small indigenous operators, this holds promise for a long-term sustainable source of financing for their upstream projects. It is interesting to note that many indigenous players in Nigeria are now listed on stock exchanges at home and abroad. This trend is expected to continue. One can only hope that Nigeria's government will divest its interests (currently averaging 55%) in the upstream JVs with IOCs and open this window of investment opportunity to all Nigerians.

The second issue pertains to ownership of assets for carrying out services in the industry. The NOGICD Act has come up with a model for effecting transfer of assets from foreign partners to local partners, just to be able to meet the requirements of the law. As ridiculous as this procedure appears, many players sincerely see it as the best way of meeting the prescriptions of the law, which emphasize ownership of assets by indigenous contractors. Again, we see an indigenization policy creeping in through the back door. The method and manner prescribed by the regulator, pejoratively referred to as "asset forcing," is seen by many as expropriation.

Asset forcing, especially in the manner envisaged in the Nigerian Content Act and regulation, is not a sustainable way of distributing oil blocks. Neither is it the best way of encouraging investment in upstream operations and maintenance of assets. Even if the policy succeeds in respect to assets already in service within the nation state, it will discourage further foreign investment in similar assets and defeat the very purpose of the policy.

Suggestions for Future Growth in Local Content Policy

Nigerian content has come a long way since 2010 when the NOGICD Act came into being. Many successes have been recorded, but there have been many challenges. There is no doubt that with necessary improvements, the law will evolve in such a way that the needs of the nation will be better met. What one would expect is that a substantial part of the act will be transferred to corresponding regulations, for example, possibly a significant portion of Part 1 of the act, starting from Section 1037. Only the high-level objectives would remain in the law, allowing the regulations to manage the finer details of implementation. With this hope in mind, I will conclude this paper with some challenging thoughts.

Domestication for value capture

The premise that domestication will translate to local value capture is not always true. It depends. The petroleum industry depends on a global supply chain. The ability to compete anywhere in the value chain is anchored on technical competence, cost competitiveness, and access to robust financing beyond the limit of someone with only a local or domestic focus. For domestication to translate to sustainable value capture, encouragement toward global positioning is key. And for the particular skills and capabilities to take root and grow, longer term contracts would be necessary. This implies that the contracting philosophy would have to be modified to support some level of job aggregation, some level of incentives to support capacity building, and some import quota restrictions to support domestic production.

Ownership of assets

It is generally believed that ownership of oil blocks translates to increased wealth for the owners of the blocks. Ordinarily that should not be a problem in any economy. The problem in Nigeria is the procedure, which many believe to be nontransparent, with too much discretion granted to the president to award blocks or to set aside the results emanating from properly conducted bidding rounds. Nigeria's current political problems are partially underpinned by this unfortunate practice. There is a need to move toward more transparent bidding rounds. Furthermore, it would be advisable to limit participation in such bidding rounds to companies that are listed on the stock exchange to allow for greater transparency, accountability, public participation, and wealth creation through shareholding. This will better guarantee access to the benefits of the commonwealth for many instead of just a few.

With respect to ownership of equipment for operations and maintenance of assets, the prescription would be same. Those who are shortlisted to be on the tender list for major industry contracts should be listed on the stock exchange.

Tool for stimulating industrial development

Local content could be a tool for stimulating industrial development. Again, one must quickly add a caveat—it depends on the policy directions. In the case of Nigeria, many opportunities have been missed and continue to be missed. One case in point is in power generation. As stated above, the possibility of more than US$30 billion annual economic activity in downstream, electricity, and transportation sectors remains untapped. The possible linkages to the steel industry, marine sector development, and legal, insurance, and other services were largely unexplored, at least until the law came into effect. A critical value-chain understanding is necessary if local content policy is to be used to unlock value and stimulate industrial development.

Linkages

Local content could be enhanced through linkages with existing sectors in the economy. Some of the most obvious areas are electricity, steel, transportation, insurance, and real estate development. When the needs and plans of the industry are known by the other sectors, it becomes obvious that possible synergies might be found that could be more cost-effective for the IOCs and still benefit the host country. IOCs should be encouraged to be active members of the local chambers of commerce, so that through such platforms, business opportunities could be shared liberally.

Another area of possible linkages worth exploring is in the various professions related to the oil industry. IOCs have no problem in most cases bringing expertise to the host country, most of the time without undergoing the legal requirements of registration with the local professional bodies. In Nigeria, for example, there is a legal requirement that for engineers, technicians, and other professionals to be registered with the Council of Registered Engineers of Nigeria (COREN). Yet in over 50 years of oil industry operations in Nigeria, few expatriates working in the oil industry have registered. And because the institutions of governance tend to be weak, these malpractices are never challenged. I believe this is one area where host nations can benefit from imported expertise, especially when experienced professionals commit to the nation's mentorship programs. Such a scheme will also assist companies in their succession planning.

Resource nationalism or business advocacy?

From our journey in Nigeria, I think it is reasonable to draw some conclusions regarding what will make a local content policy succeed, and what will not.

Where business advocacy is the bedrock of local content policy, chances are good that it will succeed. This is because the policy on its own should translate to cost competitiveness, for example between imported and locally produced inputs. Such a policy will be driven more by considerations for efficiency, sustainability, and profitability.

But where the policy is built on considerations that are fed more by resource nationalism, business rationalism tends to be downplayed, regulations tend to be prescriptive, and draconian procedures ultimately frustrate the purpose of the policy. Such policies may win cosmetic short-term victories, but in the long term, the fallacy of the whole enterprise becomes apparent, and the nations are left worse off.

For this reason, local content policies should be driven by collaboration between the business community and the government. The private sector is better placed to deliver on policies that make business sense, and getting their cooperation and ownership for the policy would produce a higher probability for success.

Regulation or law?

Where there is already a regulator for a particular industry, it would be advisable, and certainly more cost-effective, to allow the regulator to establish the rules for implementation of a local content policy. Usually, such a regulator is already empowered to make local content regulations, and no modification of existing statutes are required. If Nigeria had followed this path, effective local content regulations could have been enacted in 2000 and not 2010. It is also likely that the regulator (DPR) would have been more collaborative with the stakeholders. Issues such as target setting, measurement, and planning, among others, would have had a better platform from which to develop.

Is there any business case for a local content law? Yes, there is, but I believe local content legislation would be more successful if it was not merely sector based. For the reasons already advanced, especially concerning linkages, a local content policy must be driven at the highest level of government, and each sectoral regulator must draw inspiration from the same national policy. If intersectoral linkages and synergies are to be optimized, a central policy coordination for local content becomes imperative. Value-chain thinking, with a macro approach instead of a micro approach, would likely produce a better and more sustainable outcome.

Conclusion: A Journey, Not a Destination

In conclusion, local content in Nigeria has been a journey and has not yet reached its peak of highly functional operation. Some mistakes have been made, but some achievements have been recorded as well. The NOGICD Act might not have taken the country to the desired destination, but it has changed the paradigm. One can only expect that the right lessons would be learned, and necessary improvements brought to bear on the regulations, as implementation continues.

The central lesson from the Nigerian experience is that local content is a technology issue, a development issue, and a macroeconomic issue. Collaboration of all stakeholders across government and businesses is critical for the success of the policy. There is no one-size-fits-all approach in terms of policy prescription. It depends on the expectations of the host nation state and the willingness of businesses to work collaboratively with the government to seek the optimum way to deliver added value that is sustainable and yet profitable. Local content becomes, therefore, an iterative endeavor—a journey with all stakeholders in the same vehicle.

Notes

1. "History of the Nigerian Petroleum Industry," Nigerian National Petroleum Corporation (NNPC), http://www.nnpcgroup.com/NNPCBusiness/Businessinformation/OilGasinNigeria/IndustryHistory.aspx.
2. *Annual Statistical Bulletin 2016* (Vienna: OPEC, 2016), http://www.opec.org/opec_web/static_files_project/media/downloads/publications/ASB2016.pdf.
3. "Nigerian Crude Oil Production Data Chart," Financial Markets Dealers Association, http://fmda.com.ng/markets/crude-oil-production/.
4. Aniefiok E. Ite, Udo J. Ibok, Margaret U. Ite, and Sunday W. Petters, "Petroleum Exploration and Production: Past and Present Environmental Issues in the Nigeria's Niger Delta," *American Journal of Environmental Protection* 1, no. 4 (2013), 78–90, http://pubs.sciepub.com/env/1/4/2/.
5. Raj Kulasingam, "Outlook of the Nigerian Oil and Gas Market," *Financial Nigeria* (August 5, 2015), http://www.financialnigeria.com/outlook-of-the-nigerian-oil-and-gas-market-sustainable-69.html.
6. "Kuwait Facts and Figures," OPEC, http://www.opec.org/opec_web/en/about_us/165.htm; "Venezuela Facts and Figures," OPEC, http://www.opec.org/opec_web/en/about_us/171.htm; "Nigeria's Renewal: Delivering Inclusive Growth in Africa's Largest Economy," McKinsey Global Institute (July 2014), 11–12.
7. C. Lloyd, "Nigerian Oil and Gas," *Energy Global: Hydrocarbon Engineering* (March 11, 2015); "Shell in Nigeria: Unlocking Nigeria's Potential in Natural Gas," Shell, http://s05.static-shell.com/content/dam/shell-new/local/country/nga/downloads/pdf/unlocking-natural-gas-em.pdf.
8. "Country Overview: Nigeria, the Leading Oil and Gas Country in Africa," *Annual Report and Accounts 2014* (Seplat Petroleum Development Co., 2014), 6.
9. "Gas to the Megawatt Rescue," Heinrich Böll Stiftung Nigeria (November 6, 2014), https://ng.boell.org/2014/11/06/gas-megawatt-rescue.
10. Douglas Imaralu, "Why Nigeria Must Trade More with Africa," *Ventures Africa* (January 31, 2014), http://venturesafrica.com/why-nigeria-must-trade-more-with-africa/.
11. "Corruption Perceptions Index 2015," Transparency International, http://www.transparency.org/cpi2015.
12. B. Udo, "Fuel Subsidy Removal: Now Is Best Time to Act, World Bank Tells Buhari," *Premium Times* (December 8, 2015); Emmanuel Elebeke, "Cost of Nigeria's Subsidy Higher Than Benefits," *Vanguard News* (December 8, 2015), http://www.vanguardngr.com/2015/12/cost-of-nigerias-subsidy-higher-than-benefits-world-bank/.
13. Anthony Osae-Brown, "Ten Shocking Inefficiencies in NNPC" (March 2012), www.osae-brown-insights.blogspot.com.ng.

14. Odunewu Segun, "Nigeria Annual Gas Flare Can Generate 27000 MW: Online Tracker," *National Daily Newspaper* (December 8, 2014).

15. Jean Balouga, "Nigeria Local Content: Challenges and Prospects," *International Association for Energy Economics*, Third quarter (2012), 23–26, https://www.iaee.org/en/publications/newsletterdl.aspx?id=176.

16. Adeoye Adefulu, "Nigeria/Local Content Policy," undated presentation, Odunjinrin & Adefulu.

17. E. L. Inanga, "The First 'Indigenisation Decree' and the Dividend Policy of Nigerian Quoted Companies," *Journal of Modern African Studies*, 16, no. 2 (1978), 319.

18. "National Insurance Commission Act," Cap. N53, *Laws of the Federation of Nigeria*, 2004, http://lawnigeria.com/LawsoftheFederation/NATIONAL-INSURANCE-COMMISSION-ACT.html.

19. "Coastal and Inland Shipping Act," Cap. C51, *Laws of the Federation of Nigeria*, 2004, http://lawnigeria.com/LawsoftheFederation/Coastal-and-Inland-Shipping-%28Cabotage%29,-2003.html.

20. Although there are no sources that expressly describe the act as superprescriptive, the schedule pursuant to Sections 3(2), 11(1)–(4), and 70(d) of the act prescribes different and arbitrary percentages of local content for more than 300 services, covering the range of services provided in the industry.

21. "Who Regulates the Oil and Gas Industry?" *Practical Law*, Thomson Reuters, https://uk.practicallaw.thomsonreuters.com/5-523-4794?__lrTS=20170412002123735&transitionType=Default&contextData=(sc.Default)&firstPage=true&bhcp=1.

22. "Functions: What We Do," Department of Petroleum Resources, https://dpr.gov.ng/index/functions-of-dpr/.

23. "NAPIMS: Roles and Responsibilities," Nigerian Petroleum Investment Management Services, http://www.napims.com/oversight.html.

24. "About NCDMB," National Content Development and Monitoring Board" (May 25, 2015), http://ncdmb.gov.ng/index.php/about-ncdmb/what-we-do.

25. "About NNPC," Nigeria National Petroleum Corporation, http://nnpcgroup.com/AboutNNPC/CorporateInfo.aspx.

26. "The Nigerian Oil and Gas Industry Content Development Act 2010," Section 69, *Laws of the Federation of Nigeria*, http://lawnigeria.com/LawsoftheFederation/Nigeria-Oil-and-Gas-Industry-Content-Development-Act,-2010.html.

27. The interpretation of the phrase "first consideration" is not provided for under the act. The phrase has been explained by commentators as a scenario whereby Nigerian independent operators are considered first in all projects for which contracts will be awarded. It also goes further to include exclusive consideration to Nigerian indigenous service companies that demonstrate ownership of equipment and employ Nigerian personnel. Simply put, Nigerians have priority in bidding for contracts, provision of goods and services, and employment and training. There is preference and disposition for Nigerian companies under the act. Noteworthy is that this priority is subject to the fulfillment of conditions established by the minister.

28. Denzil Kentebe, status briefing of stakeholders on Nigerian content (July 2015), venue unknown.

29. E. Alike, "Samsung, LADOL, a Partnership That Boosts Local Content," *This Day* (October 7, 2014).

30. "Yulong's Nigeria Pipe Mill to Create 5,000 Jobs," *Energy News* (August 25, 2014).

31. K. Nwapa, "Nigerian Content Development" (keynote speech, Local Content Policies in the Oil, Gas, and Mining Sectors Conference, Vienna, Austria, September 30–October 1, 2013).

32. Jesse Ovadia, "Measurement and Implementation of Local Content in Nigeria: A Framework for Working with Stakeholders to Increase the Effectiveness of Local Content Monitoring and Development," Facility for Oil Sector Transparency in Nigeria (FOSTER) (January 2013), 10.

33. Lonadek offers technical, management and leadership, entrepreneurial, and nontechnical local content programs, while Delton Shareholding Company has set up a training center in Port Harcourt, Nigeria to service current training contracts with the SPDC for the SOKU Associated Gas Gathering project. AGIP also recently concluded training in line with the act. "Local Content Development," Lonadek, http://www.lonadek.com/modules/content&id=local_content_development; "Training," Delton Group, http://www.deltongroup.com/project-experience/training.php; "Nigerian Agip Concludes Training for 15 Youths," Nigerian Content Development and Monitoring Board (NCDMB) (June 22, 2016), http://ncdmb.gov.ng/2016/06/nigerian-agip-concludes-training-for-15-youths/.

34. "Guideline on Research and Development in the Oil and Gas Industry," NCDMB (March 1, 2012), http://www.ncdmb.gov.ng/images/GUIDELINES/RD.pdf.

35. Ovadia, "Measurement and Implementation of Local Content in Nigeria," 32.

36. "Act Gives Domestic Boost to Nigeria's Oil Services Sector," *Oxford Business Group* (July 9, 2014), http://www.oxfordbusinessgroup.com/news/act-gives-domestic-boost-nigeria%E2%80%99s-oil-services-sector.

37. The portion referenced if removed will not necessarily detract from the import and objective of the act as it provides a prescriptive to-do list for those seeking to engage in the oil and gas industry. The subsidiary legislation could be akin to those found under the Petroleum Act of 1968, which has regulations dealing with similar issues as contained in the cited part of the act.

17

NORWAY

Introduction

The Norwegian state carries out petroleum resource management based on policies implemented through laws and regulations and the exercise of administrative authority. The main characteristic of Norwegian resource management is strong host country control.[1] Norwegian authorities exercise this control through a licensing regime, combined with requirements for permits, approvals, and consents at various stages of petroleum activities. This approach has been accepted by the industry.

At the dawn of the petroleum era in Norway, the Norwegian continental shelf (NCS) was very promising. Oil companies were keen to get their share of attractive acreage. As the years passed and some large discoveries were made, interest was maintained. Norway had the benefit of being a stable, democratic state. It had capable public institutions that, albeit inexperienced with petroleum activities, were useful models to implement industrial policies. Despite political shifts in government, a relatively steady direction of petroleum politics has been maintained. From the beginning, the Norwegian authorities emphasized maintaining a close dialogue with the oil industry so as to respond swiftly to its needs. One of Norway's policy goals was to build institutional capacity and then leave resource management in the hands of competent and able institutions. These institutions functioned under a set of administrative laws enacted to safeguard the rule of law and offer predictability, transparency, and accountability in operations. Norway also had a well-functioning, impartial judicial branch. All in all, these elements contributed to an investment climate characterized by stability and predictability, despite the strong degree of host country control.

The petroleum industry contributes to the Norwegian economy through several channels. It is stated on the Norwegian government's website that "the oil and gas sector is Norway's largest measured in terms of value added, government revenues, investments and export values."[2] It is important to note that the Norwegian petroleum industry not only contributes to the Norwegian economy in the form of direct government revenues through taxation,[3] but also through "value added" and "investments and export values." The oil service industry in Norway has thrived from relatively early stages in Norwegian petroleum history. Today, many Norwegian oil service companies are operating internationally. Moreover, there are several Norwegian oil companies operating on the NCS. The national oil company Statoil[4] started as a wholly state-owned company that benefited from preferential rights to participate in petroleum activities. Statoil is now an international company that is only partially state-owned. It is a listed, integrated company that is highly specialized in offshore petroleum activities. Statoil is the dominant player on the NCS but also is active in 37 countries.

Considering the current status quo, it is interesting to note that Norway had absolutely no competence or capabilities within the petroleum industry before oil exploration activities began in the early 1960s. Decades later, Norway is well-known for its achievements related to local content and the development of its oil industry. As observed by Heum,[5]

> In fact, the way Norway has organized the petroleum activities, the political ambitions that have been pursued, and the measures taken to implement policy are in no way significantly different from what other countries with rich endowment of oil and gas have done and attempted. Nevertheless, Norway is still one of the few exceptions when it comes to really being successful in this respect.

This chapter contains an examination of the main contextual, policy, and legislative elements of Norwegian local content development over the past decades. The intent is to identify reasons for the Norwegian achievements and to discuss whether the policies and the legislative measures implemented by Norway in its local content development could be transferable to other countries. Notably, the chapter focuses on policies and legislative measures in the period from the inception of the Norwegian petroleum industry in the early 1960s until 1994, when the European Economic Area (EEA) Agreement[6] entered into force. The EEA Agreement effectively requires Norway to comply with certain parts of the European Union (EU) regulations. Among those are requirements for the free movement of goods, services, capital, and persons.[7] As a direct consequence of Norwegian accession to the EEA Agreement, the provisions aimed at promoting local content in the then-applicable petroleum law[8] were repealed. Thus, today, there are no local content requirements in the Norwegian legislative framework for petroleum activities. Norwegian players in the industry are therefore competing in the marketplace on the same terms and conditions as international players.

Brief historical overview: Interaction between the development of the Norwegian petroleum industry and petroleum policy and legislative developments

The Norwegian petroleum industry dates back some 50 years. Famously, in 1958, the Norwegian Geological Survey stated in a letter to the Norwegian Ministry of Foreign Affairs that "'one could disregard the possibility of coal, oil, or sulphur deposits being present on the continental shelf along the Norwegian coast.'"[9] The following year, however, discoveries of petroleum were made offshore of the Netherlands, in the Groningen area. International oil companies started to take interest in the North Sea and the host countries around it.

In 1962 Phillips applied for concessions over the entire NCS. The application was rejected; it was simply asking for too much, and in any event, it was too soon to commence petroleum activities on the NCS. The Norwegian authorities acknowledged that they did not have the requisite knowledge of the petroleum industry and that there was no legislation in place for such activities. As with any industry, there also was a need for competition and diversification should petroleum activities be allowed on the NCS. As the present day Norwegian petroleum authorities observe:

> The company asked for a license for the parts of the North Sea that were in Norwegian territorial waters and that were or might be designated as part of the Norwegian continental shelf,[10] and offered USD 160 000 per month. This was regarded as an attempt by the company to obtain exclusive rights. The authorities decided that it was out of the question to hand over the entire continental shelf to one company. If these areas were to be opened for exploration, more companies would need to be involved.[11]

The rejection of the Phillips application is an early demonstration of Norway's characteristically cautious approach, employing an experience- and knowledge-based, step-by-step policy for resource management and regulation. With the Phillips application, the need became apparent for Norway to define petroleum-related policies and implement them in legislation. Although they were quick to establish a legal framework so activities could commence, they did not rush to establish a comprehensive legislative framework from day one. It was agreed that regulations in law were required due to the importance of the industry, but the need to learn more about the industry and gather experience before a comprehensive law was adopted was also recognized.[12] Thus, a characteristic of Norwegian petroleum policy and legislation is that it has been developing alongside the Norwegian petroleum industry. As will be demonstrated below, the same policy was followed in the more specific regulations aimed at local content development.

In 1963, a Royal Decree[13] establishing sovereignty over what was in effect the continental shelf was adopted. Later that year, a very brief 1963 framework law was adopted. The law established the state's exclusive right to subsea natural resources, and thus the right for the king[14] to grant Norwegian and foreign persons and companies licenses to explore for and exploit subsea natural resources. It also authorized the king to issue rules on such activities.[15] In 1965 a Royal Decree[16] was adopted that implemented the 1963 framework law in more detail. It was only after the adoption of a set of main rules that the first licenses were granted in September 1965.[17] Exploration activities followed suit, and the first major discovery, the Ekofisk field, was made in 1969. Production commenced in 1971, and other major discoveries were made in the years that followed.[18] As Norwegian petroleum activities expanded both in scope and in volume, the legislative framework was further developed and refined. In 1972,[19] the 1965 Royal Decree was replaced with a more comprehensive law. A major step forward in legislative development took place in 1985, when the time was deemed ripe to introduce a new, comprehensive petroleum law and its regulations. The law, which took considerable time and effort to draft, was built on past domestic experience combined with broad international experience. The 1985 Act was later replaced by the Petroleum Activities Act No. 72 ("the Petroleum Act") on November 29, 1996. Numerous regulations have since been issued related to this law. Moreover, model production licenses and standard joint operating agreements have been developed.[20]

In addition to the main milestones in legislative development as described above, it is important to add that Norwegian authorities have always closely observed the petroleum industry and any changes in its framework conditions, whether in respect to the market, technological developments, or other areas. Early on, they established a close dialogue with the players in the industry. Despite an overall high level in continuity and predictability in policy, the history of the Norwegian petroleum policies and legislation also shows that the Norwegian authorities have been sensitive to needs. Policy makers have updated policies and legislative framework to reflect changes in the industry's conditions or to achieve new policy objectives spurred by the increasing maturity of the NCS. This is a key element that has contributed to Norway's success in maintaining attractiveness among investors while also developing local content.

Today, Norway is one of the major producers in the world. Although parts of the NCS are considered mature, new thinking has nevertheless resulted in major discoveries in mature areas over the past few years. In 2007, the Edvard Grieg field was discovered, and in 2010 the Johan Sverdrup field was discovered. Recently, exploration activities commenced in the Barents Sea for the first time. The industry's interest in these newly opened areas for petroleum activities was considerable. There is therefore reason to believe that the Norwegian petroleum industry will survive for a good few decades yet. In any event, as a result of successful Norwegian local content policies, Norwegian oil companies and the oil service industry have proven competitive on an international scale. Thus, even when Norway eventually runs out of petroleum, the Norwegian players should be capable of continuing operations globally.

Local Content Policy and Baseline

Selected policy goals related to local content development in Norway

The successful development of Norwegian local content is due in part to timing and other, possibly unique, framework conditions.[21] But it is also a result of successful modeling and implementation of local content policies and legislation. As previously discussed, there are two important features of the general petroleum policy in Norway. First, Norway employed a cautious, step-by-step approach to petroleum resource management and regulation. It also garnered experience and knowledge from the oil industry, maintaining a close dialogue with the industry in order to learn from it and adapt the regulatory environment accordingly. In the following, policies more specifically related to local content development are discussed.

The Norwegian petroleum industry dates back more than 50 years. Over this period of time, there have been numerous changes in circumstances relevant to local content development. The Norwegian local content policies and the measures to implement them have been consciously dynamic. With a dynamic approach, it is easier to ensure that the policies achieve their purpose while still benefiting Norway's industrial and economic sectors. As demonstrated in many other countries, preferential treatment of a certain sector, especially over a long period of time, may have unfortunate effects on efficiency, quality, cost levels, and the overall economy.

"Norwegianization" of the Norwegian petroleum industry through preferential treatment of Norwegian goods and services and state participation were important policy goals for Norwegian petroleum resource management from the start. In the early 1960s, Norwegian companies were not capable of carrying out petroleum activities on their own, nor could they provide the specialized services required by the petroleum industry. To get the industry started and to run it efficiently, there was a need for cooperation with experienced international companies. Petroleum activities were subject to strict state control and were allowed only under certain conditions, such as those intended to ensure the development of domestic competence and skills. Heum notes that

> politicians were in principle open to the contribution from foreign firms, while being quite determined to use the opportunity to develop domestic industrial competence. This was partly done by requiring the oil companies to set up fully operating subsidiaries in Norway, where the Norwegian authorities encouraged the recruitment of Norwegians.[22]

Hence, it was acknowledged that even from a short-term perspective, the policy goal of Norwegianization had to be balanced against other policy goals. In particular, it had to be balanced against a key policy goal for any host country with a nascent petroleum industry, namely to attract competent international companies and capture the economic rent from those activities. The following statements in a 1979 white paper[23] sum up the main policy positions on the involvement of international companies:

> It has been a clearly stated goal that the petroleum industry to the greatest extent possible shall be integrated in Norwegian business life and create economic growth and employment in Norway. At the same time it has been clear that the industry has been completely dependent on foreign capital and expertise, and that it also in the future will be necessary and appropriate with cooperation with foreign industry.[24]

The policy goal to attract foreign capital and expertise efficiently limits how far a host country can push local content requirements. Local content requirements that are too rigid or too difficult to comply with may discourage the interest of investors. Such requirements also impair efficiency and increase costs of petroleum operations, thus ultimately reducing potential state revenue from the petroleum industry.

Local content requirements on the use of Norwegian goods and services were introduced in Norwegian law by the Royal Decree of 1972.[25] In 1979, while preparing to update the petroleum law,[26] drafters contemplated the adequacy of the local content requirements embedded in the 1972 Royal Decree. As discussed in the white paper from 1979,[27]

> When modelling statutory provisions in this area, we face the question of whether the current rules on use of Norwegian goods and services shall be maintained unchanged or be changed.... One must also keep an eye on the effects that a preferential arrangement on the Norwegian continental shelf will have for the incidental cost development in the country. A preferential arrangement will affect the prices that the industry sets for its goods and services. In turn, this can lead to an increase in the salary levels in these businesses and in the industry in general. Businesses that do not enjoy preferential treatment but that are exposed for the full effects of international competition could struggle to cope with this higher salary level. It is hard to fully foresee the consequences for Norwegian economy of such a development.

The wording captures the dynamic approach to local content development established by the Norwegian authorities. It also illustrates the difficult and complex conflicts of interests and priorities policy makers must balance and the need to consider possible long-term effects. It is therefore necessary for host countries to keep a keen eye on developments in the industry so as to head off any unfortunate and unintended effects in the early stages of policy implementation.

These complex conflicts of interests and priorities render institutional cooperation among relevant ministries and other state entities important if the host country is to succeed in modeling and implementing adequate local content policies. When a host government chooses to promote local content objectives in the petroleum sector, the petroleum sector cannot be viewed in isolation. Rather, the local content policies must be in line with the other economic development policies of the host country and vice versa.[28] Information sharing and coordination among the ministries are key elements for successful implementation. Such coordination can be problematic, however, and communication and coordination between ministries can suffer from power struggles and "turf wars." This, unfortunately, is detrimental to the possibilities for success in developing local content. Luckily this has been much less of a problem for Norway than for many other countries, due in part to well-established administrative practices.[29]

It is also worth noting that the involvement of several Norwegian oil companies on the NCS contributed to increased involvement of Norwegian suppliers of goods and services. The involvement of multiple Norwegian oil companies on the NCS resulted from a policy goal set by Norwegian authorities. In 1972 it was concluded that there should be three Norwegian oil companies operating independently from one another on the NCS.[30] In the early years, they were given preferential rights in order to become established in the industry.

Another general policy goal that was key for successful local content development was the goal to develop able institutions.

Baseline and Norwegian petroleum policies on local content—two elements intertwined

Experience shows that local content policies are more likely to attain their goals if they are built on realistic expectations and on already existing industrial competencies and capabilities in the host country. In this respect, success might be easier to come by for a host country that already possesses some industrial competency and a reasonable level of education in the population. Such was the case for Norway.

To a degree it is true that in the early 1960s, Norway lacked skills and competence specific to the petroleum industry. Nonetheless, Norwegian industry generally was already developed. As Al-Kasim notes, "Before the advent of petroleum activities, Norway already possessed considerable technological expertise in several areas relevant to petroleum operations. Examples include concrete technology, dam building, marine operations, and advanced methods of numerical analysis."[31] Moreover, Norway possessed considerable skills in building and providing oil tankers to the international petroleum industry. This business, as well as shipping, was quite sizable by both domestic and international standards. Norwegians possessed considerable knowledge about offshore activities, and the country was already industrialized, with a relatively educated population and a good educational system. Norwegian businesses possessed industrial competency.

Although not directly relevant to upstream petroleum activities such as exploration, development, and production, these skills and competencies formed an excellent base to build upon. Norway's workforce and industries were able to develop the skills and competencies required for delivery of goods and services to the petroleum industry. As mentioned, the educational system was good, and most Norwegians were fairly well educated.

At the time petroleum activities commenced in Norway, the Norwegian administration was already characterized by having functional, stable democratic institutions. The Norwegian government promoted sound business practices, encouraged industrial dynamics, and modeled industrial policies.[32] Norwegian authorities were able to manage implementation of industrial policies, carrying out necessary supervision and enforcement. The task of managing the nascent petroleum sector was naturally given to the Ministry of Industry at first. Although admittedly in the beginning, the ministry had no competence within petroleum activities, it was familiar with managing industrial activities with international interfaces. Thus the ministry was able to take on this task and determine what was needed. It was also recognized that much could be learned from the international oil companies.[33]

Developing a stable, predictable environment for oil industry investors requires adequate administrative practices relating to the implementation of local content policies and the development of the petroleum industry, among other concerns. Petroleum activities commenced in Norway at a time when prudent administrative practices were already in use, characterized by a relatively high level of predictability, transparency, and accountability. The Norwegian Public Administration Act[34] was enacted in 1967. Aiming to safeguard the rule of law and thoroughness and impartiality in public administration, it established general rules on legal competence, case handling, case preparation prior to making decisions, and policies on complaints and reversals, among other regulations.[35] By and large it was a codification of previous practices for administrative case handling. The Freedom of Information Act was enacted in 1970. Subject to defined exceptions, this act established the principle of free access to information in public documents for the public in general.[36]

All in all, these baseline elements were cornerstones in enabling the Norwegian authorities to model and implement adequate local content policies for the Norwegian oil industry. It also illustrates one of the points made initially: the conditions under which Norway modeled and implemented its local content policies were geared for success. Furthermore, the offshore petroleum industry was only just beginning. There was much room for entrepreneurs to develop new technologies and establish themselves not only on the Norwegian market, but also internationally. It is hard to describe this timing as anything other than very lucky.

It is this baseline that enabled the local content development to materialize so quickly. Again reference is made to the 1979 white paper, which summed up the experiences made at that stage, a mere 14 years after petroleum activities commenced on the NCS:

> Over the 14 years that have passed, the Norwegian involvement has grown to be relatively significant. Norwegian yards have developed and built advanced drilling platforms and Norwegian rig owners have today a large fleet of semi-submersible drilling platforms, in total approximately 30 units. Norwegian yards have built a number of steel and concrete platforms for production and processing and large parts of Norwegian industry has been involved as suppliers of goods and services. The Norwegian participation in production licenses has been gradually increased. This has fortified the connections with the Norwegian business life at large.[37]

Key Players

Introduction

It was stated previously that it was a clear policy goal for the Norwegian authorities to attract and involve international companies in the nascent Norwegian petroleum industry. For the present purposes, it is sufficient to note that this effort was successful. Over the years numerous types of foreign entities, including all types of oil companies, have established business operations in Norway and blossomed. The first to arrive were the majors, and then as time passed, smaller oil companies were also established in Norway, as well as all types of oil service companies.

The Norwegian authorities had as a policy goal the development of a domestic petroleum industry. Key players in this respect were of course the Norwegian entrepreneurs, who had a lot at stake in their efforts to gain a foothold in a new industry. Norway has fostered a number of oil companies, and from the beginning, companies included not only Statoil, but also Norsk Hydro, Saga, and DNO. The oil service industry in Norway thrives, and many domestic oil service companies are also operating internationally. Without entrepreneurial driving forces in the domestic business life, however, chances of succeeding in sustainable local content development are slim. At the same time, it is difficult to say how successful the entrepreneur's initiatives would have been without the help of supportive measures and their careful implementation, supervision, and enforcement. On a global scale, there is increased regulation of local content. However, the level of compliance among the international players might not always be satisfactory. Thus the state administration plays a key role, as mere rules in and of themselves will not lead to results; compliance is a critical factor.

Acknowledging the key role that the state administration of petroleum activities plays in succeeding in local content development, the following section focuses on the state administration of the petroleum industry. It also discusses the role and evolvement of state-owned and national oil companies and how they played their respective roles in the development of local content in Norway.

State administration of the petroleum industry

As stated above, a policy goal that was key for the successful local content development in Norway was the development of able institutions. As mentioned, at the dawn of the petroleum era on the NCS, the Norwegian administration was already characterized by efficient, stable institutions that were used to promote sound business practices, encourage industrial dynamics, and model industrial policies. These institutions operated on the basis of prudent administrative practices that were characterized by relatively high levels of predictability, transparency, and accountability. This certainly formed an advantageous starting point. Nevertheless, as it became apparent that Norway's petroleum potential was sizable, the need for a specialized administrative system soon became apparent. Norway wished to maintain tight state control on the petroleum activities to be carried out on the NCS.[38] Thus, there was a need to increase specialized competence and capacity.

At first, as the natural choice, the Ministry of Industry was charged with the management of the petroleum industry. As early as 1966, a separate petroleum section was established within the ministry. Al-Kasim comments:

> In a report to the Storting [i.e., the Parliament] submitted on 12 June 1970, the Ministry was cautious regarding the future organisation of its functions within the petroleum sector. It stated amongst other things that "the Petroleum section has been established, and the personnel have been engaged, on a provisional basis. This form of organisation has been adopted

> because the Ministry does not yet know whether there will be permanent activities on the Continental Shelf.... In the longer run, the present organisation form will be undesirable. It may be necessary to establish a directorate for Continental Shelf matters, possibl[y] also a state-owned company in charge of the Government's business interests in petroleum exploration, [if] commercial finds of petroleum are made in one or more of the blocks where the Government has participation rights."[39]

Again, one can make note of the cautious, step-by-step approach to resource management. However, the perception of the petroleum potential and the sustainability of a Norwegian petroleum industry on the NCS was greatly affected by the Ekofisk field coming on stream in 1971. Thus, in June 1972, a new structure was adopted for state administration of the petroleum industry. The new regime was based on the advice of a special committee charged by the government to recommend a new organization for the administration of petroleum matters. The committee had studied examples of other, more mature petroleum countries. The new regime was based on a separation of responsibilities as follows:

- **The Ministry of Industry.** Responsibility for policy making, legislation, and licensing was given to the Ministry of Industry.
- **The Petroleum Directorate (NPD).** Subordinate to the Ministry of Industry, this directorate was responsible for technical control and regulatory and advisory functions.
- **Statoil.** A 100% state-owned company, Statoil was charged with commercial participation.

In 1978, the Ministry of Oil and Energy was established. The Ministry of Oil and Energy had authority over the functions of the NPD related to resource management, and the Ministry of Labor and Local Affairs had authority over the functions of the NPD related to health, safety, and the environment. The rationale for this split concerns the potential conflict of interests between resource management and licensing policy goals on one hand, and maintaining a high level of health, safety, and environmental protection on the other. Later, in 1994, the NPD itself was also split into the NPD and the Petroleum Safety Authority (PSA). Today, therefore, the NPD is only responsible for resource management issues, and the PSA is responsible for health, safety, and environmental issues.

Another important element of the state administration of petroleum resources with great impact on local content development was the establishment of Statoil in 1971. It was established as a 100% state-owned oil company with the objective to (singly or through other companies) explore for, exploit, transport, refine, and market petroleum and related products.[40] Over time, its roles have changed,[41] and Statoil today is a commercial company that is driven by shareholder interests, acting and reacting in accordance with the prevailing market conditions. In parallel, the state's ownership interests in Statoil have fluctuated. Notably, in 2001, the Parliament decided for Statoil to be listed on the stock exchange, and an opening was made for others to own up to one-third of Statoil's shares. In 2006, Statoil merged with Norsk Hydro ASA, a partially state-owned company. Today, Statoil is an international, integrated company with activities in 37 countries and is registered on the stock exchanges in Oslo and New York. It is also the dominant player on the NCS. Today, the state owns 67% of the total shares of Statoil.

Industry Knowledge

As stated earlier, Norwegian knowledge of the petroleum industry in the early 1960s was close to nonexistent. It was a policy goal to learn more about this new industry so as to enable competent resource management, encourage active Norwegian participation in petroleum activities, and develop local content. The focus on continued learning, through research and development, continues today.

The goal of building industry knowledge was achieved through the implementation of a variety of measures. In the 1970s, licenses and certain additional agreements established a range of requirements to build industry knowledge. For instance, licensees were required to provide for training of Norwegian personnel to enable workforce competency and training of governmental officials to build effective institutions. Licensing funds also helped to train personnel to teach petroleum-related topics in Norwegian schools and to build competence in relevant educational institutions. There were requirements to transfer technology and for licensees to conduct a certain percentage of their research and development in Norway. For example, in the 1979 licenses, it was a requirement that at least 50% of R&D activities undertaken in connection with any petroleum activities under a license had to be performed in Norway. The details were set out in separate agreements. The aim was to promote R&D activities and to develop the competence and know-how of Norwegian research institutions and contractors. This foundation served as the basis to build strong research communities in Norway, which in turn encouraged innovation in the Norwegian petroleum industry. It also was a great help in developing local content. As Norwegian petroleum authorities have noted on this topic:

> The development of new knowledge and technology is a cornerstone of Norway's resource management.... New technology has been essential for achieving the greatest possible value creation and environmentally sound utilisation of resources on the Norwegian continental shelf.... The industry's competitiveness and innovation capacity have also led to major positive spin-off effects and technological applications in other industries in Norway. Technology developed for the Norwegian shelf has given the Norwegian supplier industry a competitive advantage on the global stage.[42]

Norway has strong, independent research institutions and highly competent research communities and centers,[43] along with involvement in various companies, including Statoil. It is also connected with universities and colleges, such as Norwegian University of Science and Technology. The current plethora of projects has resulted in high levels of activity. Today, R&D activities are promoted through legislation and allocations to the Research Council of Norway.

In 2001, the Ministry of Petroleum and Energy established the Oil and Gas in the 21st Century (OG21) strategy. The aim was to address the challenges associated with efficient and responsible petroleum activities. Reportedly, the OG21 process helped oil companies, universities, research institutions, the supplier industry, and authorities to agree on a joint national technology strategy for oil and gas.[44] The strategy has since been updated.

In general, Norwegian universities and colleges provide a range of courses aimed at the petroleum industry, through relevant disciplines that include law, engineering, and geology, among others. Moreover, the Norwegian government itself offers training and capacity building to other governments. In this, the work carried out under the Oil for Development Program, as well as the training of government officials provided by the foundation Petrad, are at the forefront.

Legislative Hydrocarbon Regime Concerning Local Content

As mentioned initially, there are no "local content" requirements in the present-day legal framework for petroleum activities. In a sense, this brings us back to the starting point; in the early 1960s, there were no local content requirements in Norway's statutory law. Local content development was instead promoted and encouraged through informal agreements between the Norwegian authorities and the licensees. Later local content policies were addressed in license conditions and specific agreements between the Norwegian authorities and the licensees. As mentioned previously, training, technology transfer, and R&D support were set as requirements in license conditions and separate agreements. Moreover, state participation was gradually introduced. From the second license round in 1968, the licenses awarded were given on the condition of state participation with a carried interest.

Local content requirements on the use of Norwegian goods and services were introduced in Norwegian law by the Royal Decree of 1972.[45] Section 54 required that licensees were to make use of Norwegian goods and services to the extent that these were competitive in terms of quality, service, delivery times, and price. In order to support Norwegian suppliers of goods and services, Section 54 also required that they were to be included in invitations to tender in as far as the goods and services requirements were available in Norway. Moreover, in the evaluation of tenders, the degree of local content was to be taken into consideration. In addition, Section 54 made the licensees responsible for their contractors' and subcontractors' compliance with these requirements. Section 54 further required licensees to carry out their petroleum activities with a base in Norway. It was expected that the Norwegian organization would be of sufficient scope to perform necessary activities and to take all necessary decisions in connection with petroleum activities.[46]

Section 54 of the 1972 Royal Decree was incorporated in the Petroleum Law of 1985. Section 54 of the 1985 act stated that "in activities that are comprised by this act, competitive Norwegian suppliers shall be given real opportunities to deliver goods and services."[47]

Further requirements with regards to the procurement of goods and services were adopted in regulations. Moreover, the 1985 act established requirements as to the training of personnel and provided a legal basis for the government to adopt rules on training of public officials. This latter requirement was further implemented in regulations to encompass training of officials from the Ministry of Petroleum and Energy, the Norwegian Petroleum Directorate, or other public entity pursuant to the agreement. Moreover, it was established that the Ministry of Petroleum and Energy could direct licensees to train instructors teaching petroleum-related topics in Norwegian schools.

Not at any point did Norway have specific supervision or enforcement mechanisms for compliance with local content requirements. Although a specific office was established within the ministry for local content development, the general rules applied. However, the legal framework for petroleum activities and the manner in which it was implemented encourage both prospective and current licensees to come up with proposals as to how compliance would be ensured. For instance, how prospective licensees were planning to enhance local content would be taken into consideration as a part of the process leading up to the grant of new licenses. Moreover, if commercial discoveries were made, a local content plan was required as a part of the development plan. In this manner, the licensees were forced to be proactive.

As noted previously, local content development in Norway took place rather swiftly. Furthermore, the Norwegian authorities were sensitive to the effects of continued preferential treatment. As Norway moved towards EEA accession in the early 1990s, a white paper on repeal of the local content requirements from the Petroleum Act noted the possible consequences:[48]

> Within the petroleum sector, Norwegian oil companies have had certain reliefs and benefits compared to international companies. Thus, only the international companies have been obliged to cover exploration costs for state participation in petroleum activities. When the state has increased its shares in commercial discoveries, international companies have reduced their share, something which Norwegian companies in some cases have not been obliged to do. It is pointed out... that these forms of differential treatment now has ceased.... It is pointed out that...the purchase of goods and services will be regulated in the EEA rules for public procurement. Consequentially, purchases will as a main rule take place on the basis of competitive tenders and equal treatment.... It is pointed out that this will expand the market opportunities for Norwegian goods and services to the petroleum industry in other EEA countries. [49]

Conclusion

All in all, it is an undisputed conclusion that Norway has benefited from the local content policies implemented. Moreover, results materialized very quickly. The success is perhaps due more to framework conditions, timing, and the manner in which policies were modeled and implemented than to the specific policy goals.

An examination of Norway's local content policy illustrates how it has been carefully balanced against other interests, and how policy and requirements have been adapted in a dynamic manner. Moreover, successful local content development requires efforts from all main stakeholders in the industry: government, international companies, and local entrepreneurs. Although the latter group benefited from preferential rights, not all of them enjoyed success. There will always be an element of risk in business, and petroleum prices are volatile. As Ryggvik[50] notes:

> The history of the Norwegian contractor industry is a complex one. Not all start-up attempts were equally successful. Furthermore, the contractor industry was always far more vulnerable to economic conjunctures than the oil companies. In periods with low oil prices, the companies could save on expenses by reducing new investments (prospecting and the development of already known fields) and live off established fields. This hit many contractors directly. In the 1980s alone, the contractor industry experienced two such crises. The low oil prices at the end of the 1990s led to a similar crisis.
>
> Many Norwegian company groups ended up as parts of large foreign companies. However, there were also examples of Norwegian capital milieus which started out as parts of Norwegian companies and made comparable buy-outs abroad, eventually establishing large international companies. The common feature of the vast majority of these was that in the start-up phase, they were assisted by various forms of Norwegian protectionism.

Oil companies and their employees face economic risks from falling oil prices. One may therefore question the sustainability of ambitious local content development. As commented above by Ryggvik, an economy that is heavily dependent on one industry is vulnerable. This is especially true given the volatility of the petroleum prices. As was the case in the 1980s, the industry is again facing a low-price scenario, which hits the oil services industry hard. Oil companies are also focusing on cutting costs. Nevertheless, the Norwegian oil services industry has contributed to the domestic economy of Norway to a considerable extent for many years, and it is reasonable to assume it can continue for some time yet. As recently stated by the former Norwegian Petroleum and Energy Minister, Tord Lien, "The persons that will turn the light off on the NCS are yet to be born."[51]

Notes

1. It should be noted, though, that especially in the beginning, the Norwegian state was careful to balance this approach so as not to impair the investment climate. The main principles for resource management that were established for the first licenses in the early 1960s remain the same today. Norwegian authorities were treading cautiously in the early years so as not to lose the interest of the investors. As the number of discoveries rose, however, so did the host country's control.
2. "Economy," Norwegian Petroleum, http://www.norskpetroleum.no/en/economy/.
3. Norway has a license regime for exploration and production rights. The licensees become owners of all petroleum produced, and government revenues are collected through taxation.
4. Norway has fostered more than one national oil company over the past decades.
5. Per Heum, "Local Content Development—Experiences from Oil and Gas Activities in Norway" (SNF Working Paper No. 02/08, Institute for Research in Economics and Business Administration, Bergen, February 2008), 3, http://extrayendotransparencia.grupofaro.org/wp-content/uploads/2014/12/LC-Development-Experiences-from-oil-and-gas-in-Norway.pdf.
6. The English version is available at *The EEA Agreement*, European Free Trade Association (EFTA), http://www.efta.int/legal-texts/eea. The EEA Agreement was entered into between the then–European Community and the European Community's member states on the one hand, and the EFTA states on the other hand, comprising Iceland, the Principality of Liechtenstein, and the Kingdom of Norway.
7. "Part I: Objectives and Principles," Article 1(2), in *EEA Agreement*, http://www.efta.int/legal-texts/eea.
8. "Act of 22 March 1985 No. 11 Pertaining to Petroleum Activities" (March 22, 1985); now repealed. Also called "the 1985 Act," it has been replaced with the 1996 Act ("Act 29 November 1996 No. 72 Relating to Petroleum Activities").
9. As quoted in Morton Lund, "The Value of Flexibility in Offshore Oil Field Development Projects" (Department of Economics and Technology Management, Norwegian University of Science and Technology, Trondheim, June 1997), https://forsvaret.no/prinsix/ForsvaretDocuments/The%20value%20of%20flexibility%20in%20offshore%20projects.pdf.
10. The Norwegian state did not declare sovereignty over the NCS until 1963.
11. "Norway's Petroleum History," Norwegian Petroleum, http://www.norskpetroleum.no/en/framework/norways-petroleum-history/.

12. "Petroleum Act and Its Regulations," Official Norwegian Reports, Ministry of Petroleum and Energy, NOU 1979: 43, chapter 1.
13. "Royal Decree of 31 May 1963 Relating to the Sovereignty of Norway over the Sea-Bed and Subsoil outside the Norwegian Coast" (May 31, 1963), http://www.un.org/depts/los/LEGISLATIONANDTREATIES/PDFFILES/NOR_1963_Decree.pdf.
14. In this case, "king" refers to the government or its relevant minister.
15. "Act 21 June 1963 No. 12 on Exploration and Exploitation of Subsea Petroleum Resources" (June 21, 1963); this was more an "enabling act" for further implementation in regulations than a comprehensive petroleum law. Such regulations were made by Royal Decree on April 9, 1965, which adopted a regulation implementing the 1963 Act. The regulation established the main rules on petroleum operations, safety, and environment.
16. "Royal Decree of 9 April 1965 Relating to Exploration for and Exploitation of Petroleum Deposits in the Sea-Bed and Its Subsoil on the Norwegian Continental Shelf," http://www.un.org/depts/los/LEGISLATIONANDTREATIES/PDFFILES/NOR_1965_Decree.pdf.
17. Interestingly, and in line with the cautious, step-by-step approach to resource management and development of policies and legislation, production licenses were not granted until Norway had declared sovereignty over its continental shelf (April 1963) and agreed on delimitation with the United Kingdom and Denmark (March 1965).
18. For a more detailed overview, see "Norway's Petroleum History," http://www.norskpetroleum.no/en/framework/norways-petroleum-history/.
19. Royal Decree of December 8, 1972.
20. See "Agreement for Petroleum Activities and Model Production Licenses," Ministry of Petroleum and Energy (February 3, 2016), https://www.regjeringen.no/en/find-document/dep/OED/Laws-and-rules-2/Rules/konsesjonsverk/id748087/.
21. The framework conditions, or the baseline, will be detailed later in the text.
22. Heum, "Local Content Development," 9.
23. "Petroleum Act and Its Regulations," NOU 1979: 43.
24. Ibid., 38, official translation of relevant parts.
25. Royal Decree of December 8, 1972, Section 54.
26. "Act of 22 March 1985 No. 11 Pertaining to Petroleum Activities" (March 22, 1985), now repealed.
27. "Petroleum Act and Its Regulations," NOU 1979: 44, office translation.
28. Silvana Tordo, Michael Warner, Osmel E. Manzano, and Yahya Anouti, *Local Content Policies in the Oil and Gas Sector* (Washington, DC: International Bank for Reconstruction and Development, World Bank, 2013), xiii, doi:10.1596/978-0-8213-9931-6, or http://documents.worldbank.org/curated/en/549241468326687019/pdf/789940REVISED000Box377371B00PUBLIC0.pdf.
29. The frequent use of public hearings of new legislation is a measure that ensures that information is shared and opinions are gathered before adoption of new legislation.
30. Statoil (which was fully state-owned in the beginning), Hydro (which was part state and part privately owned), and Saga (fully privately owned).
31. Farouk Al-Kasim, *Managing Petroleum Resources: The 'Norwegian Model' In a Broad Perspective* (Alden Press, 2006) 58.

32. Heum, "Local Content Development."
33. The evolution of the administration of petroleum activities is presented in the "Brief historical overview" section and also is reviewed in the "Policy and Baseline" section.
34. Public Administration Act, February 10, 1967.
35. Norwegian administrative law is also built on certain key principles such as the principle of legality as well as principles aiming to safeguard that any discretionary decisions are properly made, i.e. the principle of proportionality et cetera.
36. Freedom of Information Act, June 19, 1970.
37. "Petroleum Act and Its Regulations," NOU 1979.
38. From the very beginning of oil industry activity, Norway has had a license-based regime for granting E&P rights. Licenses are typically granted with specific stipulations that the licensee must comply with. Moreover, in order to move forward with petroleum activities or a specific project, governmental approvals and consents are typically required.
39. Al-Kasim, *Managing Petroleum Resources*, 46.
40. Proposition No. 60 (2006–2007), 3.1, "Statoil's Development in Recent Years," *Regjeringen.no*, https://www.regjeringen.no/no/dokumenter/stprp-nr-60-2006-2007-/id461986/sec1.
41. For instance, in 2001 wholly state-owned company Petoro took over the task of management of the commercial aspects of the State's Direct Financial Interest (SDFI).
42. "Petroleum Related Research and Development," Norwegian Petroleum, http://www.norskpetroleum.no/en/environment-and-technology/petroleum-related-research-and-development/.
43. For instance, SINTEF, ARCEx (Research Centre for Arctic Petroleum Exploration), and the National IOR Centre of Norway, to mention a few.
44. "Petroleum Related Research and Development," Norwegian Petroleum.
45. Royal Decree of December 8, 1972, Section 54.
46. In addition, there were requirements to bring petroleum ashore so that processing and refinement would be carried out in Norway. This was only practical for oil, however, as gas has never been an element of significance in the Norwegian energy mix.
47. "Act of 22 March 1985 No. 11 Pertaining to Petroleum Activities" (March 22, 1985), unofficial office translation. Text can be found at www.lovdata.no (accessible only with a subscription).
48. Inntst. O. (1992–1993). Text can be found at www.lovdata.no (accessible only with a subscription).
49. Unofficial office translation. Norwegian text can be found at www.lovdata.no (accessible only with a subscription).
50. Helge Ryggvik, *The Norwegian Experience: A Toolbox for Managing Resources?* (Centre for Technology, Innovation and Culture, the University of Oslo, 2010), 61, http://www.sv.uio.no/tik/forskning/publikasjoner/tik-rapportserie/Ryggvik.pdf.
51. "De som skal slå av lyset på norsk sokkel, de er ennå ikke født," *Dagens Næringsliv* (February 2, 2016), http://www.dn.no/nyheter/energi/2016/02/02/1029/Tord-Lien/-de-som-skal-sl-av-lyset-p-norsk-sokkel-de-er-enn-ikke-fdt.

18

QATAR

Ruchdi Maalouf

Introduction

The state of Qatar is located on the northeastern coast of the Arabian Peninsula. It shares a border with Saudi Arabia and is a short distance from Bahrain and the United Arab Emirates. It is also a short distance from Iran across the Persian Gulf. Qatar is a member of OPEC (the Organization of the Petroleum Exporting Countries) and the GCC (Gulf Cooperation Council).

Qatar became an oil producer in 1947, and in 2014, it held the thirteenth largest oil reserves in the world. It had the third largest proved reserves of natural gas and was ranked as the world's third largest exporter. In 2015, Qatar was the world's largest LNG exporter. In addition to crude oil, condensate, and natural gas liquids (NGLs), Qatar also exports petrochemicals, ammonia, fertilizers, and steel.

Although its population was estimated to be in excess of 2 million people in 2014, there are fewer than 300,000 Qatari nationals. With a GDP of about US$200 billion (2013), its average earnings per capita are among the highest in the world. This is therefore a country that has no developmental problems and has strong and healthy long-term prospects. Nevertheless, the Qatari government has taken a long-term view on the exploitation of its natural resources with a moratorium on gas developments and significant downstream investments in petrochemicals, fertilizers, and power. More recently the Qatari government also has focused on healthcare, education, and tourism. These efforts include plans to involve a growing portion of its nationals in the management of the economy and to raise their education level, which is already comparatively very high.

Local Content Policy and Baseline

Policy background

Qatar has been pursuing a number of policies to ensure that its oil and gas industry benefits its state and its nationals sustainably. The long-term strategy rests on the view that its reserves of oil and gas are of strategic importance and must be maintained for the future generations and exploited to meet the needs of national security and sustainable development. The principle is the state ownership of natural resources. The long-term strategy also drove Qatar to develop a petrochemical industrial base and to develop related industries.

Qatar is also conducting an intensive exploration of its petroleum reserves with the dual objectives of expanding the lifetime of its reserves and adding to its production capacity.

Some results of that long-term strategy can already be observed. Although pursued for a long period of time, the Qatari development strategy has been formulated through a statement of principles in a document entitled "Qatar National Vision 2030." In the short term, two policies have been pursued for the past 10 to 15 years. The first is an effort to educate, train, and promote Qataris at key management positions of the national companies (see Qatarization section). The second is a moratorium on further developments in the main petroleum structure, which has been in place for the past 10 years.

Qatar National Vision 2030. Qatar's development goals were stated in a strategy paper published in July 2008, "Qatar National Vision 2030."[1] The paper identifies five challenges that Qatar needs to balance to build a sustainable future:

1. Modernization and preservation of traditions: how to preserve cultural traditions in a rapidly globalizing and increasingly interconnected world;
2. The needs of the current generation and those of future generations: how to meet today's needs without compromising the needs of future generations;
3. Managed growth and uncontrolled expansion: how to encourage Qatar's growth without depleting its resources and overstraining its economy;

4. The size and the quality of the expatriate labor force and the selected path of development: how to manage development when there is high reliance on a significant foreign workforce; and
5. Economic growth, social development and environmental management: neither economic development nor the protection of environment should be sacrificed for the sake of the other.

In order to meet these challenges, the paper identifies four pillars on which the policies must rest:

1. Human development: primarily through education and healthcare and by motivating the workforce to play a leading role in the economy;
2. Social development: modernizing the society while maintaining the traditions through social care and protection, a sound social structure, and international cooperation;
3. Economic development: through sound economic management, responsible exploitation of oil and gas, and a suitable economic diversification; and
4. Environmental development: achieving a balance between development needs and protecting the environment.

Moratorium on the North Field developments. The gas reserves that are used in the LNG developments are mainly located in Qatar's North Field, a geological structure holding up to 1,800 Tcf of gas and 50 billion barrels of condensate. Since Qatar is currently the largest LNG exporter in the world, it is notable that the bulk of its production comes from projects related to one gas field. From 2005, the Qatari government implemented a moratorium on future developments from the North Field, in order to maintain sustainable exploitation of the reservoirs. This policy is regularly reviewed but has so far remained a pillar of the development strategy of the government.

Qatarization. One concrete way the Qatari National Vision 2030 has been translated into action is the policy known as "Qatarization." Qatarization is a long-term policy through which higher education of Qataris and skills development should result in an increase of the Qatari workforce in key positions of the energy, industry, and government departments. Those positions are currently occupied by expatriate staff. The objective is to provide 50% or more of Qatari citizens with "meaningful" permanent employment. The Qatarization initiative applies both to the public and the private sector.

It should be noted that international companies participating in the oil and gas industry participate in the initiative.

The Qatar Foundation was tasked with encouraging and supporting the education of Qatari nationals and providing them relevant job opportunities. A directorate for human resources (the Qatar Foundation Human Resources Talent Sourcing and Development Directorate) supports the Qatarization policy through personal development of Qatari nationals.

The policy relies on the following activities:

- *Recruitment*. The directorate ensures that applications from candidates who have the appropriate potential are processed adequately by the relevant hiring department.
- *Succession planning*. The directorate develops Qatari candidates who have leadership potential. Qatari employees have a career development plan, which lists careers relating to their current and long-term employment.
- *Training*. Discussed in more detail in the following text.
- *Scholarships*. Top candidates are offered scholarships to study in Qatar and abroad.
- *Internships*. Internship opportunities are encouraged.

Overview of key prevailing local conditions

The oil and gas industry is the largest sector of the economy and generates a significant portion of the government revenue. However, strong efforts of diversification can be observed, thanks to the development of an industrial base around petrochemicals and fertilizers, and also with construction materials (through Qatar Steel and Qatar Primary Material Company [QPMC]). The government supported those choices due to the proximity to the oil and gas industry through feedstock (petrochemicals and fertilizers) and the requirement for primary materials in the associated services (petroleum operations, construction, etc.).

The development of an industry, particularly a petrochemical industry, is central to the government's plans to diversify the economy and sustainably use the petroleum reserves. Central planning has consisted of growing the economy through the development of an export industry centered around the ports of Ras Laffan and Mesaieed (which are also oil and gas ports). This has allowed a steady growth of Industries Qatar (IQ) in petrochemicals, fertilizers, and steel, and it is now one of the larger chemical producers in the Middle East.

In addition to developing an export industry, Qatar has also supported the development of modern health facilities, tourism infrastructure, and western-style education institutions.

The development of higher education and the policy of raising the education level of the local workforce to prepare it for highly skilled roles are also essential to Qatarization. The government provides every citizen free education and scholarships to study abroad. The Higher Education Institute (HEI) awards full scholarships to Qatari nationals admitted into a variety of designated undergraduate and graduate schools in the United States and other countries.

Key Players

Host governments and their agencies

The executive and legislative authorities rest with the emir, who appoints the government (the Council of Ministers). The minister of Energy and Industries is in charge of the energy policy of Qatar, along with the chairman of Qatar Petroleum (as well as RasGas and Qatargas). Qatar Petroleum is also the government's arm for oil and natural gas activities.

The group of national oil companies

Qatar Petroleum. Formerly Qatar General Petroleum Corporation, Qatar Petroleum is a 100% state-owned company established by statute in 1974 to replace the Department of Petroleum. It is responsible for the oil and gas industry in Qatar and is the government's agent in oil and natural gas activities. Qatar Petroleum may participate either directly (it operates some of the petroleum fields) or indirectly through subsidiaries or in joint ventures with foreign companies through PSAs or development and fiscal agreements. Its activities include exploration, production, storage, transportation, and sales of crude oil, petroleum products, natural gas, and fertilizers.

Qatar Petroleum's activities extend beyond the realm of petroleum developments. It is also involved in industrial cities in Qatar, in the metals industry, and in international E&P and LNG terminals through Qatar Petroleum International (QPI).

Qatar Petroleum has also been granted regulatory prerogatives on behalf of the government (as discussed further later in the text).

Other Qatar Petroleum companies. In the Qatar Petroleum group of companies, RasGas and Qatargas stand out in that they are the primary vehicles for the exploitation of the natural gas reserves. Together they make Qatar the largest LNG producer in the world, with 14 LNG trains with the capacity to produce 77 million tons of LNG annually.

Other Qatar Petroleum companies of note include the following:

- Refineries: Qatar Petroleum Refinery and Laffan Refinery
- Gas-to-liquid: Oryx GTL and Pearl GTL
- Petrochemicals and fertilizers: Qatar Fertilizer Company (ammonia and urea), Qatar Petrochemical Company (ethylene, low-density polyethylene, sulfur), Qatar Fuel Additives Company (methanol, methyl tertiary butyl ether), Qatar Vinyl Company (vinyl chloride monomer, caustic soda), Qatar Chemical Company (high-density polyethylene, alpha olefin), Q-Chem 2 (high-density polyethylene, alpha olefin), Qatofin (linear low-density polyethylene), Qatar Melamine Company (melamine), Ras Laffan Olefins Company (ethylene)
- Metal: Qatar Aluminum, Qatar Steel

Other companies. Tasweeq (Qatar International Petroleum Marketing Company) is another state-owned company, created by statute, with the aim of marketing certain petroleum products and by-products such as LPG, sulfur, condensates, naphtha, motor gasoline, gasoil, and jet fuel.

Nakilat (Qatar Gas Transport Company) is a Qatar-listed company that transports LNG and LPG.

Woqod (Qatar Fuel) is a Qatar-listed company that is the exclusive distributer of fuel in Qatar. This includes diesel, gasoline, aviation fuel, bunker fuels, ship-to-ship bunkering within Qatari waters, bitumen importation and distribution, LPG for cooking, and lubricants.

Oil and gas companies operating in the country

Most large oil and gas companies have a foothold in Qatar. Many are in partnership with Qatar Petroleum.

A significant partnership exists in LNG between Qatar Petroleum and ExxonMobil, and there are also partnerships with Total, ConocoPhillips, and Shell. A list of the partnerships in the LNG industry is given in table 18–1.

Table 18–1. LNG industry partnerships

Company	Trains	Shareholders
RasGas	1, 2	Qatar Petroleum 63%, ExxonMobil 25%, KORAS 5%, Itochu 4%, LNG Japan 3 %
RasGas II	3, 4, 5	Qatar Petroleum 68.8 %, ExxonMobil 29.5%, CPC 1.7%
RasGas III	6, 7	Qatar Petroleum 70%, ExxonMobil 30%
QatarGas	1, 2, 3	Qatar Petroleum 65%, ExxonMobil 10%, Total 10%, Marubeni 2.5%, Mitsui 2.5%
QatarGas 2	4, 5	Train 4: Qatar Petroleum 70%, ExxonMobil 30% Train 5: Qatar Petroleum 58%, ExxonMobil 25%, Total 17%
QatarGas 3	6	Qatar Petroleum 68.5%, ConocoPhillips 30%, Mitsui 1.5%
QatarGas 4	7	Qatar Petroleum 70%, Shell 30%

In upstream oil and gas, Maersk Oil, Occidental Petroleum, and Total operate offshore fields.

In petrochemicals, notable joint ventures include the following:

- Laffan Refinery (Qatar Petroleum, 51%; ExxonMobil, 10%; Total, 10%; Idemitsu, 10%; Cosmo, 10%; Mitsui, 4.5%; and Marubeni, 4.5%)
- Oryx, GTL (Qatar Petroleum, 51%, and Sasol, 49%)
- Qatar Chemical Company (Mesaieed Petrochemical Holding [100% Qatar Petroleum subsidiary], 4%; Chevron Phillips Chemical, 49%; and Qatar Petroleum, 2%)
- Q-Chem 2 (Qatar Petroleum, 51%, and Chevron Phillips Chemical, 49%)
- Pearl GTL (Qatar Petroleum and Shell, under a development and production sharing agreement, 100% funded by Shell)

Industry Knowledge

The government of Qatar implements a policy of developing a diverse knowledge-based economy to build the country's long-term prosperity. Qatar is trying to learn from the failures of the educational systems of its Gulf neighbors by using the example of the US primary and secondary educational system and by establishing partnerships with reputable US universities.

Qatar Foundation

Qatar Foundation is the principal vehicle used by the government for education. Qatar Foundation started a flagship project with a 2,500-acre campus in Doha hosting six US universities: Weill-Cornell Medical College (medical studies); Carnegie Mellon University (computer science, business studies); Georgetown School of Foreign Service (political science and international affairs); Virginia Commonwealth University (design); Texas A&M University (engineering); and Northwestern University (journalism). To these, a number of recognized institutions have been added, including University College London (archaeology) and HEC Paris (executive MBA). Furthermore, Qatar Foundation has developed the local university offering with programs at the Hamad Bin Khalifa University (law, science, engineering, and translation). This higher education framework supports the Qatarization policy through the development of appropriate skills.

Qatar Petroleum Learning and Development Department

The Qatar Petroleum Learning and Development Department organizes training and career development for Qatari trainees and Qatar Petroleum staff. This also includes educational support to the energy and industry sectors of Qatar.

The Qatar Petroleum Learning and Development Department provides the following programs in Qatar:

- Technical and vocational education and training programs
- Professional and development training programs
- University placement and scholarships aligned with the higher education requirements of the energy and industry sectors
- Educational and various other resources via the Qatar Petroleum Library and e-learning facilities
- Training in English language competency to enhance the skills of Qatari employees

It should be noted that the local content requirement for foreign companies operating in Qatar appears to be limited to supporting the government effort to increase the number of Qatari nationals in managerial positions and does not include involvement with the Qatarization policy.

Legislative Hydrocarbon Regime Concerning Local Content

Regulatory bodies

The regulatory structure of Qatar is a direct consequence of its constitutional structure. All legislative and executive powers are vested in the emir. The emir appoints the government (the Council of Ministers), including the minister of Energy and Industry. The minister of Energy and Industry is the chairman of Qatar Petroleum, RasGas, and Qatargas. Qatar Petroleum is also the government's arm for oil and gas activities. Consequently, the regulatory authority is exercised by the minister, while remaining subject to the emir's ultimate control.

Relevant policies, legislative framework, and key provisions

Although the Qatari economy grew from oil and gas activities, there is no overall petroleum regime. Certain principles are laid out in the laws, but particular activities tend to be regulated through contractual arrangements between Qatar and the IOCs, typically through Qatar Petroleum.

Petroleum legislation. The Permanent Constitution of the State of Qatar came into effect in June 2005. Under Article 29 of the Constitution, "Natural wealth and its resources including oil and gas are the property of the State, and the State shall preserve and exploit the same in the best manner in accordance with the provisions of law."

Decree-Law 10/1974 established Qatar Petroleum and granted certain rights to the company, including two of notable interest:

- The exclusive right to represent the state in negotiating and contracting all matters relating to exploration and drilling of oil and gas (Article 4)
- The right to establish and participate in companies related to its activities (Article 5).

Decree-Law 4/1977 on preserving oil wealth set out safety obligations and international standards for petroleum operations. It also included the requirement for persons engaged in petroleum operations to submit their plans to Qatar Petroleum and the Ministry of Energy and Industry for approval. Furthermore, they must inform Qatar Petroleum periodically of the execution of their project.

Other relevant legislation includes Law 8/2004, Protection of the Maritime Facilities of Petrol and Gas; Decree-Law 30/2002, Environmental Protection Law (and its regulations); and Decree-Law 15/2007, Organization of Marketing and Sale of Regulated Products Outside the State of Qatar (also known as the Tasweeq Law).

Law 3/2007 regarding the exploitation of natural resources sets out the rules for natural resources other than petroleum and reaffirms Qatar Petroleum's "general and exclusive privilege to explore, prospect and produce oil and natural gas and other hydrocarbon products and derivatives" (Article 4).

Foreign investment. Foreign investment is primarily governed by Investment Law 13/2000. Foreign investment is normally allowed but is limited to 49% of the capital for most business activities, with a Qatari partner or partners holding at least 51%. Foreign investment is not allowed in commercial agencies and real estate. The following activities allow 100% foreign ownership:

- Agriculture, industry, health, education, tourism, development and exploitation of natural resources, energy, or mining (subject to government approval)
- Foreign financial services (at the Qatar Financial Center)
- Business consultancy and technical services, information and communication services, cultural services, sports services, entertainment services, and distribution services.

It should be noted that Article 68 of Law 5/2002, the Commercial Companies Law, allows the 51% Qatari government/49% foreign JV vehicles to contract out of the law (for instance, by creating different classes of shares). As a result, "Article 68" companies are a common feature of foreign investments in oil and gas in Qatar.

Relevant contracts

Qatar follows an original model of petroleum licensing whereby the government exercises a high level of state participation and risk-sharing for the most critical developments (predominantly relating to gas) and a PSC system for other developments (predominantly relating to crude oil). The terms and conditions of these petroleum developments are governed by private contracts where Qatar Petroleum represents the government and may also be the equity holder. There is no detailed petroleum legislation that regulates those contracts.

The most important government contracts used in Qatar are the following:

- Exploration and production sharing agreements (EPSAs) and development and production sharing agreements (DPSAs) are used when the government is not a participant.
- Development and fiscal agreements (DFAs) are used when the government is a participant.

The contracts are not publicly known and no templates have been published. Only a limited overview of their contents is available, as detailed below.

Development and fiscal agreements (DFAs). When the government takes equity in a project, the agreement of choice is the DFA. DFAs are used in LNG and related developments. An agreement will become effective after approval through a decree of the emir of Qatar.

Parties. The parties to a DFA are Qatar Petroleum, representing the Qatari government, and the foreign partner (the contractor). Qatar Petroleum normally takes a majority shareholding of 70% or more.

Scope. Under the DFA, Qatar Petroleum and the contractor will undertake to enter into a joint venture agreement in relation to a joint venture company that will conduct certain petroleum operations (LNG liquefaction and sales, but also helium and pipeline gas). The joint venture company will be granted the right to explore, appraise, develop, and produce nonassociated gas and liquids in the area defined in the DFA. (This will include the right to liquefy, store, and sell the nonassociated gas.) Qatar Petroleum will reserve certain rights as specified in the agreement.

Term. The term of the DFA is generally 20 years or more and is renewable.

Conduct of operations. The joint venture company must abide by good industry practices and local laws and maintain insurance. The joint venture company must provide all information to the government and provide access to the operations.

Royalty. The joint venture company will pay a royalty based on the gas delivered based on the calorific value and on the condensate sold based on a percentage of the proceeds.

Taxes. The joint venture company will be subject to Qatar income tax law. The government may grant a tax holiday of up to 10 years. Assets may be depreciated linearly over 10 years.

Title to assets. The joint venture company will have title to the assets.

Other relevant terms. The government will enjoy an audit right on the accounts. The contractor will be free to export any currency and funds. There will be limited termination rights. The law of the contract will be Qatari law, and dispute resolution will either be before the local courts or under international arbitration rules. Contracts will contain stabilization provisions. There are no local content obligations other than in relation to Qatari participation and Qatarization. In particular, the DFA does not seem to contain obligations relating to the purchase of locally produced goods, transfer of technology, or training obligations. Nevertheless, under the legislative framework, the parties to the DFA retain a broad discretion to include additional commitments on an ad hoc basis. There are no published DFAs, and it is not possible to assess how diverse the terms of existing agreements are.

Exploration and production sharing agreements (EPSAs) and development and production sharing agreements (DPSAs). Where there are no proven reserves, the contract of choice is the EPSA. Where there are proven reserves, the DPSA is used. Their terms and conditions vary. An agreement only becomes effective after approval through a decree of the emir of Qatar.

Parties. The parties to an EPSA or DPSA are Qatar Petroleum, representing the Qatari government, and the foreign partner (the contractor).

Scope. The contractor will be granted the right to explore, appraise, develop, and produce petroleum in the area and field as defined in the EPSA or DPSA. The contractor will have the right to store, transport, and sell the petroleum, although Qatar Petroleum will reserve certain rights, such as the right to purchase the gas for LNG production.

Term. The term of the EPSA or DPSA will generally be 25 to 30 years, which can be extended by 5-year increments.

Exploration. The exploration phase is 5 to 8 years and may be divided into two periods. The contractor will have to relinquish at least 20% of the contract area after a first exploration period. At the end of the exploration, any area that is not designated as a producing, development, or appraisal area will also have to be relinquished.

Minimum obligations. The EPSA or DPSA will contain minimum work obligations and minimum expenditures. If there are two periods, the contractor must fulfill the minimum obligations of the first period before a second period can start. The content of the minimum work obligations are typical and include seismic data acquisition and analysis and well drilling.

Appraisal and development plans. After a discovery has been made, the contractor will have to prepare an appraisal program for the government's approval. The government will have a limited time to review the program. If approved, the contractor will have a period of six months to determine the commerciality of the discovery. If the discovery is commercial, the contractor will prepare a development plan and submit it to the government.

Conduct of operations. During the production phase, the contractor will have to prepare and submit an annual work program and budget. In the conduct of operations, the contractor will have to abide by good industry practices and local laws and must maintain insurance. It will have to provide all information to the government and provide access to the operations. Any flaring of associated gas will have to follow best industry practices and will be subject to government permission.

Management committee. The management committee will consist of an equal number of members between the contractor and Qatar Petroleum. The management committee will be responsible for the exploration, appraisal, and development, as well as the work program and budget. It will also establish certain procedures, such as lifting.

Price. Crude oil is typically priced at market level. Associated gas that is not needed for the petroleum operations must be delivered to the government.

Profit sharing. Profit sharing is determined on a case-by-case basis according to a sliding scale.

Cost recovery. A portion of crude oil or nonassociated gas will be allocated to cost recovery on the basis of an agreed maximum annual rate (typically 40%). The balance will be carried forward, and any crude oil or nonassociated gas in excess of the costs will be apportioned between the government and the contractor at a 9 to 1 ratio.

Other contractor costs. A bonus payment is due on the first export of hydrocarbons and as agreed-upon production thresholds are reached. The contractor must also pay rentals for the contract area. No royalties are due under an EPSA or DPSA.

Taxes. The contractor will be subject to Qatar income tax law, although the government is understood to assume the tax liability on behalf of the contractor. The assets can be depreciated linearly over 10 years.

Title to assets. The title to the fixed and movable assets will pass to the government upon cost recovery. The contractor will continue to enjoy the exclusive use of those assets for the duration of the EPSA or DPSA.

Other relevant terms. The EPSA or DPSA will also grant the government rights to audit the accounts. The contractor will be free to export any currency and funds. Limited termination rights are normally included. The law of the contract is Qatari law, and dispute resolution is either before the local courts or under international arbitration rules. Stabilization provisions are also normally included in the EPSA or DPSA. There are no local content obligations other than in relation to Qatari participation and Qatarization. In particular, the EPSA and DPSA do not seem to contain obligations relating to the purchase of locally produced goods, transfer of technology, or training obligations. Nevertheless, under the legislative framework, the parties to the EPSA and DPSA retain a broad discretion to include additional commitments on an ad hoc basis. There are no published EPSAs and DPSAs, and it is not possible to assess how diverse the terms of existing agreements are.

Current Issues and Challenges

Qatar does not face specific issues and challenges other than those issues that affect the global oil and gas industry. The current framework appears stable, and there is a track record of implementing long-term strategies. The current expectation would be that the government will maintain stable and reliable institutions and continue the policies in place.

From a geopolitical point of view, one point worth noting is that the North Field belongs to a geological structure shared with Iran and locally known as South Pars. While the moratorium is in the long-term interest of a sustainable long-term Qatari development, competition for the reserves of the giant field casts a shadow on the ability of Qatar to control the exploitation of North Field/South Pars. Iran's stated plans to restart their previously frozen LNG developments will have an impact on Qatar, but the results are difficult to foresee.

From the domestic point of view, it should be noted that the Qatarization policy has suffered some criticism (see further reading below). The pace of Qatarization has disappointed some. The difficulty of motivating a workforce that is sheltered from hardship and competition has also been highlighted. While the objectives and time line were very ambitious, the execution has also been criticized for being too systematic and emphasized too late in the curriculum. It has also been said that quantity was favored over quality, which would defeat the goal of the policy. Overall, the intentions of Qatarization have been perceived by some players as measurable objectives, whereas the success depends on building up a process within the education system and the national companies, as discussed in the concluding section.

Conclusion

Being one of the richest states in the world per capita, Qatar has demonstrated an original approach to local content in that the government has taken the development of the economy (including that of an advanced petroleum industry) in its hands. Rather than making petroleum licensing conditional on an IOC developing local content, Qatar Petroleum leads the developments and manages investments in the oil and gas industry. Qatar's development strategy seems sensible, as its population is small and its natural resources are essentially hydrocarbons, predominantly natural gas. The approach to local content in the wider sense is thus far reaching: state ownership of the natural resources, majority state participation and operatorship of the large projects, and government-led industrial expansion. From a narrower perspective, for measures favoring local companies, industries, and employees, local content obligations are less extensive.

The Qatari example provides a direction that can be followed to build a sustainable and diversified economy out of abundant natural resources. Lessons include the following:

- Building long-term partnerships with the various IOCs, which is key to obtaining fair terms and the necessary technology transfer
- A stable and steady policy from the central government (with the benefit of stability in Qatar's case), which can direct the development of industries derived from petroleum exploitation
- Sustainability, which involves taking a long-term view of development and profit, and may include implementing a moratorium on developments if that best serves the long-term interests of the country.

Qatar's successful development is a valuable precedent of the transformation of a petroleum-based economy into a diversified economy where local content is at the fore. This is valid for the development of the local population through the distribution of wealth and through the education and promotion of local employees in the management and control of the national companies. It is also valid for the diversification of national companies that was achieved through Qatar Petroleum with the government's support under long-term policies. It is noteworthy that sustainability is a key pillar of the Qatari policy, as it should be for any long-term approach to development.

Naturally, one has to take into account the facts that Qatar has a modest number of nationals and considerable gas reserves. But this would limit the measure of success to the wealth statistics. Wealth may be relative, but the effort to achieve some degree of diversification, to plan development sustainably, and to raise the education level of the population are the objective markers of success.

The Qatarization policy may provide another measure of Qatar's success. Although the ambitious target of 50% of Qataris has not been met, the national oil and gas employees of the Qatari companies are gaining ground in filling the upper management tiers. This success is increasingly being recognized outside Qatar in the international oil and gas industry.

Note

1. "Qatar National Vision 2030," General Secretariat for Development Planning (July 2008), http://www.mdps.gov.qa/en/qnv/Documents/QNV2030_English_v2.pdf.

19

TIMOR-LESTE

Manuel Santos Vitor, Catarina Cortez Gamito, and Bernardo Correia Barradas

Introduction[1]

The Constitution of the Democratic Republic of Timor-Leste establishes that all resources of the soil, subsoil, territorial waters, continental shelf, and the exclusive economic zone (EEZ), which are essential for the economy, are owned by the state.[2]

Timor-Leste has two main areas for the purposes of the development of petroleum activities: the Timor-Leste Exclusive Area (TLEA) and the Joint Petroleum Development Area (JPDA). While the former refers to petroleum activities carried out within the territory of Timor-Leste[3] (its territorial sea, EEZ, and continental shelf), the latter is the result of decades of territorial disputes between Timor-Leste, the Commonwealth of Australia, Portugal, and Indonesia.[4] It currently encompasses a shared zone—the Timor Gap—between the Timor Zone and the Australia Zone. The JPDA was established by the Timor Sea Treaty entered into between the small Southeast Asian half-island nation and Australia. Due to lack of delimitation of sovereignty and maritime boundaries, both countries jointly manage and develop petroleum activities.[5]

Thus, in order to evaluate local content in Timor-Leste's petroleum industry, we need to assess not only the assemblage of contractual instruments agreed between the governments of Timor-Leste and Australia, but also the existent national legislation and related documents.

Industry Overview

Timor-Leste does not have any local O&G industry, lacking both essential infrastructures and human resources to fully operate the petroleum sector. Moreover, all of the oil exploration is undertaken offshore of the country's south coast (in the Timor Sea).[6] The companies operating there do not have, with the natural exception of the national oil company Timor Gap, any relevant presence onshore, since all the E&P activities have been conducted either offshore or in Australia.[7]

Nonetheless, since the petroleum sector is considered one of the key pillars of the future development of the country,[8] Timor-Leste's authorities are undertaking relevant efforts to attract extraction players to the country. A key project is the Southern Coast Project, known locally as the Tasi Mane project,[9] comprising three clusters of development on the south coast and managed by Timor Gap on behalf of the Ministry of Petroleum and Mineral Resources.

The Tasi Mane project is still in its early stages and includes an LNG plant in Beaço,[10] a refinery and petrochemical plant in Betano,[11] a supply base in Suai,[12] and support infrastructures, such as a highway and an airport.

This project encompasses the government's medium- and long-term goals for the local oil and gas industry.

Apart from the Tasi Mane project, the Timorese government's goal is to develop the country's petroleum industry. The government particularly wants to promote maximum participation by Timorese citizens and national companies in the O&G industry through the development of national training programs.[13] Such programs include training on geology, petroleum and chemical engineering, and project management, among other disciplines.

Key Players

The key players of the Timor-Leste's petroleum industry include several regulatory bodies, the national oil company of the country (Timor Gap or *Timor Gás & Petróleo EP*), and O&G companies and service providers, as detailed in the following text.

Ministry of Petroleum and Mineral Resources

Pursuant to the Organic Law of the VI Constitutional Government (GOL),[14] the Ministry of Petroleum and Mineral Resources (MPMR), or *Ministério do Petróleo e Recursos Minerais*, is the

> central Government body responsible for the design and execution of the energy policy and of the management of mineral resources, including petroleum and other strategic minerals, approved by the Council of Ministers, as well as the licensing and regulation of the extraction activity, and industrial activity of improvement of oil and minerals, including petrochemical and refining.[15]

There are no specific provisions for local content in relation to the MPMR's functions. Nonetheless, local content policy is implicitly included through the

> guarantee, through the proper legal, administrative and technical instruments, [of] the maximum involvement of Timor-Leste in the oil and mineral resources sector, and the promotion of national opportunities in the sector in order to attract and retain related foreign investment.[16]

Even so, the GOL could have gone further and establish a more comprehensive and specific set of guidelines and obligations on local content for the MPMR to oversee.

Several entities come under the purview of the MPMR: the National Petroleum and Minerals Authority, Timor Gap, the Institute of Petroleum and Geology, and Timor Mines.

National Petroleum and Minerals Authority

The National Petroleum and Minerals Authority, or *Autoridade Nacional do Petróleo e Minerais* (ANPM), was established under Decree-Law No. 20/2008, on June 19, 2008, which was last amended by Decree-Law No. 1/2016 on February 9, 2016. The ANPM is Timor-Leste's entity responsible for the management and regulation of petroleum activities in the country's exclusive jurisdictional area (TLEA) and in the JPDA pursuant to the Decree-Law No. 20/2008,[17] the Petroleum Activities Law,[18] and the Timor Sea Treaty of 2002.

Within the JPDA, the ANPM is the designated authority.[19] The designated authority is the executive body within the three-tiered joint administrative structure of the JPDA, responsible before the Joint Commission and entrusted with carrying out the daily regulation and management of petroleum activities in the JPDA.[20]

Under Decree-Law No. 20/2008, as last amended,[21] the ANPM has the responsibility to prepare, enact, and implement regulations and administrative decisions that impose (i) certain levels of national contribution (local content) to the oil sector, such as local employment creation,[22] and (ii) the maximization of the use of local goods and services by the industry, within the entire petroleum activities sector. This responsibility was further detailed in ANPM's "Institutional Strategic Plans of 2009–2012"[23] *and of 2016-2018* wherein "maximising employment opportunities for Timor-Leste nationals" was established as one of ANPM's strategic goals. Also, in ANPM's Objectives and Programs,[24] we may find the revision of the draft of the JPDA Regulation (including the insertion of local content and commercial provisions). The latter objective was achieved with the enactment of the Petroleum Mining Code,[25] which repealed the Interim Petroleum Mining Code. The first objective was established[26] to "ensure the employment of a minimum of 250 TL (Timor-Leste) nationals in the upstream petroleum sector" by 2010, apart from generally promoting the capacity and qualification of the Timorese national workforce. This was further enhanced by the ANPM's *Institutional Strategic Plan of 2016–2018* which currently provides in its Objective 1 for the "maximization of employment of Timorese nationals, and trained in the upstream sector."

As such, and according to ANPM's *Annual Report 2013*,[27] a total of 413 Timorese nationals were employed either directly or indirectly to support the Bayu-Undan and Kitan oilfield projects at the JPDA. The major employers at the Bayu-Undan project were ESS (catering services), CAJV/Caltech (operation and maintenance), and CHC Helicopters for offshore personnel transportation. At the Kitan contract area, the major employers were Eni (the PSC operator), SDV (now Bolloré Logistics), GAP-MHS, ESS, and Bluewater.

In 2014, there were a total of 659 Timor-Leste nationals employed in the Bayu-Undan and Kitan projects, an increase of 37% from 2013.

Concerning goods and services, combined procurement amounts for Bayu-Undan and Kitan local[28] goods and services are presented for the years 2012 to 2014 in table 19–1.[29]

In 2012, the total expenditure for goods and services (produced and/or supplied by providers in Timor-Leste) for the JPDA was approximately US$20 million.

Table 19–1. Goods and services procurement, 2012–2014

Year	Amount in US$	Increase from previous year
2012	$47,117,980.00	n/a
2013	$58,568,987.00	+12%
2014	$60,906,781.24	+4%

Institute of Petroleum and Geology

The Institute of Petroleum and Geology (IPG), or Instituto de Petróleo e Geologia, is a public institute created by Decree-Law No. 33/2012, July 18, 2012. Its purpose is to record, produce, manage, maintain, and disclose geological information, primarily related to oil and gas resources. This information serves as the basis for the exploration, production, and development of Timor's national mineral resources. Additionally, the IPG (i) coordinates and promotes activities for the production of geological information and knowledge related to Timor-Leste, including offshore areas; and (ii) establishes a national registry of geological information and natural resources of Timor-Leste, among other functions.[30]

Although recently established, the IPG already occupies a central role in creating local knowledge and increasing the employment qualifications of Timor's national citizens. Nevertheless, the IPG is still taking its first steps, namely by promoting frequent training activities abroad for its staff and seeking to attract foreign know-how. The IPG is already a relevant source of knowledge for the industry and is taking significant steps toward becoming a well-known center for reliable and up-to-date geological information.

Timor Gap (Timor Gás & Petróleo)

Timor Gap (Timor Gás e Petróleo E.P.) is Timor-Leste's national oil company, established under Decree-Law No. 31/2011, July 27, 2011. Timor Gap's corporate purpose is to participate in the exploration and production of petroleum resources in the national territory (and in the JPDA), both onshore and offshore.

In 2013, Timor Gap signed its first production sharing contract (PSC), through its wholly owned subsidiary Timor Gap, PSC JPDA 11-106, to begin exploration offshore in the JPDA. Operations are conducted through a joint venture partnership with Eni (the PSC operator, with a 40.53% share) and INPEX (with a 35.47% share).

Timor Gap is also responsible for the Tasi Mane[31] project. As mentioned previously, this project includes the Suai supply base, the Suai airport upgrade project, the refinery and petrochemical plant at Betano, the Beaço LNG plant, and construction of a highway. This ambitious governmental project is aimed at creating a national hydrocarbons industry.

Timor Gap also joined the MHS Aviation Berhad (Malaysian Helicopter Services) in 2012 to create GAP-MHS Aviation (in which Timor Gap has a share of 60%). GAP-MHS provides aviation support services to offshore petroleum operations in the Timor Sea. Finally, in 2014, Timor Gap Oil and Gas Marine and Logistics was established to provide general services to the marine industry and to render logistics and support services for the marine industry operating in the Timor Sea.

Oil and gas companies operating in Timor-Leste and relevant operations

TLEA. The sole contract area in Timor-Leste's offshore area (S-06-04-E) was granted under a PSC-PSA contract that was in force from November 3, 2006 to February 11, 2013. The PSC operator was Eni Timor-Leste (from Italy), having as PSC JV partners Galp (a Portuguese company) and the Korea Gas Corporation (a South Korean company).

Unfortunately no significant discoveries were made, and most of the exploration areas were relinquished in 2013.

JPDA. This joint production area, as agreed upon by Timor-Leste and Australia, currently includes 8 contract areas, 11 fields (including Troubadour, Bayu-Undan, Greater Sunrise, and Kitan), and 60 wells.

ConocoPhillips (JPDA 03-12) is the PSC operator in two of the contract areas (03-12 and 03-13), with Woodside Energy Limited in two others (03-19 and 03-20). The remaining contract areas are operated by Oilex, Eni, and Minza Limited, with many PSC JV partners, including some of their own subsidiaries or fully owned companies (such as ConocoPhillips Timor Sea and ConocoPhillips Emet). Other JV partners include Santos, Inpex Sahul, Tokyo Timor Sea Resources, Shell Development, OG ZOCA, Videocon, GSPC, Bharat PetroResources JPDA, Pan Pacific Petroleum, Japan Energy E&P JPDA, INPEX Timor Sea, and Talisman Resources.

From these companies, only ConocoPhillips, Woodside Petroleum, and Eni have a physical presence onshore, with country managers (Timorese nationals), facilities, and local staff.

Petroleum services companies operating in Timor-Leste and relevant operations

Petroleum services companies represent approximately 85% of all employment of national citizens (659 in 2014) due to oil and gas industry activities at Bayu-Undan and Kitan.[32] The remaining 15% is divided between the two main PSC operators, ConocoPhillips and Eni.

The main petroleum services companies with relevant presence in Timor-Leste are listed in table 19–2, with reference to their respective areas of activity.

Table 19–2. Service providers in Timor-Leste

Company	Activity	Country of Origin
Air North	Aviation	Australia
Worley Parsons	Resources and energy	United States
SDV/Bolloré Logistics	Logistics	France
Bluewater	Floating, production, storage and offloading	The Netherlands
ESS Support Services	Food and support services	Australia
CHC Helicopters	Aviation services	Canada
CAJV/CALTECH (Clough AMEC)	Operations and maintenance	Australia
Gap MHS	Aviation services	Malaysia/Timor-Leste
Svitzer Asia Pte Ltd	Safety and support at sea	Denmark

From the companies listed in table 19–2, Air North, SDV, and Worley Parsons have relevant commercial activity both onshore and outside the oil and gas industry. SDV's parent company, Bolloré, was recently awarded with the construction and management of Timor-Leste's new international port and container terminal at Tibar, near the capital city of Dili.

Industry Knowledge

Unfortunately, there is a lack of industry knowledge in Timor-Leste. This results from an underdeveloped educational system and the fact that Timor-Leste is still an industrially young nation, lacking skilled personnel.

To overcome these challenges, entities such as the MPMR, the ANPM, and Timor Gap promote training and qualification programs abroad for their staff, either in international educational institutions or by means of secondments with international oil and gas companies. During 2014, and in addition to other training programs and conferences, Timor Gap had seven employees involved in secondments and on-the-job-training programs. Participating companies included PTT Global Chemical PC (Thailand), Toyo-Thai and TT (Thailand), Shell, and Worley Parsons.

Legislative Hydrocarbon Regime Concerning Local Content

Usually, local content requirements are established in the bidding and negotiation phase and subsequently remain until the end of the petroleum operations.

As referred to previously, Timor-Leste's petroleum activities are carried out within two different areas, the TLEA and the JPDA. As such, different regulatory frameworks are applicable to both areas. Local content requirements applicable in Timor-Leste are established in the laws, treaties, and relevant contracts relating to each area of development.

TLEA

No significant discoveries have been made in the TLEA, and most of the exploration areas were relinquished in 2013. Nonetheless, there is a specific legal framework for the development of petroleum within the area. The main applicable legislation consists of the Petroleum Activities Law and the Public Tenders in Respect of Petroleum Contract Awards Decree (No. 7/2005), also called the Public Tendering Decree.[33]

According to the Petroleum Activities Law and pursuant to the Constitution of the country, "the title to and control over petroleum in the territory of Timor-Leste are vested in Timor-Leste." As such, it depends on the government and, more specifically, on the MPMR to oversee the granting of authorizations or licenses to develop and carry out petroleum activities within the TLEA.

Such authorizations may be granted through direct negotiation but usually are granted by public tender. The bids to obtain an authorization should include proposals for "training of, and giving preference in employment in the petroleum operations to nationals of Timor-Leste"[34] and for "the procurement of goods and services from persons based in Timor-Leste."[35] This is further enhanced by the provisions of the Public Tendering Decree, which establish that the "assessment of the bids shall be made by reference to the procurement of local goods and services, and training of nationals of Timor-Leste."[36] Thus, according to the wording of both the Petroleum Activities Law and the Public Tendering Decree, although nonbinding, these local content provisions are key factors of valorization of the bids.

According to Law No. 13/2005, authorizations for petroleum activity must stipulate that Timor-Leste may participate in authorized petroleum operations up to a maximum equity of 20%.[37] The decision is made by the Council of Ministers or by the PM in this case pursuant to delegated powers from the Council of Ministers as per article 22(1) of the Petroleum Activities Law. The participation of the state may occur during any phase of petroleum operations in accordance with the provisions of the contract.[38]

Furthermore, the Petroleum Activities Law grants to the government the authority to enact regulations on compliance reporting by authorized persons with obligations set out in the law and authorizations. Such reports include those relating to (i) the training and employment of Timorese nationals and (ii) procurement of Timor-Leste goods and services.[39] Even though permitted, no such regulations have been enacted.

Also, the Public Tendering Decree sets forth that whenever there is a foreign company participating in the bidding, it must produce "a written commitment to establish a company in accordance with the laws of Timor-Leste, with an administrative office to carry on operations in Timor-Leste."[40] Signature of the contract is conditional on the fulfilment of this commitment.

Finally, the Public Tendering Decree sets forth two mandatory terms and conditions to be included in a PSC. These are:

1. the existence on an obligation regarding the training and capacity building of, and giving preference in employment in petroleum operations to, nationals of Timor-Leste, and
2. the existence of an obligation to procure local goods and services, if mentioned by the State.[41]

In line with the above, the PSC model provided by the ANPM also has relevant provisions on local content applicable in Timor-Leste.[42] As such, the PSC model contains the following:

1. A mandatory requirement for the operator to carry out its petroleum operations from an office located in Timor-Leste.[43]

2. If not subject to the consent of the Ministry, the need to include in the development plan to be delivered to the MPMR the contractors' proposal for
   (a) the use of Timor-Leste goods and services; and for
   (b) training and employment of Timor-Leste nationals.[44]
3. The need to comply with the proposal for the application of an authorization regarding training, employment, and the acquisition of goods and services and to
   (a) give preference to the acquisition of goods and services from persons based in Timor-Leste, provided they are offered on competitive terms and conditions,
   (b) give preference in employment in petroleum operations to nationals of Timor-Leste (moreover, a report demonstrating compliance with the above obligations shall be submitted to the MPMR within a certain periods of time).[45]
4. The obligation of the contractors to publicize, hence drawing the attention of suppliers based in Timor-Leste, the tenders that shall take place for the provision of goods and services for petroleum operations (except by consent of the ministry).

Thus, except for the provisions both in the Petroleum Activities Law and the PSC model establishing the state's participation in petroleum operations, there are no applicable mandatory local content rules. Furthermore, and as stated above, rules on granting preference to goods and services provided by Timorese companies, as well as the hiring and training of Timorese nationals, are only key factors of valorization of the bids and are thus nonbinding.

JPDA

All of Timor-Leste's currently extracted oil comes from fields in the JPDA. The JPDA is regulated by a set of international treaties and agreements between Timor-Leste and Australia that commenced on May 20, 2002 with the signing of the Timor Sea Treaty.

As such, the legal framework that regulates the activities within the JPDA is, namely, the Timor Sea Treaty,[46] which established the JPDA in the Timor Sea, the Interim Petroleum Mining Code,[47] the Petroleum Mining Code,[48] the Interim Regulations Issued under Article 37 of the Interim Petroleum Mining Code,[49] the Interim Directions Issued under Article 37 of the Interim Petroleum Mining Code,[50] the Interim Administrative

Guidelines for the JPDA,[51] the Treaty on Certain Maritime Arrangements in the Timor Sea (CMATS), [52] the Memorandum of Understanding on the Greater Sunrise,[53] and, finally, the International Unitization Agreement on the Greater Sunrise.[54]

In relation to local content provisions, the Timor Sea Treaty established that both countries shall take appropriate measures to (i) give preference to the hiring of Timorese nationals or residents in the country, and to (ii) promote training and opportunities for such nationals and residents.[55] Also, provisions on local content may be found in the Petroleum Mining Code for the JPDA (PMC),[56] adopted pursuant to Article 7(a) of the Timor Sea Treaty and repealing the Interim Petroleum Mining Code. The provisions state that the Application for an Authorization[57] shall include proposals for training Timorese nationals and will give preference in employment for authorized activities[58] to national and permanent residents of Timor-Leste, along with the acquisition of goods and services from persons based in the country.[59]

This provision on local content was inserted within the revision of the draft of the JPDA regulations and reflects Timor-Leste's position on the negotiation process. This provision merely establishes a preference for hiring of Timorese nationals within each authorization. Nonetheless, it is worth noticing that the reference to the acquisition of goods and services from persons based in Timor-Leste does not include this preference requirement.

Thus, the local requirement contents in each authorization must be negotiated on a case-by-case basis, which shall be reflected in the relevant contract that is then submitted to the Joint Commission's approval. Such is the case with the PSC for Contract Area JPDA 11-106, entered into between the ANPM (in its role as "Designated Authority") and Eni JPDA 11-106 B.V. (as PSC operator), INPEX Offshore Timor-Leste, and Timor Gap's subsidiary TIMOR GAP PSC 11-106, Unipessoal.

As an example, we may mention section 5(4) of PSC 11-106, which specifies that

> the contractor shall take reasonable steps to comply with the proposals submitted in respect of training, employment and acquisition of goods and services, and shall: (a) give persons based in Timor-Leste a real opportunity to compete for delivery of goods and services, provided they are offered on competitive terms and conditions; (b) with due regard to occupational health and safety requirements, give preference in Petroleum Operations to national and permanent residents of

Timor-Leste, and (c) within thirty (30) days of the end of each calendar year, submit to the Designated Authority a report demonstrating compliance with the above obligations.

In light of the above, we may state that the terms used in the wording of the agreement, such as "reasonable steps," "real opportunity," and "preference," do not create more than a so-called soft obligation on the part of the contractors. The final compliance and actual outcome are naturally difficult to assess.

Another case that may be mentioned is the provision in Annex D, where the contractor [60] proposes (note the nonbinding wording):

1. to ensure that local content provisions are included in the contractor's sub-contracts and, in particular, for the drilling rig, rig services, supply, logistics and the petroleum infrastructures' project;
2. to provide the resources to actively ensure that sub-contractors are in compliance with their local content provisions;
3. to contribute to the establishment of petroleum infrastructure in Timor-Leste in the form of the Data Tape and Core Storage Facility at Hera, capped at US$1,9 million, and, finally,
4. to provide six months of Dili-based training in English, IT, HSE and Petroleum Geoscience and Engineering to ten (10) qualified Timorese graduates, secondment to five (5) qualified Timorese graduates during the Exploration Period and additional secondment to five (5) qualified Timorese graduates in the event that one or more contingent wells is drilled.

Thus, in light of the above, it is possible to conclude that nonbinding provisions on local content are also applicable to the JPDA. This is mainly due to the lack of locally available qualified personnel and goods and services providers.

Downstream sector

Downstream activities in Timor-Leste are regulated by the Downstream Sector Decree, Decree-Law No. 1/2012, February 1, 2012; ANP Regulation No. 1/2012, October 24, 2012, On Administrative Procedures, Requirements, and Fees for the Granting, Renewal, and Modification of Downstream Activities' Licenses; and ANP Regulation No. 2/2012, September 3, 2012, On Downstream Activities' Inspection Division.[61] The entity responsible for the regulation and supervision of the downstream activities in Timor-Leste is the ANPM.[62]

The above-mentioned legislation and regulations also set forth mandatory local content requirements in relation to foreign companies operating in the downstream sector in Timor-Leste. Decree-Law No. 1/2012 established rules to maximize "Timor-Leste's participation" in downstream activities.[63] However, no further regulations were enacted, so this rule has the same nonbinding nature as the other rules mentioned previously in this text.

Conclusion

Timor-Leste, as a developing nation, needs to overcome several hurdles in the country's petroleum sector. Challenges that must be overcome in order to compete with developed oil-producing countries include the lack of infrastructure, inadequate industry knowledge, and an insufficiently qualified workforce. In addition, the division of the exploration of the petroleum resources into two areas (and two countries)—the TLEA and the JPDA—results in a duality of applicable legislation, creating additional challenges.

Nonetheless, and as stated throughout this text, the government of Timor-Leste is promoting and enforcing innovative, proper, and necessary efforts to develop and improve the petroleum sector of the country. Furthermore, the government understands that through the implementation of local content requirements, the goals of promoting and developing the local industry connected to the petroleum sector, as well as the training of Timorese nationals, can be accomplished.

Although the applicable guidelines and rules for local content are currently nonbinding, it is our understanding that in the future, mandatory provisions on local content in the petroleum sector shall be enacted and implemented in the country.

Notes

1. This chapter was prepared through a collaborative effort between PLMJ Law Firm (Manuel Santos Vítor/Catarina Cortez Gamito) in Portugal and Bernardo Correia Barradas, an attorney with CBMS Law Firm in Timor-Leste.
2. "Natural Resources," Article 139, Constitution of the Democratic Republic of Timor-Leste (March 22, 2002).

3. "Territorial Scope of Application," Article 3, Law No. 13/2005 (September 2, 2005); also known as the Petroleum Activities Law, http://timor-leste.gov.tl/wp-content/uploads/2010/03/Law_2005_13_petroleum_activities_.pdf.
4. Mats Lundahl and Fredrik Sjöholm, "The Oil Resources of Timor-Leste: Curse or Blessing?" (Working Paper 229, Stockholm School of Economics, Stockholm, October 2006), http://swopec.hhs.se/eijswp/papers/eijswp0229.pdf.
5. Francisco Pereira Coutinho and Francisco Briosa e Gala, "David and Goliath Revisited: A Tale About the Timor-Leste/Australia Timor Sea Agreements," in *Texas Journal of Oil, Gas and Energy Law* 10, no. 2 (2015), http://tjogel.org/journalarchive/Issue10/10.2.DavidGoliath_Francisco.pdf.
6. Timor-Leste is bordered by the Banda Sea to the north and the Timor Sea to the south.
7. As a rule, Timor-Leste's code for commercial companies and its tax laws do not require any local incorporation (the exceptions include banking and insurance activities). A company can operate in the country as nonresident or by means of a permanent establishment, and it can also be registered and have a representative without the need to have physical facilities. Moreover, and as regards O&G activities, it should be noted that the JPDA is neither Timorese nor Australian territory.
8. "Petroleum and Minerals," Section 3.5, Program of the Sixth Constitutional Government, http://timor-leste.gov.tl/?cat=39&lang=en#prog3.5.
9. *Tasi Mani* translates as "Male Sea" in the local language, Tétum, which is a reference to the south sea. The south sea is rough in comparison to the North Sea, locally known by the name of *Tasi Feto* or "Female Sea."
10. A village on the south coast of Timor-Leste.
11. Ibid.
12. Ibid.
13. See Timor-Leste's "Program of the Sixth Constitutional Government and Timor-Leste's Strategic Development Plan 2011–2030" (Dili, Timor-Leste: RDTL, 2011), http://timor-leste.gov.tl/wp-content/uploads/2012/02/Strategic-Development-Plan_EN.pdf.
14. Decree-Law No. 6/2015, March 11, 2015; one unofficial translation is available at https://www.laohamutuk.org/econ/OGE15/OR/VIGovOrgDL6-2015en.pdf.
15. "Ministry of Petroleum and Mineral Resources," Article 28(1), Decree-Law No. 6/2015, March 11, 2015.
16. Ibid., Article 28(1)(c) and (d).
17. "Nature," Article 1, and "Powers and Functions," Article 3, in Decree-Law No. 20/2008, "National Petroleum Authority," June 19, 2008, as last amended, http://web.anpm.tl/webs/anptlweb.nsf/vwAll/Resource-ANP%20Decree%20Law%20No.20-2008%20/$File/Decree-Law-2008-20-ANP%20decree%20law.pdf?openelement.
18. "On Petroleum Activities," Law No. 13/2005, September 2, 2005; also called the Petroleum Activities Law.
19. "Regulatory Bodies," Article 6, Timor Sea Treaty, http://timor-leste.gov.tl/wp-content/uploads/2010/03/R_2003_2-Timor-Treaty.pdf.
20. Ibid., Article 6(b).

21. Decree-Law No. 20/2008, Article 3(6).
22. Ibid., Article 22(2):

    The ANP shall establish technical, commercial and socio-economic requirements, such as the level of local employment creation and of purchase/use of local goods and services, that shall be mandatory for bidders; and project requirements in terms of environmental quality and protection, industrial safety and the safety of the public at large.

23. Autoridade Nacional do Petróleo e Minerais, "Institutional Strategic Plan 2009–2012," http://www.anp-tl.org/webs/anptlweb.nsf/vwAll/Resource-ANP%20Institutional%20Strategic%20Plan%202009%20-%202012/$File/ANP%20strategic%20Plan-External.pdf?openelement and *Institutional Strategic Plan* for 2016–2018, http://www.anpm.tl/anpm-strategic-action-plan-period-2016-2018 at http://www.anpm.tl/anpm-strategic-action-plan-period-2016-2018/.
24. Objective 10, Program 1, in ANP's "Objectives and Programs," included in the Institutional Strategic Plan for 2009–2012. This is further included in the Institutional Strategic Plan for 2016–2018 in Objective 3.
25. The Petroleum Mining Code is one of the main legislative texts that constitute the legal framework established for the administration of petroleum activities in the JPDA. The legal framework applicable to the JPDA is addressed in more detail in the "Legislative Hydrocarbon Regime Concerning Local Content" section of the text.
26. Goal 3, Objective 1, Program 2 in ANP's "Objectives and Programs," included in the *Institutional Strategic Plan for 2009–2012*.
27. ANP Annual Report 2013, available from http://www.anpm.tl/.
28. Local for Timor-Leste; the local content requirements at the JPDA are applicable to both countries.
29. There is a pressing need for a comprehensive legal and regulatory definition of the term "local" in relation to local content policies.
30. Article 3(1) and 3(2), Decree-Law No. 33/2012, July 18, 2012.
31. See the "Industry Overview" section of the text.
32. This figure disregards employment created onshore at the regulatory bodies and within Timor Gap, which in 2014 reached about 250. Most of these were employed at ANP and Timor Gap, with 78 and 101 employees, respectively.
33. "Public Tenders in Respect of Petroleum Contract Awards," Decree No. 7/2005 (October 5, 2005), http://mj.gov.tl/jornal/lawsTL/RDTL-Law/RDTL-Decree-Laws/Decree-Law-2005-7.pdf.
34. "Invitation to Apply," Article 13(3)(a)(iii), Law No. 13/2005, September 2, 2005.
35. Ibid., Article 13(3)(a)(iv).
36. "Bid Assessment," Article 9(3), Decree No. 7/2005 (October 5, 2005).
37. "Participation by the State in Petroleum Activities," Article 22(3), Law No. 13/2005, September 2, 2005. This mandatory term is also provided for in the Annex of the Public Tendering Decree (Decree No. 7/2005, October 5, 2005).
38. "Participation by the State in Petroleum Activities," Article 22(1) and (4), Law No. 13/2005, September 2, 2005.
39. Ibid.

40. "Contents," Article 7(3)(d), Decree No. 7/2005 (October 5, 2005). Moreover, when the contract is awarded to a group of companies, the commitment refers to the leading company (the operator).
41. "Annex," Decree No. 7/2005 (October 5, 2005).
42. "TLEA: Production Sharing Contracts," ANPM, http://www.anp-tl.org/webs/anptlweb.nsf/vwAll/TLEA.
43. "Interpretation," Article 1(6)(d) of the PSC Model, regulated under the Petroleum Activities Law.
44. Ibid., "Work Programmes and Budget," Article 4(11)(d)(vii)(aa) and (bb).
45. Ibid., "Conduct of Work," Article 5(4).
46. Resolution No. 2/2003, April 1, 2003, which ratified the Timor Sea Treaty. The treaty is available at http://www.anp-tl.org/webs/anptlweb.nsf/vwAll/JPDA.
47. The Interim Petroleum Mining Code was adopted by the Joint Commission on April 2, 2003, in accordance with Article 7(b) of the Timor Sea Treaty.
48. "Petroleum Mining Code for the Joint Petroleum Development Area" (PMC), available at http://www.anp-tl.org/webs/anptlweb.nsf/vwAll/JPDA.
49. "Interim Regulations Issued under Article 37 of the Interim Petroleum Mining Code" (June 2003), available at http://www.anp-tl.org/webs/anptlweb.nsf/vwAll/JPDA.
50. "Interim Directions Issued under Article 37 of the Interim Petroleum Mining Code" (June 2003), available at http://www.anp-tl.org/webs/anptlweb.nsf/vwAll/JPDA.
51. "The Interim Administrative Guidelines for the JPDA," available at http://www.anp-tl.org/webs/anptlweb.nsf/vwAll/JPDA.
52. See Resolution No. 4/2007, dated 8 March 8, 2007, which ratified the Treaty on Certain Maritime Arrangements in the Timor Sea (CMATS). Text of the CMATS is available at http://www.anp-tl.org/webs/anptlweb.nsf/vwAll/JPDA. Please note that on January 9, 2017, the Permanent Court of Arbitration issued a trilateral joint statement confirming the termination of CMATS. In the case of Timor-Leste, the Timorese Parliament adopted on 10 January 2017, the Parliament Resolution no. 01/2017, published on 16 January at the Official Gazette, in which it unilaterally terminated the CMATS between the two countries.
53. "Greater Sunrise—Memorandum of Understanding," available at http://www.anp-tl.org/webs/anptlweb.nsf/vwAll/JPDA.
54. "Greater Sunrise—International Unitization Agreement," available at http://www.anp-tl.org/webs/anptlweb.nsf/vwAll/JPDA.
55. "Employment," Article 11, Timor Sea Treaty.
56. "Applications," Section (5)(4), in "Petroleum Mining Code for the Joint Petroleum Development Area" (PMC), http://www.anp-tl.org/webs/anptlweb.nsf/vwAll/Resource-Petroleum%20Mining%20Code/$File/PMCtoCoM0602.pdf?openelement.
57. Pursuant to "Interpretation," Section 1, PMC,

    "Authorisation" means an Access Authorisation, a Production Sharing Contract or a Prospecting Authorisation, or an agreement (if any) made in respect of such an Authorisation or Production Sharing Contract....

"Authorised Person" means: (a) in respect of a Production Sharing Contract, the Contractor; and (b) in respect of any other Authorisation, the person to whom the Authorisation has been granted.

Section 9 of the PMC addresses access authorization.

58. As provided by Section 1 of the PMC.

59. Except in relation to PSC numbers 03-12, 03-13, and 03-20.

60. Eni, TIMOR GAP, and INPEX.

61. In addition, there are Regulations No. 1/2013, September 18, 2013, on Installation and Operation of Fuel Filling Stations; and No. 1/2014, January 15, 2014, on Fuel, Biofuel and Lubricant Quality Standards and Specifications.

62. "Powers and Functions," Article 3(4), Decree-Law No. 20/2008, June 19, 2008, as last amended.

63. Article 5(b), Downstream Sector Decree, Decree-Law No. 01/2012, January 15, 2012.

20

CONCLUSION

Eduardo G. Pereira and Tonje Gormley

Introduction: The Importance of Local Content Development

This book provides a detailed analysis of one of the most topical issues in the international petroleum industry today: how to create added value from the petroleum industry through developing the local content in petroleum activities. It presents an overview of how a different host countries have approached local content development. We have seen examples of what a host country could or should take into consideration to model adequate local content policies and have examined how such policies may be implemented through statutory, regulatory, or contractual requirements. Host governments may receive considerable revenues from sales of their share of petroleum produced or from relevant taxes, service fees, or other dues collected from companies carrying out upstream petroleum activities. Nonetheless, collection of funds will not secure maximum contribution to the domestic economy, nor will it further development of the economy of the host country as a whole.

We have seen that all the countries analyzed in this book have in one form or another modeled policies to promote local content development and have implemented such policies in laws, regulations, and contracts. Interestingly, we note local content development is topical in nearly all host countries, irrespective of their context. It is not just developing countries that focus on local content development; local content development has also been on the agenda in long-established, industrialized countries like Norway and Canada. In Norway, the era of mandatory local content requirements is over. Today, Norwegian companies compete for business on the same terms and conditions as international companies. This was initially spurred by Norway's need to comply with EU and EEA regulations. Comments in white papers published as early as 1979, however, express concern for the overall effects on the Norwegian economy of continued preferential treatment of an industry. In Canada, we have seen that local content is still continuously developed at different governmental levels, and that local content policies are pursued. Yet it is noted that there is considerable diversity in the local content requirements of the subjurisdictions under consideration in Canada. This is related to differences in the maturity of the oil and gas sector in each subjurisdiction. We have also seen how local content development is as pertinent in host countries where the petroleum industry is nascent, such as Lebanon and Kenya, as it is in countries that can be characterized as mature petroleum countries, such as Angola, Brazil, Iran, Libya, Mexico, Nigeria, and Russia.

We have seen that a host country that maintains a dynamic approach to local content development can use that responsiveness to redirect policies to tackle current problems. For instance in Libya, which has a petroleum industry that dates back decades, renewed focus on local content development can function as a means to contribute to a postconflict economy that is in dire need of a boost. In Canada, local content development is pursued hand in hand with policy goals of promoting indigenous people's rights. In Qatar the issues to tackle are of a different nature. Local content policy in Qatar is a policy goal whereby the number of Qatari employees in the petroleum industry can be increased, since expatriate employees have a dominant role in the Qatari petroleum industry. The Qatarization policy is a long-term policy through which higher education of Qataris and skills development should result in an increase of Qatari employees in key positions in the energy industry and in government departments. The objective is to provide 50% or more of Qatari citizens with "meaningful" permanent employment. By this we may deduce that although the local content development policies in themselves might be more or less similar throughout the world, the objectives for modeling them are specific to the host country.

Local content development—challenges and pitfalls

We note that local content development in the form of business opportunities is often an expectation of the people when reports of a budding petroleum industry emerge in new petroleum countries. However, the international competition that a new domestic industry faces represents a major challenge. Today, the international petroleum industry is a well-established industry. There will always be a need for technical improvements as frontier areas challenge oil and gas companies to find new solutions. Even so, there are numerous experienced oil and gas services companies that operate globally. It is difficult for host countries to build a full-fledged national industry from scratch that will be able to compete against a well-developed network of global goods and services providers. These competitors are often strong players both in a financial and technical sense. Adding to that, many international oil and gas companies have framework contracts with a pool of suppliers that can offer their goods and services globally on favorable terms and conditions for established customers.

Competition issues may be addressed by way of prioritizing certain types of national business activities to build up an industrial or services base. If a gap in the market is identified, then the host country should focus on this. If the host country possesses relevant industrial capacity that can easily be transferred into use in the petroleum industry, then the host country can use this as a stepping stone. Again, Norway is a good example. There was a need to develop offshore technologies, and Norwegian industry possessed relevant industrial skills and competencies to capitalize on this opportunity, with the assistance of the Norwegian government.

Another common expectation is that the petroleum industry will lead to job creation. In reality, the industry does not generate a plethora of jobs in all of its phases, and when some phases are finished, the jobs are no longer available. There is also a trend toward increased mechanization of petroleum activities. The need for unskilled manpower in modern petroleum activities has diminished over the years, although there will always be room for unskilled labor and the need for regular goods and services, such as food, hotels, catering, etc. The biggest challenge, however, is that many of the jobs that the industry creates require specialized skills and competence. Government leaders in host countries hope to develop a more skilled national labor force to build domestic economic strength. Unfortunately, many countries lack an effective basic educational system, much less institutions of higher learning for degrees in petroleum-related fields. So the optimal desire for a host government is for requirements to hire qualified nationals. Such requirements are usually not effective, however, unless they are combined with requirements to establish efficient and meaningful training programs to secure the transfer of expertise and knowledge.

Drafting the local content requirements in an adequate manner also represents a challenge. The basic recipe for implementing local content policies is rather similar in most jurisdictions. Host countries will implement the policy by establishing requirements in laws, regulations, and contracts. The requirements typically establish preferential rights for state, or sometimes privately owned, oil companies. For instance, in Timor-Leste, the petroleum law establishes that authorizations must stipulate the right of Timor-Leste to participate in the petroleum operations up to a maximum equity of 20%.[1] Moreover, it is common to establish requirements that give preferential rights to the procurement of local goods and services and to employment of nationals of the host country. Ancillary to this, one will typically find requirements to transfer technology, carry out research and development, and train the national workforce. The regulatory framework in Ghana is a good example of the typical approach taken to implement policies on the latter issues. Section 23 of the Petroleum Exploration and Production Act (PNDCL 84) of Ghana was passed in 1984, and it contains local content provisions under Part III, "Rights and Obligations of Contractors and Sub-Contractors." Section 23, as quoted in chapter 10, requires training for Ghana's national workforce, as well as transfer of technology.[2]

Broadly speaking, though, there are two main models for drafting rules on giving domestic providers of goods and services preferential rights. Some countries have adopted flexible rules, such as those that give preferential rights to domestic goods and services of similar price and quality when available. Other countries, such as Nigeria and Ghana, have set very prescriptive, detailed requirements concerning mandatory local content percentages in given areas of business within a specified period. Each host country must evaluate which model is more suitable in its given context. Irrespective of which model is chosen, however, it is a challenge to find a realistic level for thresholds and local content levels.

It is also a challenge to establish adequate tools for compliance in the form of legal basis for supervision and enforcement. However, even if such tools are adequate, we see very often that practically speaking, authorities are not equipped with capacity and competence to perform their administrative responsibilities. It is important that the host country also implements policies for institutional capacity building alongside, or as a part of, local content development. Supervision and enforcement requires human resources as well as technical and commercial competencies.

All in all, we note that the most challenging aspect of local content development probably is to strike the right balance with other policy goals. Importantly, most host countries would not wish for local content policies to negatively impact the investment climate or to be detrimental to other development policy goals. When evaluating the impact of local content requirements, they should be viewed as a whole. There are many factors to consider when evaluating the impact that local content requirements may have on the investment climate and the domestic economy. Considerations include requirements for state participation, use of local goods and services, training and employment, transfer of technology, research and development, joint venture requirements, reporting, and tender requirements. All of these must be considered, as well as the manner in which the requirements are drafted and enforced. In addition to the practical challenges in making such complex evaluations, the host government might experience political challenges if local content development takes place to a lesser extent or at a slower pace than what people expect. Having the political strength to think long term and choose the right priorities might be a challenge. In certain host countries, it might not even be politically viable.

Success criteria?

We note that the host countries that can be said to have succeeded in their efforts might have had timing and unique framework conditions on their side. Again, Norway serves as an example. At the dawn of the Norwegian petroleum industry, the North Sea was attractive to international oil companies and benefited from several favorable framework conditions, such as experience within the offshore industry, a high level of education, a stable democracy, and a generally competent administration.

Local content efforts are being instigated by regulatory requirements and contractual obligations. Experience shows, however, that it is very hard to achieve the goal of creating added value from petroleum activities unless the measures are based on an understanding of the likely longevity of the natural resource base. It is important to consider what the host country wishes to achieve with its petroleum industry. Framework conditions such as existing human capital, existing infrastructure and industrial capacity, and the possible effect on the economy as a whole must also be considered. Local content requirements that are not developed based on a policy that considers the prevailing local conditions as well as the overall short- and long-term industrial and development goals of the host country can easily fail. Conversely, if the context of the host country is carefully considered and policies are adapted to realistically fit the specific context, the chances for success are higher.

Another success criterion is cooperation. In modeling local content policies, it is important that all relevant ministries work together, agreeing on goals and approach, and balancing and adjusting sector policies when and if required. Host countries must maintain a dynamic approach to local content development. The petroleum industry changes rapidly, both with respect to what is needed in goods and services as well as to market conditions. If compliance for local content requirements becomes too expensive, the oil companies may be hesitant to invest in the country, and discoveries and E&P efforts may decrease. The competition to attract international oil companies, with their capital, know-how, and infrastructure benefits, may defeat the main policy goal of most host countries to carry out petroleum activities and capture the maximum feasible economic rent. On the other hand, poor or absent local content policies and regulations might affect the development of the entire nation and result in the "resource curse."

A realistic view on timing and possible areas for development is an important success criterion. The following questions should be asked: Is there a need for a full-service domestic oil services industry? Can the host country facilitate the development of a domestic oil services industry in time for the companies to comply with mandatory requirements? Can the domestic oil services industry match the pace of the development of the petroleum industry? Experience shows that there is a risk for decreased efficiency in petroleum operations and increased corruption with the wrong type of government intervention. Nigeria, and to some extent, Brazil, are good practical examples of those risks. Such problems with local content development have the capability of creating long-term detrimental effects, including lower total government take in the future. It is for this reason that some commentators state that local content requirements can be a trade-off for profitability. This is indeed the challenge that both host countries and IOCs face.

Lessons to Be Learned?

How might emerging petroleum countries benefit from past experiences, both from a regulatory and a compliance point of view?

As we mentioned before, we analyzed nearly 20 jurisdictions from all continents, with some of the most relevant producers as well as less-advanced petroleum provinces, including most of the countries studied in this text.

The idea behind this book is that practical examples from experienced jurisdictions (i.e., Group A) could provide valuable inputs for other petroleum provinces (such as Groups B and C) (table 20–1).

Table 20–1. Groups of host countries

	Background	Jurisdiction – Examples
Group A	This group is the core of our book as they represent relevant producers of oil and/or gas and they should have higher experience with local content to certain extent.	We consider the following countries as part of Group A: Angola, Algeria, Brazil, Canada, China, Egypt, Iran, Libya, Nigeria, Norway, and Qatar.
Group B	This group represent emerging countries with recent discoveries and they are yet to test their local content measures.	We consider the following countries as part of Group B: Ghana and Mozambique.
Group C	This group represent petroleum provinces with higher hope for large hydrocarbon discoveries but they are yet to find and/or produce large-scale hydrocarbon discoveries.	We consider the following countries as part of Group C: Cyprus, Kenya, Lebanon, and East-Timor.[1]

Each author was given an outline for chapter contents in order to ensure similar parameters for reasonable comparison of the jurisdictions. The outcome of our research was interesting, as it was fairly easy to spot similarities and distinctions between these groups.

Group A has a large variety of oil and gas countries with decades or even a century or more of hydrocarbon E&P activities. There is no consistency among this group. Some of these countries failed to achieve a sustainable and enduring progress with their local content policies over past decades, such as Brazil and Nigeria. Others countries enjoyed fairly rapid success, such as Norway. Others took more time to develop their path, such as Angola and Libya.

Group B has similarities between these two countries, as both have recently made large hydrocarbon discoveries and they are trying to reap the most benefit from those resources. They are eager to increase local content, but they do not yet have extensive practices and experiences to evaluate the success or failure of their policies.

Group C is probably the one with lowest concern for local content up to this point. They do not have any relevant discoveries and production, so they do not have any immediate concerns. However, they are awakening to local content concerns and trying to build new legislation based on international experiences.

Implementation of policies is often carried out through regulatory requirements or contractual obligations, or both. One of the main dilemmas that any government faces is to strike a reasonable balance between adding value and maintaining a generally positive investment environment.

We can use requirements for preferential treatment of local goods and services to illustrate this point. On one hand, a typical approach is to give preference to local companies provided they have similar quality, price, and delivery times as those offered by international companies. To develop that kind of competitive capacity and competency from scratch is, as pointed out, challenging. However, there are good arguments to use this model initially, as it sets a threshold for domestic companies to participate, and it is less likely to impair the investment climate notably. Realistically, developing local content takes time. This might not be a politically attractive model for a sitting government facing unreasonably high expectations from its populace.

On the other hand, many host countries have gone from one stance to another in an attempt to find an adequate model for implementing local content policy goals in a manner presumed to force compliance. Some host countries have chosen to adopt a stronger approach through establishing fixed local content percentages within a given time frame, with penalties for companies that fail to perform as expected. There are reports that this approach works; the level of local content is increased. However, this approach can be detrimental to the investment climate and to the policy goals of the host country seen as a whole, especially if there is no realistic chance that international companies can achieve compliance. This in turn may lead to noncompliance (which fails to achieve local content goals), corruption, or a reduced participation interest from international investors. If the requirements lead to a forced delivery of noncompetitive goods and services, or if they are impossible to comply with and there are no chances for exceptions, then investors might search for another hydrocarbon province in which to invest.

The chapter on Brazil shows examples of how some of these exceptions to noncompliance with local content rules could work as follows (see table 4–4):[3]

It is important to find a balance between the reality of the host country and its expectations. It is important also to try to foresee the overall effects of short- and long-term local content development policy goals in a given host country. Some might argue that unrealistic obligations of local content might force the industry to move forward. Nigeria and Brazil are good examples of lessons learned in this regard. Strong local content requirements have proven unrealistic in the past few years. At one time, however, they helped to encourage international services providers and the local industry to invest heavily in domestic capacity during a time when the petroleum industry was growing. This would be a quite different scenario in times of low oil and gas prices or if there were no relevant discoveries in the country.

On a more positive note, an interesting example from the chapter on Equatorial Guinea may help new oil and gas nations to develop local content from scratch:[4]

> Equatorial Guinea's National Content Regulation, Article 17, mentions the ministerial order that establishes that "companies in the hydrocarbon sector engaged in exploration and production, as well as contractors and subcontractors, shall offer split contracts, with the intention of creating smaller contracts, and to the extent possible, invite domestic firms to participate, in accordance with their capabilities and resources."

This is relevant for nascent oil and gas provinces, as it is fairly difficult to build an oil and gas industry from scratch. Quite often the governing bodies of countries tend to be unreasonable or unrealistic. They might be draconian in their regulations, or they might establish a loose regulatory regime that makes investors wary. Instead, they need to be innovative in the way they address important obstacles to early local content development.

In addition, there is little point to enacting rules or ensuring that companies are subjected to contractual obligations unless compliance is supervised and enforced. Nigeria is a good example of this situation. Local content requirements have existed there for several decades, but Nigeria still faces a variety of issues with its local content policies and regulations. Some host countries may be challenged with restraints on capacity and competence to carry out these tasks. Institutional capacity building should, in such instances, be an integral part of local content policies and their implementation. Norway is an example of how this was identified and pursued as a policy goal, resulting in competent institutions that were able to carry out the roles and responsibilities allocated to them in terms of supervision and enforcement. Moreover, we note that often when host governments are empowered with measures to enforce local content requirements and obligations, they take a negative approach. In some instances, such as by offering bonuses for exceeding local content in different areas, a positive approach toward compliance can encourage higher performance.

Host countries must ensure that there are adequate mechanisms to audit and to follow up on compliance with the given contractual and regulatory framework. Some countries might provide a loose provision requiring preference to local providers within similar quality and price guidelines. Others might provide a more rigid approach, obligating the award of contracts or acquisition of goods with local companies within a minimum percentage. But how do host governments or national oil companies actually verify and control the compliance of these provisions? The reality is that a large number of countries fail to supervise and enforce the local content requirements, as they do not have adequate mechanisms to verify compliance. In Brazil, this was tackled by creating a plan for an authorized third party to certify the local content rules. This third party is duly approved and inspected by the national petroleum agency. This model might not fit all host countries, however, as this effectively implies delegation of public authority.

Moreover, the common rule appears to be that penalties are imposed on noncompliant players. We have seen that this may lead to oil and gas companies entering into back-to-back arrangements with their subcontractors. The result can be quite negative, as it increases the costs of the services and does not guarantee compliance for any party involved. If the host government provided a more constructive approach, the level of compliance could be increased. For example, an oil company could gain (i) an extension to complete its local content requirement, or (ii) tax incentives if it matches or exceeds the local content commitment. In this sense oil and gas companies and services providers could share the economic benefit to invest in local content, not only to secure contracts but also to receive other advantages from the host government. It would be a win-win situation for all parties involved. But not many countries are willing to follow this path. Brazil at least acknowledges that there is a need for exceptions and waivers when the noncompliance is due to lack of capability to source the services or goods locally, as previously discussed. However, even this solution might not be entirely satisfactory; applying for exceptions could entail delays.

In addition, it is important to have a clear understanding on what is local content and what is considered as an added value. For example, hiring a local broker to bring the goods from an international market should not be considered as fulfillment of a local content requirement; there was no real added value to the country but rather to a local broker. Another example is to acquire the entire materials abroad and then assemble them in the host country. This is better than using a broker or bringing "goods" ready for use from abroad. At least assembling the goods in the host country will bring higher employment and some local expertise to the country.

The definition of a "local" company can vary significantly from country to country. For example, in Brazil the criterion to be considered a Brazilian company is fairly straightforward. All that is required is the incorporation of a company in Brazil, regardless of the ownership. In this sense, a foreign investor could incorporate a company in Brazil with 100% foreign equity and still be considered a local company. The same situation is very different in other countries like Angola and Nigeria, as their local laws require at least 51% of local national ownership to be considered a local company.

A red flag relating to local content policies is the likelihood of increased corruption. Special care should be taken in places where the ownership of a local company must be composed of a certain percentage of local nationals or whenever preferred local players are indicated or recommended by any governmental authority. In some cases, politicians and other individuals might abuse the system to force international companies to partner with local companies for political or financial gain.

In short, large discoveries of oil and gas, or prospects for such discoveries, are likely to spark higher local content expectations from the host government and civil society. But it is important to adopt a step-by-step, analytical, and fact-based approach to the progression in the relevant policies, laws, and contracts. The key essence of local content is to add value to the industries and the development of the host country. The chapter on Iran provided an interesting graph showing the value creation of transfer of technology (fig. 20–1):[5]

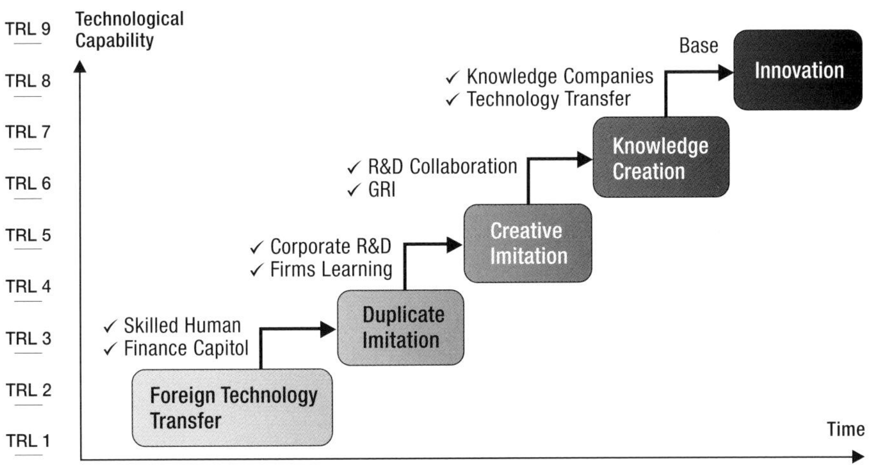

Fig. 20–1. Technological catch-up strategy implementation road map

Some might argue that a local content policy is successful if the local industry is capable of meeting international demand, and Norway is a good example of this concept. However, as already mentioned, Norway was lucky with respect to timing and framework conditions, such as industrial capacity and competence in the government and the private sector alike. It had stable, competent institutions to further its policy goals. It was acknowledged early on that for the domestic oil services companies to favorably and sustainably impact Norway's GPD, the companies would have to be able to compete not only domestically, but internationally as well. This is what has brought added value to the Norwegian government, economy, and society over time.

The above illustrates one important point: learning from best practices elsewhere is not a simple exercise. The "copy and paste" method of borrowing a foreign model is unlikely to bring the expected success in a given jurisdiction. As observed by Heum, Norway's petroleum activities, political pursuits, and policy choices do not differ greatly from those of other countries. "Nevertheless, Norway is still one of the few exceptions when it comes to really being successful in this respect."[6] It is important to understand the context and progression for each country in terms of human capacity, infrastructure, economic development, and community and political support, among other factors.

An important role of local content policies that is often overlooked by host governments is the transfer of technology. The Iran chapter provided interesting graphs on the relevance of transfer of technology, innovation, and the local content value chain creation (fig. 20–2 and fig. 20–3).[7]

It was not the objective of this book to provide empirical research on the differences in terms of added value in each country analyzed in this book. That would require an extensive and international collaboration to evaluate a large variety of indices and statistics (employment, oil and gas industry growth, growth of related industries, GPD, exports levels, etc.) in each jurisdiction. However, our book strives to provide valuable inputs from relevant oil and gas jurisdictions with regard to their local content background, policies, and regulations.

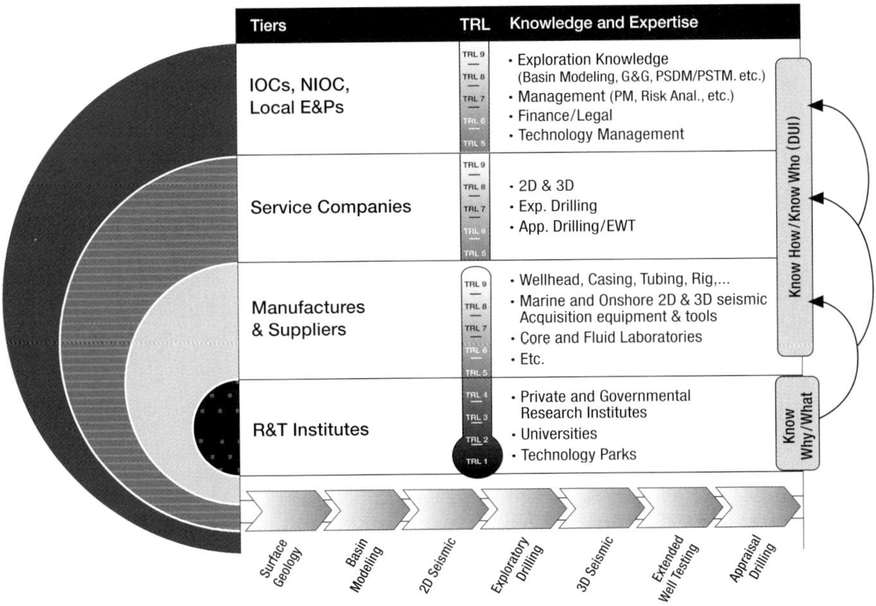

Fig. 20–2. Technology transfer and local content enhancement in exploration phase

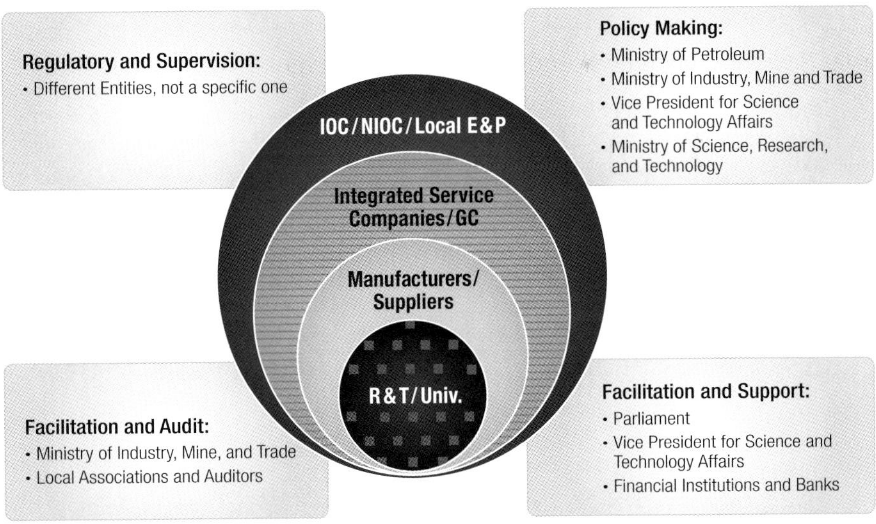

Fig. 20–3. Network of environmental entities together with key players of the E&P supply chain

We have seen that there is no secret formula to follow for any petroleum province to achieve a sustainable local content policy. But there are key principles and concerns host countries should be aware of before they implement their policies, and that is exactly what this book addresses. The goal of this book was to examine the overall policy goals and regulatory regimes in a large variety of oil and gas jurisdictions. In this way, the different levels of research and training investments, the main concerns with policy implementation, and possible ways forward for successful local content development could be compared and discussed. Hopefully it can serve as an inspiration for regulators and other players alike.

Notes

1. The decision to participate in petroleum operations is made by the Council of Ministers (or by the prime minister if so designated), and the participation of the state may occur during any phase of petroleum operations in accordance with the provisions of the contract. For more information, see chapter 19.
2. See chapter 10, "Ghana."
3. See chapter 4, "Brazil."
4. See chapter 9, "Equatorial Guinea."
5. See chapter 11, "Iran."
6. Per Heum, "Local Content Development—Experiences from Oil and Gas Activities in Norway" (SNF Working Paper No. 02/08, Institute for Research in Economics and Business Administration, Bergen, February 2008), 3, http://extrayendotransparencia.grupofaro.org/wp-content/uploads/2014/12/LC-Development-Experiences-from-oil-and-gas-in-Norway.pdf..
7. See chapter 11, "Iran."

INDEX

A

B

C

D

E

F

G

H

I

J

K

L

M

N

Q

R

S

T

U

V

W

X

Y

Z

ABOUT THE AUTHORS

Dr. Theophilus Acheampong is a petroleum economist and political risk analyst with experience in providing economic and investment analysis as well as research and strategic advisory for projects in the global energy industry. His work focusses on fiscal and other tax-related issues in upstream oil and gas and downstream energy markets. He has extensive consultancy experience in building fiscal models, performing macroeconomic analysis, and working with governments and international institutions in developing countries. Theo's country risk analysis experience covers political and business risks in sub-Saharan Africa, which includes providing risk forecasts and analysis of the impact of business risks, namely tax, legal, and operational risks on industry sectors and companies, including war, terrorism, and political unrest. He has worked as a senior analyst and consultant for firms such as IHS Markit, Crystol Energy, and Kina Advisory Limited, among others. Theo has a PhD in petroleum economics and an MSc in petroleum, energy economics, and finance from the University of Aberdeen, Scotland. He has a BSc in chemical engineering from the Kwame Nkrumah University of Science and Technology, Kumasi, Ghana.

Sayed El Araby, petroleum geologist, has held senior executive positions including deputy chairman for exploration and agreements with Egyptian national oil companies including the EGPC, EGAS, and Tharwa Petroleum Co.

Berryl Claire Asiago is an energy lawyer who specializes in the emerging energy upstream subsectors in East Africa (Kenya, Tanzania, Uganda, Ethiopia, and Djibouti). Asiago is a doctoral researcher at the University of Eastern Finland (UEF), the Center for Climate Change, Energy, and Environmental Law, where she is involved in relevant research projects regarding legal, regulatory, and institutional frameworks associated with petroleum sectors. Asiago is an "of counsel" for Cabinet Z-K International, a law firm incorporated and registered under the laws of the Republic of Djibouti. She has the mandate to draft, review, and evaluate energy-related contracts and other general contracts associated with the energy sector, such as PPPs, in Djibouti. Asiago handles and provides general legal advice in aspects of the energy industry, including upstream and downstream oil and gas operations, power purchasing and related concession agreements, and energy trading. Prior to joining UEF, Asiago worked for several years in South Sudan, Kenya, the United Kingdom, and Finland, building hands-on experience in legal research that has been acquired from working with different organizations in different countries, thus exposing her to different cultural environments and hastening her legal research decisions. Asiago holds an LLM (Dip) in international and European energy law (Finland), An LLM masters in international banking and insurance law (UK), and an LLB (India).

Nigel Bankes is professor of law and holder of the chair of natural resources law at the University of Calgary and an adjunct professor of law with the KG Jebsen Centre for the Law of the Sea at UiT, the Arctic University of Norway. He posts on developments in Canadian oil and gas and energy law at ABlawg: http://ablawg.ca/.

Bernardo Correia Barradas is a lawyer and founding partner of the CBMS Law Firm in Timor-Leste. Bernardo is currently finalizing his LLM in petroleum law and policy from Dundee University and is a graduate of the Lisbon University Law School. He is a consultant for the World Bank in the Finance and Markets group. He worked as senior legal adviser for the Central Bank and Petroleum Fund of Timor-Leste from 2013 to 2017. He also acted as lawyer, adviser, and company secretary for Heineken Asia Pacific and then Heineken Timor. He was legal adviser to the president of the Republic of Timor-Leste, Taur Matan Ruak, and, earlier in his career, worked in the Portuguese law firms Abreu Advogados and Vieira de Almeida.

Loïc Conan, a Paris and New York qualified lawyer, has been based in Cairo, Egypt, since 2010, where he is a senior legal counsel at the Egyptian energy group PICO International Petroleum/Cheiron Petroleum Corporation. He has also recently joined the Africa Projects Team at the global law firm Eversheds-Sutherland.

Paolo Esposito holds a master's in law (Naples) and an LLM (London) He is an avvocato (Italy), a solicitor of the Supreme Court of England and Wales, and a fellow of the Chartered Institute of Arbitrators (London). Esposito has broad experience in both domestic and international disputes and specialises in the resolution of a wide range of commercial and investment disputes through complex and multiparty arbitrations, including matters under bilateral investment treaties and the Energy Charter Treaty. He has also experience as a mediator and a member of arbitration panels. Esposito focuses on offshore oil and gas and mining issues and is special counsel and business consultant for several mining companies and contractors operating in the Sub-Saharan region. He teaches international arbitration at the Chartered Institute of Arbitrators in London and has published articles on arbitration topics.

Kamran Fatahi has 17 years of experience in project and portfolio management in industrial and upstream oil and gas projects, field development planning of oil and gas fields, economic and risk analysis of oil exploration and development, cost estimation, business strategy and development, and strategic alliance formation. He was involved in preparation of field development plans (FDP) for more than of 20 oil and gas fields in Iran as team leader or economic assessment specialist. He is member of several advisory committees for the Ministry of Petroleum related to IPC contracts and its technology transfer and local content appendices. He was project manager of the E&P Organizational Competency and Pre-Qualification Model for Potential Iranian Interested Companies for the Ministry of Petroleum. Fatahi serves as project portfolio manager at Tehran Energy Consultants and holds an MSc in industrial engineering from TMU (2002) and is a PhD candidate in industrial and systems engineering from AUT. He has been a visiting scholar at the University of Florida. Recently he founded Mehr Energy Exploration and Production based in Iran and acts as CEO. He is a member of the Society of Petroleum Engineers (SPE) and the Iranian Project Management Association and is the author of several papers in international journals and conferences.

Catarina Cortez Gamito graduated from the law school of Universidade Católica Portuguesa (2010) and has an LLM in commercial and corporate law from the Queen Mary University of London (2012). Catarina joined PLMJ as an associate in 2015 and was admitted to the Portuguese Bar Association in 2012. Since the early days of her internship at another law firm, and now in her third year as an associate at PLMJ, Catarina has been working on all aspects of energy and natural resources in Portugal, Timor-Leste, and across several African jurisdictions, such as Angola, Mozambique, Guinea-Bissau, and Equatorial Guinea.

Waniss A. Otman is a petroleum economist and analyst who specializes in the area of risk and uncertainty analysis. He holds a PhD in risk analysis of investment in the upstream oil and gas sector (Aberdeen University), as well as related MSc, LLM, MSc, and BSc degrees. This perspective has strengthened his background in areas of risk and uncertainty analysis of the petroleum industry. He works in the oil and gas industry as an advisor and consultant to NOCs and IOCs. He has contributed immensely through several published books and writes numerous articles for professional journals as well presentations in world forums and to IOCs.

Patrícia Rosário is a qualified Portuguese lawyer and managing partner of Rosario Consulting Group. She holds a JD from Catholic University of Lisbon and an LLM in comparative and international dispute resolution from the University of London, Queen Mary. Her practice focuses on foreign investment/public international law, international construction law, natural resources, and international arbitration. Patrícia has professional expertise in international contract management (in particular FIDIC) and international arbitration and adjudication. Her experience also includes structuring investments in Africa and Europe. Patrícia has acted for international developers, employers, and contractors in some of the largest and most complex building and engineering projects undertaken in Europe and Africa and was involved in legislative drafting and public affairs advice. She acts as an arbitrator, adjudicator, and counsel, advising parties with respect to arbitration proceedings in different countries and under all the major arbitration systems including in English, Spanish, and Portuguese languages.

Thomas Kojo Stephens is a lawyer, consultant, and lecturer in petroleum law, policy, regulation, and revenue management, and is a member of the Ghana bar association. He attended Emory university as a Bill Gates Scholar, graduating in 2003 with a BA in political science and philosophy (highest honors, Phi Beta Kappa, and summa cum laude). He obtained an LL.B from the University of Ghana, Legon, in 2006, a BL from the Ghana School of Law in 2008, and an LL.M from Cornell University in 2011. He attended the University of Aberdeen, Scotland, as a Commonwealth Scholar and obtained his PhD in petroleum law and policy in 2014. He has published in the area of law, particularly petroleum law, and currently engages in consultancy services in the petroleum industry to a number of high profile entities. He is the advisor/consultant for the Ghana jurisdiction for the International Energy Law Advisory Group (IELAG). He is a lawyer with the law firm Akufo-Addo, Prempeh and Co., and a lecturer on petroleum law and natural resources law, among other subjects, at the University of Ghana Law School. He is also a member of Ghana's Public Interest and Accountability Committee (PIAC), a statutory body with oversight over the use of Ghana's petroleum revenue, where he chairs its Legal subcommittee and sits on the Technical subcommittee.

Daniel Szyfman leads the infrastructure and energy practice in the Rio de Janeiro unit of the law firm Machado Meyer, with a focus in oil, gas, and energy industries. Szyfman also participates in the structuring of public projects (PPP), project finance operations, and matters related to maritime law. A relevant part of his practice is dedicated to complex operations, often cross-border, such as in M&A operations and complex contracts structuring for projects construction. Szyfman also provides clients with legal assistance in general regulatory and corporate matters, supporting the client in strategic matters concerning public law issues. Szyfman has also practiced for international groups willing to establish their business in Brazil, whether in the assets acquisition, participation in auctions, or organic growth.

Manuel Santos Vítor graduated from the law school of Universidade de Lisboa (1988) and has an LLM in EU law from the College of Europe in Bruges (1989). He joined PLMJ as a trainee in 1989 and was admitted to the Portuguese Bar Association in 1991. With more than 25 years of experience in legal practice, Manuel was PLMJ's comanaging partner and then managing partner between 2008–2015 and currently heads one of PLMJ's M&A practices and PLMJ Energy, the energy and natural resources team of the various jurisdictions where PLMJ is present. Manuel also dedicates a considerable amount of his time to insurance and aviation law. His name has been associated with important projects in these areas, both in Portugal and internationally. An AIPN and IBA member, Manuel Santos Vitor has been closely involved in the internationalization of PLMJ's work. He is recommended by the Chambers Europe, Chambers Global, Global Law Experts and Best Lawyers directories.

Leslie Zhang Weihua (张伟华), vice president and general counsel, United Energy Group, has nearly 16 years of experience in the oil and gas business. Before joining United Energy Group, he was the director of the Project Management Division of the legal department of CNOOC Group. Other previous positions at CNOOC included head of the Consulting Team in the legal department and acting head of the Merger/Acquisition Team in their legal department. He was a core member of the legal team representing CNOOC on its $15.1 billion acquisition of Nexen Inc. He was recognized by Legal500 as one of Asia Pacific's top corporate counsels and one of the best general counsels in China Mainland and the Hongkong area. He currently serves as a committee member for the AIPN Model Contracts Drafting Committee and as a member of the AIPN China Steering Committee. Before CNOOC Zhang held corporate counsel positions in two Fortune 500 companies: CITIC and PetroChina. Zhang has sophisticated work experience on large-scale cross-border transactions with multinational companies. He is familiar with international oil and gas industry practices and has advanced business negotiation skills. Zhang has published four Chinese books on Mergers and Acquisitions and he is a guest professor at the Beijing Foreign Studies University and a researcher fellow appointed by the Renmin University of China.